VOLUME EIGHTY THREE

CURRENT TOPICS IN
DEVELOPMENTAL BIOLOGY

Sex Determination and
Sexual Development

Series Editor

Paul M. Wassarman
Department of Developmental and Regenerative Biology
Mount Sinai School of Medicine
New York, NY 10029-6574
USA

Olivier Pourquié
Investigator Howard Hughes Medical Institute
Stowers Institute for Medical Research
Kansas City, MO, USA

Editorial Board

Blanche Capel
Duke University Medical Center
Durham, USA

B. Denis Duboule
Department of Zoology and Animal Biology
NCCR 'Frontiers in Genetics'
Geneva, Switzerland

Anne Ephrussi
European Molecular Biology Laboratory
Heidelberg, Germany

Janet Heasman
Cincinnati Children's Hospital Medical Center
Department of Pediatrics
Cincinnati, USA

Julian Lewis
Vertebrate Development Laboratory
Cancer Research UK London Research Institute
London WC2A 3PX, UK

Yoshiki Sasai
Director of the Neurogenesis and Organogenesis Group
RIKEN Center for Developmental Biology
Chuo, Japan

Cliff Tabin
Harvard Medical School
Department of Genetics
Boston, USA

Founding Editors

A. A. Moscona
Alberto Monroy

VOLUME EIGHTY THREE

Current Topics in
DEVELOPMENTAL BIOLOGY

Sex Determination and Sexual Development

Edited by

MARK VAN DOREN
Department of Biology
Johns Hopkins University
Baltimore, Maryland

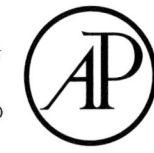

AMSTERDAM • BOSTON • HEIDELBERG • LONDON
NEW YORK • OXFORD • PARIS • SAN DIEGO
SAN FRANCISCO • SINGAPORE • SYDNEY • TOKYO
Academic Press is an imprint of Elsevier

ELSEVIER

Academic Press is an imprint of Elsevier
525 B Street, Suite 1900, San Diego, CA 92101-4495, USA
30 Corporate Drive, Suite 400, Burlington, MA 01803, USA
Linacre House, Jordan Hill, Oxford OX2 8DP, UK
32 Jamestown Road, London, NW1 7BY, UK

First edition 2008

Copyright © 2008, Elsevier Inc. All rights reserved.

No part of this publication may be reproduced, stored in a retrieval system or transmitted in any form or by any means electronic, mechanical, photocopying, recording or otherwise without the prior written permission of the publisher

Permissions may be sought directly from Elsevier's Science & Technology Rights Department in Oxford, UK: phone (+44) (0) 1865 843830; fax (+44) (0) 1865 853333; email: permissions@elsevier.com. Alternatively you can submit your request online by visiting the Elsevier web site at http://elsevier.com/locate/permissions, and selecting *Obtaining permission to use Elsevier material*

Notice

No responsibility is assumed by the publisher for any injury and/or damage to persons or property as a matter of products liability, negligence or otherwise, or from any use or operation of any methods, products, instructions or ideas contained in the material herein. Because of rapid advances in the medical sciences, in particular, independent verification of diagnoses and drug dosages should be made

ISBN: 978-0-12-374496-8
ISSN: 0070-2153

For information on all Academic Press publications visit our website at elsevierdirect.com

Printed and bound in USA
08 09 10 11 12   10 9 8 7 6 5 4 3 2 1

Working together to grow libraries in developing countries

www.elsevier.com | www.bookaid.org | www.sabre.org

ELSEVIER   BOOK AID International   Sabre Foundation

Contents

Contributors ix
Preface xi

1. Somatic Sexual Differentiation in *Caenorhabditis elegans*
Jennifer Ross Wolff and David Zarkower

1. Background 2
2. Telling One from Two: The Primary Signal 4
3. The Global Sex Regulatory Pathway 5
4. Attaining Sexual Dimorphism 10
5. The Interface Between Sex Determination and Sex Differentiation 26
6. Something Old, Something New: Conservation of Nematode Sexual Regulators 29
7. The Future 30
Acknowledgments 30
References 31

2. Sex Determination in the *Caenorhabditis elegans* Germ Line
Ronald E. Ellis

1. Nematode Gender: Hermaphrodites, Females, and Males 42
2. The Core Sex-Determination Pathway 44
3. Transcriptional Control of Germ Cell Fates by TRA-1 47
4. Modulating the Sex-Determination Pathway to Create Hermaphrodites 50
5. Maintaining a Balance Between TRA-2 and FEM-3 Levels 53
6. Do FOG-1 Levels Control Germ Cell Fates? 55
7. Mysteries 55
8. Conclusion 58
Acknowledgments 59
References 59

3. The Creation of Sexual Dimorphism in the *Drosophila* Soma
Nicole Camara, Cale Whitworth, and Mark Van Doren

1. Sex Determination 66
2. The Creation of Sexual Dimorphism 74

v

	3. A New Look at an Old Problem	93
	References	96

4. *Drosophila* Germline Sex Determination: Integration of Germline Autonomous Cues and Somatic Signals

Leonie U. Hempel, Rasika Kalamegham, John E. Smith III, and Brian Oliver

1.	Introduction	110
2.	Sexually Dimorphic Niches Regulate Germ Cells	110
3.	Sex Determination in the Germ Cells	122
4.	Germ Cells in a Sex-Mismatched Soma	128
5.	Concluding Remarks	142
	Acknowledgments	143
	References	143

5. Sexual Development of the Soma in the Mouse

Danielle M. Maatouk and Blanche Capel

1.	History of Sex Determination in Mammals	152
2.	Molecular Control of Sex Determination	159
3.	Secondary Sex Differentiation	170
4.	Summary	176
	References	177

6. Development of Germ Cells in the Mouse

Gabriela Durcova-Hills and Blanche Capel

1.	Origin of the Germ Cell Lineage	185
2.	Specification of Primordial Germ Cells in Mice	186
3.	Migration and Population of the Urogenital Ridge	190
4.	Maintaining Pluripotency in the Germ Cell Lineage	192
5.	Sex-Specific Development of Germ Cells in XX and XY Gonads	193
6.	The Sexually Dimorphic Cell Cycle	195
7.	Epigenetic Regulation in the Mouse Germ Line	198
8.	Sexually Dimorphic Epigenetic Regulation	200
9.	Germ Cells to Stem Cells and Back	202
10.	Conclusions	205
	Acknowledgments	206
	References	206

7. **The Neuroendocrine Control of Sex-Specific Behavior in Vertebrates: Lessons from Mammals and Birds**

Margaret M. McCarthy and Gregory F. Ball

1. Introduction: General Principles and the Scope of the Chapter	214
2. Historical Overview	216
3. Reproductive Behavior in Mammals	222
4. Reproductive Behaviors in Birds	231
5. Unanswered Questions and Future Directions	241
References	242

Index	249
Contents of Previous Volumes	255

Contributors

Gregory F. Ball
Department of Psychological and Brain Sciences, Johns Hopkins University, Baltimore, Maryland 21218

Nicole Camara
Department of Biology, Johns Hopkins University, Baltimore, MD 21218

Blanche Capel
Department of Cell Biology, Duke University Medical Center, Durham, North Carolina 27710

Gabriela Durcova-Hills
Wellcome Trust/Cancer Research UK Gurdon Institute of Cancer and Developmental Biology, University of Cambridge, Cambridge CB2 1QN, United Kingdom

Ronald E. Ellis
Department of Molecular Biology, School of Osteopathic Medicine, B303 Science Center, The University of Medicine and Dentistry of New Jersey, 2 Medical Center Drive, Stratford, New Jersey 08084

Leonie U. Hempel
Laboratory of Cellular and Developmental Biology, National Institute of Diabetes and Digestive and Kidney Diseases, National Institutes of Health, 50 South Drive, Bethesda, Maryland 20892

Rasika Kalamegham
Laboratory of Cellular and Developmental Biology, National Institute of Diabetes and Digestive and Kidney Diseases, National Institutes of Health, 50 South Drive, Bethesda, Maryland 20892

Danielle M. Maatouk
Department of Cell Biology, Duke University Medical Center, Durham, North Carolina 27710

Margaret M. McCarthy
Departments of Physiology and Psychiatry, University of Maryland School of Medicine, Baltimore, Maryland 21201

John E. Smith III
Laboratory of Cellular and Developmental Biology, National Institute of Diabetes and Digestive and Kidney Diseases, National Institutes of Health, 50 South Drive, Bethesda, Maryland 20892

Brian Oliver
Laboratory of Cellular and Developmental Biology, National Institute of Diabetes and Digestive and Kidney Diseases, National Institutes of Health, 50 South Drive, Bethesda, Maryland 20892

Mark Van Doren
Department of Biology, Johns Hopkins University, Baltimore, MD 21218

Cale Whitworth
Department of Biology, Johns Hopkins University, Baltimore, MD 21218

Jennifer Ross Wolff
Department of Biology, Carleton College, Northfield, Minnesota 55057

David Zarkower
Department of Genetics, Cell Biology, and Development, University of Minnesota 6-160 Jackson Hall, Minneapolis, Minnesota 55455

Preface

Vive la difference!

Back when we were embryos a genetic switch was thrown, causing us to head down either a male or female path with profound consequences on the development of our bodies and our brains. One of the most fascinating events of animal development is the creation of sexual dimorphism, the often dramatically different forms taken on by males and females of a species. Many of these differences center around the process of sexual reproduction; the production of sperm vs. eggs, and the mechanisms for attracting a mate, courting a mate and getting the gametes together. This volume focuses on our understanding of how sex-specific development is regulated and leads to sex-specific traits and behaviors.

In addition to summarizing the current knowledge in the field, there are two additional goals we have set for this volume. The first is to discuss not just how sexual identity is established, but how it is used to create sexual dimorphism. While we have made considerable progress in understanding how the primary sex determination switch works in some well-studied model systems, we are only beginning to understand how sexual identity is used to control the sex-specific development of tissues and organs, including the brain. Thus, in addition to discussing the process of sex determination in different organisms, we also focus on how sex-specific development is regulated to create sexual dimorphism and sex-specific behavior. The second goal is to compare how sexual dimorphism is created in the soma vs. the germline. The most universal aspect of sexual development is the production of the sex-specific gametes. Sperm and egg are recognizable across diverse species, even if other sex-specific traits are less conserved. One central theme for gamete development is the interaction between the germline and soma that controls this process. Therefore, we include chapters addressing how sexual dimorphism is created in the soma and the germline for three well-studied model systems. In deciding how to focus this volume, many interesting topics were necessarily not included. Thus, discussion of the wide range of sex determination mechanisms, sex-specific traits and sexual behaviors in the animal kingdom, and how they evolve, must be left for future volumes.

Lastly, I would like to thank all the authors who have contributed to this volume. In addition, I want to particularly acknowledge one author who was unable to contribute. Anne McLaren was originally planning to write the chapter on mouse germline development, before she was tragically killed in an accident last year. We are deeply indebted to Blanche Capel and Gabriela Durcova-Hills for stepping in to write this chapter. Fittingly, they were able to do so using some of Anne's own notes and figures. It is clear that, while Anne's presence in the field of Germ Cell Biology will be dearly missed, her influence will live on for years to come.

CHAPTER ONE

SOMATIC SEXUAL DIFFERENTIATION IN *CAENORHABDITIS ELEGANS*

Jennifer Ross Wolff* *and* David Zarkower[†]

Contents

1. Background	2
2. Telling One from Two: The Primary Signal	4
3. The Global Sex Regulatory Pathway	5
3.1. Overview of sex determination pathway	5
3.2. XOL-1 and the SDC proteins control sex determination and dosage compensation	6
3.3. Committed regulators of sex: Controlling TRA-1	7
4. Attaining Sexual Dimorphism	10
4.1. The somatic gonad	12
4.2. The sex muscles	14
4.3. The ventral hypodermis	16
4.4. The male tail	18
4.5. The nervous system and sex-specific behavior	23
4.6. Sex-specific cell death	25
4.7. Covert sexual specialization	26
5. The Interface Between Sex Determination and Sex Differentiation	26
5.1. Transcriptional regulation by TRA-1	27
5.2. Translating global decisions into local action	28
6. Something Old, Something New: Conservation of Nematode Sexual Regulators	29
7. The Future	30
Acknowledgments	30
References	31

Abstract

The two sexes of the nematode *Caenorhabditis elegans* are the self-fertile hermaphrodite (essentially a female with a mixed germ line) and the male, and these differ extensively in anatomy, physiology, and behavior. At hatching,

---

* Department of Biology, Carleton College, Northfield, Minnesota 55057
[†] Department of Genetics, Cell Biology, and Development, University of Minnesota 6-160 Jackson Hall, Minneapolis, Minnesota 55455

C. elegans larvae of each sex are nearly indistinguishable, differing mainly in the sex-specific death of a handful of neurons. After birth, however, a number of blast cells undergo radically different lineages and differentiation programs in the two sexes, leading to adults in which about one-third of cells are overtly dimorphic. The first C. elegans mutants causing discordance between genetic and phenotypic sex were isolated more than 30 years ago. Since then much progress has been made in uncovering the chromosomal elements and downstream regulatory pathways that control sex determination and sexual differentiation in the worm. The primary signal for sex determination is the ratio of X chromosomes to sets of autosomes, with hermaphrodites normally having two X chromosomes (XX) and males one (XO). The X:A signal is exquisitely dose-sensitive and operates via a group of X-linked regulators acting in opposition to a group of autosomal regulators that compete for the control of the master sex regulator *xol-1*. The activity of *xol-1* coordinately regulates the formation of an active X chromosome dosage compensation complex and the activity of a sex determination regulatory cascade. The sex determination pathway globally controls all sexually dimorphic features by conferring sex specificity on downstream regulatory modules, largely via the action of TRA-1, a Ci/GLI family transcription factor with high activity in hermaphrodites and low activity in males. Much of this regulation involves the imposition of sex-specific activity on general developmental regulators in specific cell lineages. Recent work has answered long-standing questions about the molecular mechanisms controlling the sex determination pathway and shown that some C. elegans sexual regulators have counterparts regulating sexual development in other phyla.

1. BACKGROUND

The nematode *Caenorhabditis elegans* has two naturally occurring sexes, the male and the hermaphrodite (a self-fertile female). The two sexes differ extensively in somatic tissues, with many anatomical, physiological, and behavioral dimorphisms that are required for efficient self- and cross-fertility (Fig. 1.1). Hermaphrodites are larger than males and have 959 somatic nuclei, whereas males have 1031 somatic nuclei, largely on account of male-specific neurons that mediate copulation (Sulston and Horvitz, 1977). Sex is genetically determined in *C. elegans*: hermaphrodites have two X chromosomes (XX) and males have one (XO) (Nigon, 1949). Males can arise either spontaneously, due to infrequent meiotic loss of an X chromosome, or through male/hermaphrodite mating, and their presence allows exchange of genetic material between individuals. Overall, one third or more of somatic cells show overt sexual specialization and these sex

Somatic Sexual Differentiation in *C. elegans*   3

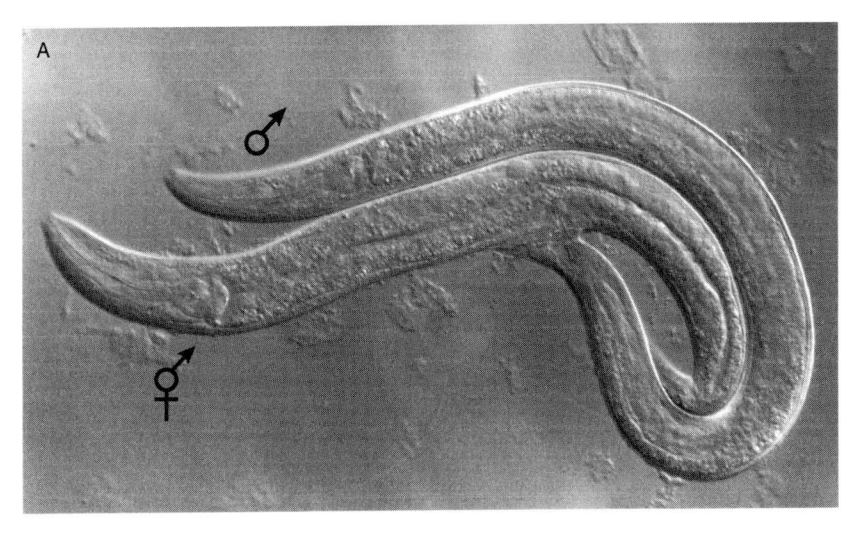

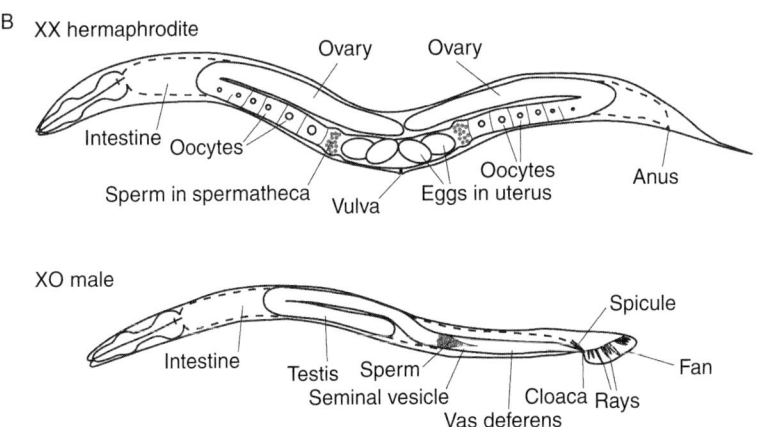

Figure 1.1 The *C. elegans* male and hermaphrodite. (A) A male and hermaphrodite *in flagrante delicto*. Both worms harbor the *dpy-28(y1)* mutation, which selectively reduces hermaphrodite body size. Image source: B. J. Meyer. (B) The two sexes differ in body size and in tissues that support reproduction, including the gonad, nervous system, and tail structures.

differences affect all tissue types (Sulston and Horvitz, 1977). Among the most pronounced dimorphisms are those directly involved in reproduction, including copulatory structures, portions of the nervous system controlling mating and egg laying, and the somatic gonad (Fig. 1.1).

C. elegans is one of a troika of animals (the others being the fruit fly and the mouse) in which intensive study of sex determination has been possible. The genetic pathway controlling sex determination was the first regulatory pathway to be described in *C. elegans* (Hodgkin and Brenner, 1977).

Three subsequent decades of genetic analysis, paired more recently with molecular studies, have led to a detailed understanding of the mechanisms that specify the sexual path of a *C. elegans* embryo and direct the execution of the ensuing developmental program.

This chapter provides an overview of major sexual dimorphisms in *C. elegans* and how their establishment is controlled during development. It also describes five concepts that have emerged from the study of sexual development in *C. elegans*. First, the primary signal for sex determination, the ratio of X chromosomes to sets of autosomes, or X:A ratio, exhibits remarkable sensitivity to chromosomal composition and acts via a number of *cis*- and *trans*-acting signals, both autosomal and sex-linked. Second, the proximal target of the X:A signal, the master regulator *xol-1*, couples anatomic sex to sex chromosome dosage compensation by coordinately regulating the formation of an X chromosome dosage compensation complex and the activity of a sex determination regulatory pathway. Third, the sex determination pathway controls all aspects of somatic sexual differentiation, including anatomy, physiology, and behavior, by regulating the Ci/GLI family transcription factor TRA-1. XOL-1 and TRA-1 can be regarded as the upstream and downstream master regulators, respectively, of sex determination. Additional regulators have evolved to permit mixed-sex gametogenesis in the hermaphrodite (see chapter by R. Ellis). Fourth, downstream of the "global" sex determination pathway are specialized sexual regulators that control limited aspects of sexual differentiation as well as general developmental regulators whose activity is made sex-specific in certain cells by the global pathway. Fifth, comparison of the nematode sex determination pathway with those of other animals has revealed that the process is fluid and fast-evolving but appears to involve some relatively ancient regulatory factors with analogous functions.

2. Telling One from Two: The Primary Signal

Sex is genetically determined in *C. elegans*. Nigon established by cytogenetic analysis that hermaphrodites have five pairs of autosomes and a pair of sex chromosomes (2A;2X, or XX), while males have five autosomal pairs and a single sex chromosome (2A;1X, or XO) (Nigon, 1949). In principle, sex might be determined solely by the number of X chromosomes, but analysis of tetraploids demonstrated that 4A;4X (X:A = 0.5) animals are hermaphrodites while 4A;2X animals (X:A = 1.0) are males, and thus it is the ratio of sex chromosomes to autosomes that constitutes the sex determining signal (Nigon, 1951). This conclusion was confirmed and extended using polyploids that showed the worm can reliably distinguish the remarkably small difference between an X:A ratio of 0.67 (3A;2X = male) and one of 0.75 (4A;3X = hermaphrodite) (Madl and Herman, 1979). Dissection of the X chromosome with partial duplications further indicated that it must contain at least three dose-sensitive signal elements (Madl and Herman, 1979).

Some of the signal elements on the X chromosome (XSEs), as well as several autosomal signal elements (ASEs), have been identified by genetic analysis. These signal elements compete for control of *xol-1*, the master upstream regulator of sex determination (see following section). In XX animals, the higher dose of X-linked elements allows them to prevail and inactivate *xol-1*, whereas in XO animals the single XSE dose is insufficient. The XSEs include the RNA binding protein FOX-1 and the nuclear hormone receptor transcription factor SEX-1 (Akerib and Meyer, 1994; Carmi and Meyer, 1999; Carmi *et al.*, 1998; Hodgkin *et al.*, 1994; Nicoll *et al.*, 1997; Skipper *et al.*, 1999). Overexpression of FOX-1 can post-transcriptionally repress a *xol-1* reporter, and SEX-1 can repress *xol-1* transcription, probably via direct binding to the *xol-1* promoter, confirming that *xol-1* regulation involves multiple molecular mechanisms (Carmi *et al.*, 1998; Nicoll *et al.*, 1997). So far two ASEs have been isolated, encoding the putative transcription factors SEA-1, a T-box protein, and SEA-2, a zinc finger protein (Meyer, 2005; Powell *et al.*, 2005). How the XSEs and ASEs compete to regulate *xol-1* is unknown.

3. THE GLOBAL SEX REGULATORY PATHWAY

3.1. Overview of sex determination pathway

Sexual fate throughout the body is determined by a "global" pathway controlled by *xol-1* (Fig. 1.2). More specialized "local" pathways act downstream of the global pathway to control sexual differentiation of specific tissues and cell types. The global pathway was formulated based on extensive genetic epistasis tests between mutations in sex-determining genes and subsequent molecular analysis has provided strong support (Hodgkin, 1987b; Villeneuve and Meyer, 1990). The pathway in Fig. 1.2 is somewhat simplified; a number of ancillary genetic interactions also occur and these probably serve to fine-tune the pathway (Hodgkin, 2002).

The global pathway can be divided into two segments. The upstream portion (from the X:A signal to *sdc-2*) has a dual function, coordinately controlling sex determination and X chromosome dosage compensation, the process that equalizes expression of X-linked genes between the two sexes. The downstream portion (*her-1* to *tra-1*) is dedicated exclusively to sex determination. Joint control of dosage compensation and sex determination by the X:A ratio ensures that both processes adopt the same sex-specific activity state in a given individual. This is essential, as inappropriate dosage compensation in either sex can be lethal. To accommodate the

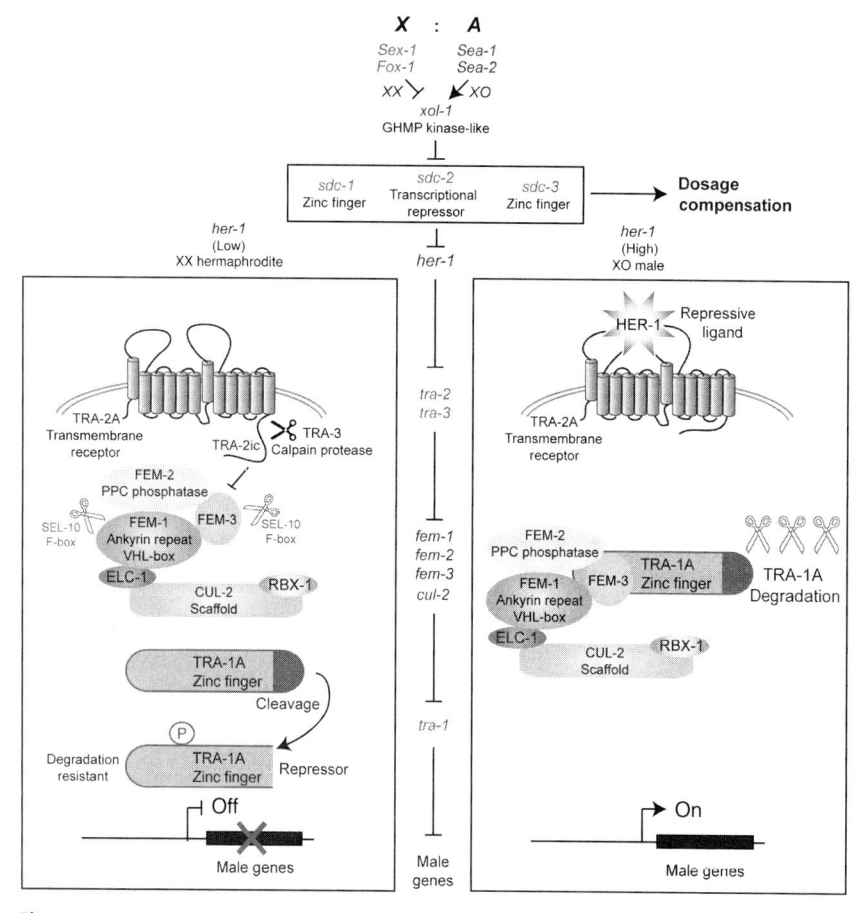

Figure 1.2 Pathway controlling somatic sex and dosage compensation in *C. elegans*. Masculinizing genes are colored blue and feminizing genes are red. Regulatory interactions shown are described in main text. Figure adapted from Kuwabara (2007) with permission.

mixed-sex germ line of the hermaphrodite, additional regulators are required in germ cells to transiently reverse the sex determination switch during larval development, as described in Chapter 2.

3.2. XOL-1 and the SDC proteins control sex determination and dosage compensation

Sex determination and dosage compensation are linked by the three SDC (sex determination and dosage compensation) proteins. SDC-1 and -3 have zinc fingers, while SDC-2 is novel. These proteins function in XX animals to control sex determination by repressing transcription of the *her-1* (hermaphroditization of XO) gene and to control dosage compensation by chromosome-wide repression of X-linked gene expression (reviewed by Meyer, 2005). The sex-specificity of the SDC complex stems from XX-specific expression of SDC-2. SDC-2 is somehow repressed in XO animals by XOL-1, a protein related to GHMP small molecule kinases but without ATP binding (Dawes et al., 1999; Luz et al., 2003). Association of a complex containing the SDC proteins with the *her-1* promoter represses its transcription by 20-fold in XX animals (Chu et al., 2002; Perry et al., 1994). Potential SDC binding sites have been identified in *her-1* (Perry et al., 1994). Intriguingly, two of these elements contain a 15-bp sequence that is absent from X, and this may allow formation of a complex specific to sex determination (Chu et al., 2002).

Association of a distinct SDC-containing complex (the dosage compensation complex, or DCC) with the X chromosomes of hermaphrodites early in embryogenesis results in repression of most X-linked genes. Unlike regulation of *her-1*, the DCC only represses X-linked transcription by one-half (Meyer, 2005; Meyer and Casson, 1986). Also, unlike the candidate SDC binding sites that mediate sex determination, those mediating DCC recruitment are not unique to X; instead it appears that a high density of these sequences in specific sites on X is critical for DCC recruitment (McDonel et al., 2006).

While both sex determination and dosage compensation involve all three SDC proteins, there are clear differences in how they function. First, the magnitude of transcriptional repression (20-fold vs 2-fold) differs, and so may the *cis* elements that mediate repression. Second, the two complexes differ in composition, with the DCC containing a unique component, DPY-21 (Yonker and Meyer, 2003). Third, SDC-3 is critical for *her-1* recognition, while SDC-2 is the key factor in X chromosome regulation (Yonker and Meyer, 2003). Fourth, autosomal repression by the SDC proteins is apparently restricted to *her-1*, while X chromosome repression is nucleated at specific sites but then spreads to affect most X-linked genes (Csankovszki et al., 2004; McDonel et al., 2006). Finally, the two roles of SDC-3 can be separated genetically, with mutations in an ATP binding motif disrupting *her-1* regulation and mutations in the zinc fingers disrupting dosage compensation (Klein and Meyer, 1993).

3.3. Committed regulators of sex: Controlling TRA-1

3.3.1. HER-1 inactivates TRA-2 cell nonautonomously

The sex determination-specific segment of the global pathway is controlled by *her-1* and culminates in *tra-1* (transformer 1) (Fig. 1.2). HER-1 is a small secreted protein that inhibits activity of the patched-related transmembrane protein TRA-2A in XO animals, thereby promoting male development

(Perry et al., 1993). Genetic mosaic analysis has shown that *her-1* acts cell nonautonomously, consistent with its molecular identity (Hunter and Wood, 1992). HER-1 appears to bind directly to TRA-2A based on genetic analysis and transfection studies in cultured cells (Hamaoka et al., 2004; Kuwabara, 1996a).

The secretion of HER-1 provides a mechanism for regulation of germ line sexual fate by somatic cells (Hunter and Wood, 1992). Having a cell nonautonomous step in the regulatory pathway also has been suggested to provide a means for coordinating sexual fates among groups of lineally unrelated somatic cells. This may be particularly important in an animal like *C. elegans* with its largely invariant cell lineage and limited capacity for regulative development.

In addition to inhibition by HER-1, *tra-2* also is regulated by translational repression, particularly in the germ line. This regulation is mediated by elements in the 3′-UTR of *tra-2* mRNA, and the disruption of these elements can cause weak somatic feminization of XO animals and increased translation of reporter genes (Doniach, 1986; Goodwin et al., 1993).

3.3.2. TRA-2 inhibits the FEM proteins with help from TRA-3 and SEL-10

TRA-2A inactivates the three FEM (feminization) proteins in XX animals to allow female development. Regulation of the FEMs by TRA-2A appears not to be at the level of transcription or mRNA stability, as *fem* mRNA levels are similar in the two sexes (Ahringer et al., 1992; Gaudet et al., 1996). TRA-2A and FEM-3 can interact *in vitro* and in yeast two hybrid assays (Mehra et al., 1999), and this interaction may render the FEM proteins inactive, possibly by sequestering FEM-3. Overexpression of FEM-3, but not FEM-1 or FEM-2, can masculinize XX animals, consistent with the idea that FEM-3 levels are critical to FEM activity (Mehra et al., 1999).

The TRA-3 and SEL-10 proteins enhance the feminizing activity of TRA-2A. TRA-3 is a calcium-dependent protease that can cleave TRA-2A *in vitro* and appears to serve as a critical cofactor for TRA-2, releasing an intracellular peptide (TRA-2ic) that may have enhanced FEM-inhibiting activity (Barnes and Hodgkin, 1996; Sokol and Kuwabara, 2000). SEL-10 is an F-box containing protein, one of a large family of proteins that direct E3 ubiquitin-protein ligase complexes to substrates. SEL-10 can interact with FEM-1 and FEM-3 and loss of *sel-10* weakly masculinizes the soma (Jager et al., 2004). A simple model for this segment of the pathway is that interaction of FEM-3 with the intracellular portion of TRA-2A blocks FEM activity, and this inhibition is overcome in XO animals by binding of HER-1 to the extracellular domain of TRA-2A. TRA-3 and SEL-10 may boost the efficiency of this regulatory step by enhancing the activity of TRA-2A and limiting the level of the FEMs, respectively.

3.3.3. The FEM proteins regulate TRA-1 by sex-specific proteolysis

The final step in the global pathway is the regulation of the terminal gene *tra-1*. *tra-1* is alternatively spliced to encode two proteins with C2H2 zinc fingers closely related to those of the *Drosophila* Ci and vertebrate GLI proteins, and it is the only *C. elegans* Ci/GLI homolog (Zarkower and Hodgkin, 1992). The larger protein, TRA-1A, can perform all known functions of *tra-1*, whereas no activity has been assigned to TRA-1B (Schvarzstein and Spence, 2006; Zarkower and Hodgkin, 1993). Ci and GLI proteins are transcriptional regulators that transduce hedgehog signaling. Nematodes lack a Hedgehog pathway, but similarity of TRA-2A to the patched receptor and TRA-1 to Ci/GLI suggests the possibility that a hedgehog-like signaling module may have played an ancient role in nematode sex determination (Kuwabara *et al.*, 2000).

Regulation of *tra-1* is mainly post-transcriptional (Zarkower and Hodgkin, 1992), and involves two regions of the protein. The first, known as the GF region, is a short segment near the amino terminus that is altered by a large number of gain-of-function mutations and overlaps a GSK3 phosphorylation consensus sequence (de Bono *et al.*, 1995; Hodgkin, 1980, 1987a). The second region, in the carboxy terminal portion of TRA-1A, interacts with TRA-2 and is important for XX spermatogenesis (Doniach, 1986; Lum *et al.*, 2000; Wang and Kimble, 2001). Genetic evidence is extensive and compelling that the FEM proteins are essential negative regulators of *tra-1*, acting to reduce its activity in XO animals and thereby permit male development (Hodgkin, 1987b). Molecular cloning of the *fem* genes did disappointingly little to illuminate the molecular mechanism by which the FEMs function: FEM-1 has ankyrin repeats, FEM-2 is a protein phosphatase whose substrate awaits identification, and FEM-3 is a novel protein (Ahringer *et al.*, 1992; Chin-Sang and Spence, 1996; Pilgrim *et al.*, 1995; Spence *et al.*, 1990).

Difficulties in obtaining effective TRA-1 antibodies delayed the investigation of its regulatory mechanisms. Recently, however, this problem was overcome and TRA-1A was shown to be cleaved in both sexes to produce a shorter phosphorylated active isoform denoted TRA-1^{100}, which is regulated by FEM-dependent male-specific protein degradation (Schvarzstein and Spence, 2006). This is counter to an earlier report that TRA-1 is regulated by sex-specific nuclear export (Segal *et al.*, 2001).

The mystery of the FEM protein regulatory mechanism was further unraveled following the identification of FEM-1 in a proteomic analysis of Cullin-2 (CUL-2) interacting proteins (Starostina *et al.*, 2007). CUL-2 is a component of E3 ubiquitin ligase complexes of the CBC (CUL2, Elongin B, and Elongin C) class. Within this large family of complexes, CUL-2 acts as a scaffold, associating at one end with an Rbx/Roc1 RING finger protein, and at the other an Elongin C adaptor protein and a substrate recognition subunit (SRS) of variable composition. Closer examination revealed that FEM-1 has a VHL box, a motif that mediates binding of the SRS to elongin C in CBC E3 ligase complexes.

Further experiments demonstrated that the CBC^{FEM-1} complex plays a role in sex determination. As expected, *cul-2* loss of function causes elevated TRA-1 levels and partial feminization in XO animals. TRA-1 expressed in human cells is cleaved to generate TRA-1^{100} (actually ~135 kDa), suggesting that the processing mechanism is conserved, and FEM-1 expression in these cells results in proteasome-dependent degradation of TRA-1. This degradation is enhanced by expression of FEM-2 and FEM-3. The three FEM proteins bind each other in transfected cells, and in the presence of a proteasome inhibitor, the three FEMs and TRA-1 can associate with human CUL-2. TRA-1 with a missense mutation in the GF domain is resistant to CUL-2/FEM mediated degradation although, unexpectedly, the GF mutant TRA-1 protein associates equally well with the FEM proteins. Finally, transfection of FEM-2 and FEM-3 stimulates CBC^{FEM-1} mediated ubiquitination of TRA-1 in human cells.

Taken together, these results strongly suggest a model in which a CBC complex containing the three FEM proteins assembles in XO animals (where TRA-2 is inactive) and efficiently targets TRA-1A and possibly TRA-1^{100} for proteosomal degradation. This is likely the primary mechanism by which *tra-1* activity is rendered sex-specific. A number of questions remain. What is the cleavage mechanism that creates TRA-1^{100} and is it analogous to cleavage of Ci/GLI? Why is GF TRA-1 protein resistant to processing? Is the regulatory mechanism the same in soma and germ line? Does TRA-1 escape degradation in the male somatic gonad primordium (see below)? What is the role of TRA-1 phosphorylation and of FEM-2 phosphatase activity?

4. Attaining Sexual Dimorphism

Despite the dramatic sexual dimorphisms of the adult, at hatching there are few overt differences between the two sexes (Sulston and Horvitz, 1977; Sulston *et al.*, 1980; Figs. 1.3 and 1.4). Hermaphrodites have two neurons, the HSN cells, that will later control egg laying, whereas males have four CEM neurons in the head, which serve a sensory function (Chasnov *et al.*, 2007; White *et al.*, 2007). The two sexes also differ in the position of scavenger cells called coelomocytes, which are anterior to the gonad in hermaphrodites and flank it in males. Finally, the blast cell B is larger in males, presaging their much more elaborate B lineage, which will give rise to posterior copulatory structures not found in hermaphrodites.

For the adult soma to become sexually dimorphic, the male and hermaphrodite lineages must diverge in favor of sexually distinct programs of development. Sexually specialized cells begin to differ in the extent, timing, and polarity of divisions as well as in their response to axial patterning cues.

Somatic Sexual Differentiation in *C. elegans*

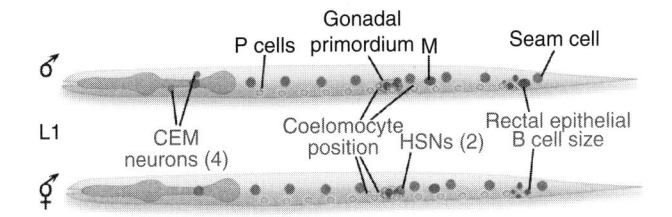

Figure 1.3 Comparison of the newly hatched L1 male and hermaphrodite. Blue labels indicate sex-specific differences established in the embryo. Black labels indicate cells that express different post-embryonic fates or lineages in the male and hermaphrodite. (Lints and Hall, 2005. Adapted with permission from WormAtlas, http://wormatlas.org/). (See Color Insert.)

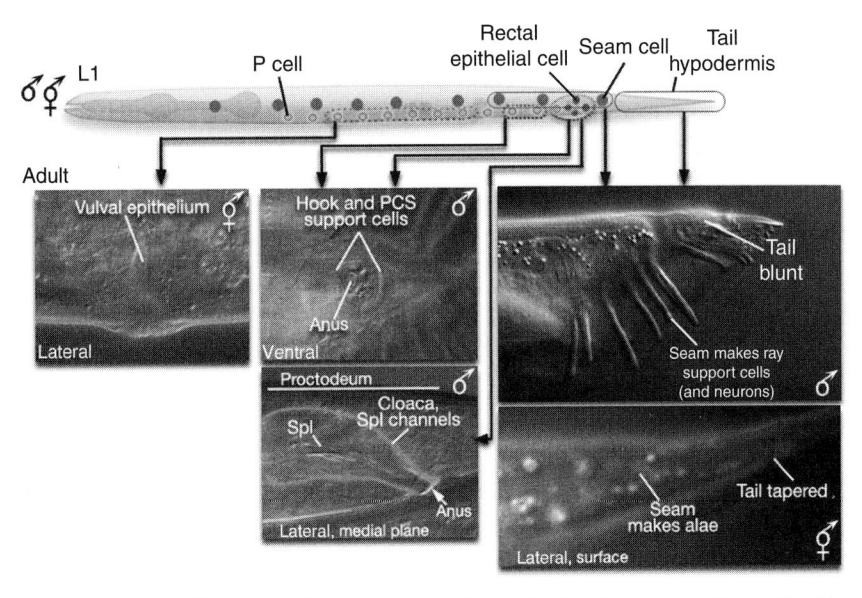

Figure 1.4 Sexually dimorphic structures of the adult arise from larval blast cells. The hermaphrodite vulva and male hook are descendants of the lateral hypodermal P cells. The proctodeum, including spicules and cloaca, is derived from rectal epithelial blast cells. The sensory rays of the male and the alae of the hermaphrodite are descendants of the lateral hypodermal seam cells. PCS, post-cloacal sensilla; Spl, spicules. (Lints and Hall, 2005. Adapted with permission from WormAtlas, http://wormatlas.org/).

In some cases, equivalent progenitors produce sex-specific organs of equal complexity in each sex. In others, structures occur sex-specifically as a result of blast cell division or cell death occurring only in one sex. Below, we highlight some of the more significant sexual dimorphisms and discuss the role that regulators of pattern, the cell cycle, asymmetric division, and apoptosis play in their development; comprehensive reviews of sex-specific development and behavior in the worm are available elsewhere (Barr and

Garcia, 2006; Emmons, 2005; Herman, 2006; Schafer, 2005; Sternberg, 2005). How the global sex determination system exerts overall control over the sexual differentiation of these features is further discussed in Section 5.

4.1. The somatic gonad

The gonads of the adult male and hermaphrodite are highly distinct in structure and function. The J-shaped testis of the male produces and delivers sperm into a posterior cloaca; the two U-shaped ovotestes of the hermaphrodite produce, store, and unite sperm and eggs, delivering fertilized embryos through a central uterus and vulva (Fig. 1.5). In spite of these differences, the gonads are derived from a fundamentally similar cell lineage that is modified sex-specifically at key points in development (Kimble and Hirsh, 1979; Sulston and Horvitz, 1977).

4.1.1. Division of the somatic gonad precursors defines the proximal–distal gonad axis

At hatching, the four-cell gonad appears the same in both sexes, and is arranged with the somatic gonad precursors (SGPs), Z1 and Z4, flanking the two germline precursors, Z2 and Z3. The first division of the SGPs is asymmetric, and defines a gonad axis in which the fates of central daughters (Z1.p and Z4.a) are defined as proximal, and those of the flanking daughters (Z1.a and Z4.p) are defined as distal. The asymmetry of the SGP division depends on Wnt and MAPK signaling factors that specify the distal fate of Z1.a and Z4.p (Siegfried and Kimble, 2002). Activation of the Wnt/MAPK pathway results in reduced nuclear levels of the terminal Wnt pathway regulator POP-1/TCF in distal SGP daughters (Siegfried *et al.*, 2004). When the Wnt/MAPK pathway is inactive, POP-1 is distributed symmetrically to both SGP daughters and both assume a proximal fate (Siegfried and Kimble, 2002; Siegfried *et al.*, 2004). Three genes, *hnd-1*, *ehn-1*, and *ehn-3* are necessary for proper positioning and survival of the SGPs (Mathies *et al.*, 2003). TRA-1 is required sex nonspecifically for proper positioning of the SGPs and symmetry of the SGP divisions, and also functions redundantly with *ehn-3* in SGP division and maturation (Hodgkin, 1987a; Mathies *et al.*, 2004; Schedl *et al.*, 1989).

4.1.2. The SGP division is sexually dimorphic

In both sexes, the mature gonad is oriented such that mitotic germ cells proliferate at the distal pole and the mature gametes exit the gonad through structures derived from the original proximal pole. However, the anatomical constraints of the central vulva in hermaphrodites and the posterior cloaca in males dictate that the shapes of the gonads, and thus the morphogenetic processes that create them, differ dramatically (Kimble and Hirsh, 1979). Indeed, sexual dimorphism is apparent from the onset of SGP division.

Somatic Sexual Differentiation in *C. elegans*

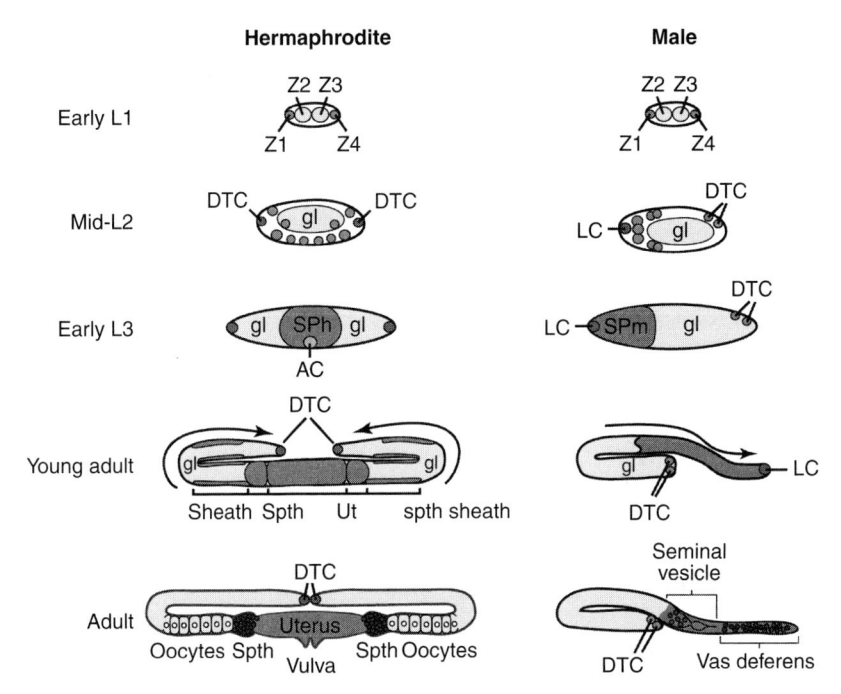

Figure 1.5 Sexually dimorphic gonadal development. At hatching, the four cell gonadal primordia of the male and hermaphrodite appear identical, but undergo sexually dimorphic division, migration, and fate specification to create dissimilar adult structures. The asymmetrical, J-shaped testis of the male includes an anterior mitotic germ line (gl) and posterior vas deferens and seminal vesicle (sv). The bilaterally symmetrical hermaphrodite ovotestes are surrounded by sheath cells, and include a distal mitotic germline (gl) and proximal spermatheca (spth) and uterus (ut). DTC, distal tip cell; LC, linker cell; AC, anchor cell. After Miskowski *et al.* (2001) with permission. (See Color Insert.)

In hermaphrodites, the daughters of the SGPs are of roughly equal size and remain on either end of the gonad. Subsequent divisions and morphogenesis generate a bilaterally symmetrical, two-armed gonad centered on the uterus and vulva. The shape of the gonad, thus, mirrors the geometry of the original proximal–distal axis.

In males, the initial symmetry must be broken to generate the J-shaped testis. This break in symmetry is reflected in male-specific features of the SGP divisions: the proximal SGP daughters are larger than the distal ones and they migrate anteriorly, displacing the distal daughters posteriorly (Kimble and Hirsh, 1979). Subsequent divisions and migrations extend the male somatic gonad toward its posterior connection with the cloaca. The sexually dimorphic aspects of the Z1/Z4 division depend on the forkhead transcription factor FKH-6, which acts downstream of TRA-1 and the sex determination pathway (Chang *et al.*, 2004). In males lacking

FKH-6, the proximal daughters of the SGPs divide symmetrically and do not migrate, mimicking the hermaphrodite pattern of division. Subsequently, FKH-6 and TRA-1 act in parallel to promote proliferation of the male Z1 and Z4 lineages. Both aspects of SGP division asymmetry are linked to the cell cycle: CYD-1/Cyclin D promotes both the FKH-6 mediated male-specific asymmetry in size and WNT/MAPK-mediated asymmetry of SGP fates (Tilmann and Kimble, 2005).

4.1.3. Proximal and distal SGP daughters act as gonad organizers

The establishment of the proximal–distal axis defines cells with distinct roles in organizing the gonad. In both sexes, each distal daughter of Z1 and Z4 produces a distal tip cell (DTC), which maintains the proliferative state of the germ line (Kimble, 1981). In hermaphrodites, the DTCs also lead the outgrowth of the gonad arms along their U-shaped trajectory (Kimble, 1981). The migration of the DTCs requires a network of regulators of cell motility, including global positional cues such as UNC-6/netrin, UNC-129/TGFβ, integrin receptors, and the metalloprotease GON-1 (Blelloch *et al.*, 1999; Colavita *et al.*, 1998; Cram *et al.*, 2006; Hedgecock *et al.*, 1990; Meighan and Schwarzbauer, 2007).

In males, gonadal leader cell function instead resides in a proximal daughter of the SGPs, the linker cell, which guides the gonad in its J-shaped path to the posterior. Although not as well understood as the hermaphrodite DTC migration, linker cell migration is regulated by some of the same molecules, including UNC-6/netrin and GON-1, that guide the DTCs (Blelloch *et al.*, 1999; Hedgecock *et al.*, 1990). Additionally, the Hox transcription factors MAB-5 and LIN-39, which are expressed in both sexes, are required male-specifically for the posterior migration of the linker cell (Clark *et al.*, 1993; Maloof and Kenyon, 1998; Salser *et al.*, 1993; Wang *et al.*, 1993). In *mab-5; lin-39* double mutants, the linker cell migrates to the anterior in a significant proportion of animals (Clark *et al.*, 1993). How these general cues are interpreted sex-specifically is not known.

4.2. The sex muscles

The sexually dimorphic musculature of the worm reflects the unique behaviors performed by each sex (Fig. 1.6). Hermaphrodites use uterine and vulval muscles to expel embryos, whereas males employ a highly specialized set of sex muscles to control movements of copulatory structures and to aid in the locomotory behaviors of mating.

Like the gonads, the sex muscles stem from fundamentally similar lineages that are modified sex-specifically (Sulston and Horvitz, 1977; Sulston *et al.*, 1980). The sex muscles in both sexes are derived from the mesoblast (M). In hermaphrodites, M divides during the first larval stage (L1) to produce two sex myoblasts (SMs), which are precursors of the eight

Somatic Sexual Differentiation in C. elegans

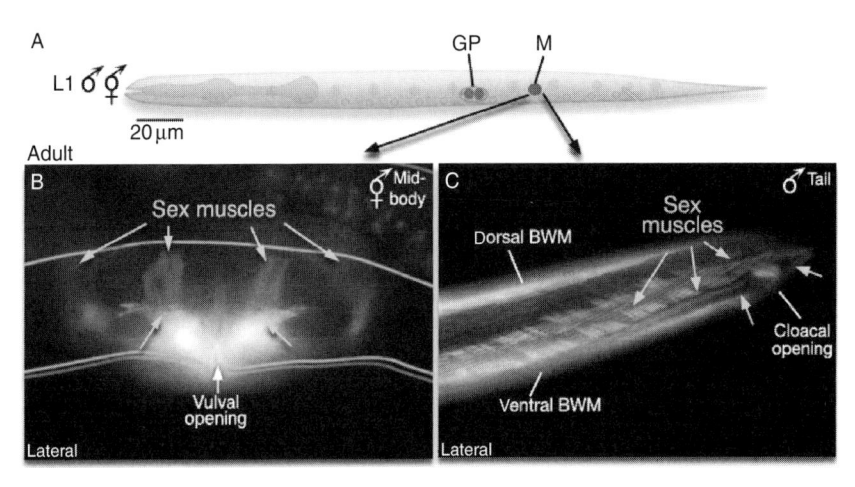

Figure 1.6 Sex muscles. (A) The sex muscles of both sexes are derived from the mesoblast M, which generates two sex myoblasts (SMs) in hermaphrodites and six SMs in males. (B) Vulval and uterine muscles of the adult hermaphrodite, expressing *hlh-1/ hlh-2::gfp* (Strain source: B. Harfe, M. Krause, A. Fire). (C) Diagonal sex muscles of the adult male, expressing *unc-27::gfp* (Strain source: L. Jia and S. Emmons). BWM, body wall muscle; GP gonadal primordium. (Lints and Hall, 2005. Adapted with permission from WormAtlas, http://wormatlas.org/). (See Color Insert.)

uterine and eight vulval muscles, as well as 14 body wall muscles, and two coelomocytes (Sulston and Horvitz, 1977). These early fate decisions among the descendants of M depend on the Hox proteins MAB-5 and LIN-39, which act, in part, by regulating the expression of the myogenic bHLH protein HLH-8 (Harfe et al., 1998; Liu and Fire, 2000). Additionally, a LIN-12/Notch signal is required to distinguish the fates of coelomocytes and SMs (Greenwald et al., 1983).

The hermaphrodite SMs are produced in the posterior and migrate anteriorly during the second larval stage (L2). This migration is guided by attractive and repulsive cues originating in the gonad, as well as a gonad-independent attractive cue (Stern and Horvitz, 1991; Thomas et al., 1990). The attractive cue produced by the gonad is the EGL-17 fibroblast growth factor (FGF), which acts through an FGF receptor (EGL-15) expressed by the SMs (Branda and Stern, 2000; Burdine et al., 1997, 1998; DeVore et al., 1995). The gonad-independent attractive mechanism depends on other conserved mediators of migration, including the distintegrin and metalloprotease UNC-71, the guanine nucleotide exchange factor UNC-73, and UNC-53, a homolog of vertebrate neuronal navigator (NAV) proteins (Chen et al., 1997; Huang et al., 2003; Steven et al., 1998; Stringham et al., 2002).

Male sex muscle development differs from that of hermaphrodites at the point of fate specification of the daughters of M, in the direction of the SM migration, in timing of SM division, and in the structure and function of the sex muscles produced. The initial divisions of M in males are similar to those of the hermaphrodite. However, the two cells that would become coelomocytes in hermaphrodites each give rise to two SMs. As a result, males produce six SMs that are the precursors of 41 sex muscles, including those that control movements of the spicules and gubernaculum and specialized body wall muscles (diagonals, longitudinals, and obliques) (Sulston and Horvitz, 1977; Sulston et al., 1980). In contrast to the hermaphrodite SMs, the male SMs postpone their migration until the L3 larval stage, migrate posteriorly rather than anteriorly, and divide as they migrate.

Regulation of the male M lineage is not well characterized, making a direct comparison of the molecules mediating sex muscle development in the two sexes difficult. In males, as in hermaphrodites, descendants of M are mis-specified in *mab-5* mutants. Thus MAB-5 is important for fate specification among descendants of M in males. MAB-5 additionally regulates SM migration in males; in the absence of *mab-5*, SMs migrate anteriorly (Kenyon, 1986). This finding suggests that at least one cue mediating anterior SM migration is present in males, but is overcome by the activity of MAB-5, the sex determination pathway, or both.

4.3. The ventral hypodermis

The P blast cells (P1–P12) of the ventrolateral hypodermis produce a lineage that diverges into two branches, each producing sex-specific structures. In the posterior lineage, analogous signals mediate development of very different structures, the vulva in hermaphrodites and the copulatory hook in the male tail. In the anterior lineage, subtly different patterns of division and cell death result in motor neurons that innervate sex-specific but analogous muscles.

4.3.1. The posterior P (Pn.p) lineage produces the vulva and the hook

At hatching, the six pairs of P blast cells compose the ventral epidermis of the worm. During L1, these cells migrate ventrally, intercalating into the ventral nerve cord, and divide to produce the Pn.a neuroblasts (see below), and the Pn.p cells, which retain their epidermal character (Sulston and Horvitz, 1977). Soon after this division, Pn.p cells either differentiate and fuse with the hypodermal syncytium, or remain unfused and undifferentiated. In hermaphrodites, unfused cells in the midbody (P3.p–P8.p) produce the vulva (Sulston and Horvitz, 1977). In males, unfused cells in the posterior (P9.p–P11.p) contribute to the sensory hook, a sclerotized copulatory structure that houses two sensory neurons and their supporting socket and sheath cells (Sulston et al., 1980) (Fig. 1.7).

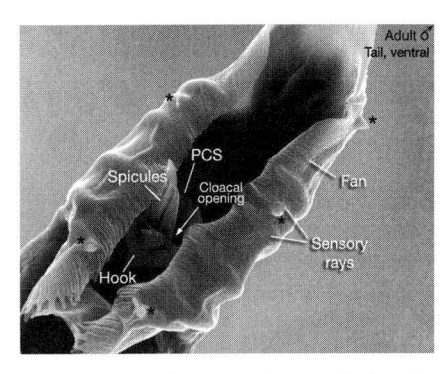

Figure 1.7 The male tail. Scanning electron micrograph showing a ventral view of the copulatory structures of the adult male tail. PCS, postcloacal sensilla; *, Sensory ray opening. (Lints and Hall, 2005. Adapted with permission from WormAtlas, http://wormatlas.org/).

4.3.2. Sex-specific activity of Hox proteins regulates competence in Pn.p cells

In both sexes, the decision to fuse or not to fuse, and thus whether to remain competent to differentiate into vulva or hook, depends on the Hox proteins LIN-39 and MAB-5. In both sexes, LIN-39 is expressed in P3–8.p, and MAB-5 is expressed in P7–11.p (Maloof and Kenyon, 1998; Salser et al., 1993). In hermaphrodites, LIN-39 prevents fusion in its expression domain, whereas MAB-5 is inactive (Clark et al., 1993; Wang et al., 1993). This hermaphrodite-specific inhibition of MAB-5 depends on the chromatin regulator EGL-27 and the bHLH transcription factor REF-1 (Alper and Kenyon, 2001; Ch'ng and Kenyon, 1999). In males, both MAB-5 and LIN-39 are active, and either protein can prevent fusion of Pn.p cells in which it is expressed alone (P3–6.p and P9–11.p) (Salser et al., 1993; Wang et al., 1993). However, in cells where both LIN-39 and MAB-5 are present, each inhibits the other, allowing fusion. Thus, the dimorphism of the cell fusion pattern arises from sex-specific activity of the Hox genes, rather than sex-specific expression.

4.3.3. A LIN-12/Notch signal specifies cell fates within the vulva and hook competence groups

The development of the vulva has served as a seminal model for the definition of fates among a group of equally competent cells (for a detailed review, see Sternberg, 2005). Classic cell ablation studies confirmed that all six unfused Pn.p cells (P3.p–P8.p) are competent to form the vulva; however, vulval fates are normally limited to the descendants of the primary cell, P6.p, and the secondary cells, P5.p and P7.p (Sulston and Horvitz, 1977; Sternberg and Horvitz, 1986). An EGF/MAPK-mediated signal from the gonadal anchor cell defines the primary fate of P6.p (Kimble, 1981;

Sternberg, 2005). The secondary fates of P5.p and P7.p are defined, in turn, by LIN-12/Notch signaling between P6.p and its neighbors. The remaining unfused cells do not normally participate in this inductive cascade, and thus assume a tertiary, nonvulval fate.

In males, the fates of the hook competence group are determined by signaling that is, in part, analogous to the vulva-inducing signal. The primary cell, P11.p, produces hook-associated hypodermal cells, whereas the secondary cell, P10.p, produces the hook sensillum (neurons, socket, and sheath) and a hypodermal cell that secretes the hook's sclerotized cuticle (Sulston *et al.*, 1980). The tertiary cell P9.p is not normally specified as a hook cell. As in the vulva, the delineation of primary and secondary roles depends on a LIN-12/Notch-mediated lateral signal (Greenwald *et al.*, 1983). However, no EGF signal establishing primary fate has been described.

4.3.4. Hox proteins regulate motor neuron fates in the anterior P (Pn.a) lineage

In both sexes, the Pn.a neuroblasts divide during the L1 larval stage to give rise to the postembryonic motor neurons of the ventral nerve cord (Sulston, 1977). These neuroblasts are also responsible for the production of neurons that innervate the sex muscles. In hermaphrodites, six VC motor neurons innervate the vulval muscles. In males, nine CA and ten CP motor neurons innervate muscles used for mating-associated movements (White, 1988).

VC and CA/CP motor neurons arise from homologous cells, the posterior great-granddaughters of the Pn.a neuroblast (Pn.aap, e.g., P3aap). Sex-specificity of cell death, timing, and number of divisions contribute to their sexual dimorphism (Sulston and Horvitz, 1977; Sulston *et al.*, 1980). In hermaphrodites, the Hox protein LIN-39 is required for survival of the six VC precursors (P3–P8.aap) during L1 (Clark *et al.*, 1993). Pn.aap cells distal to the vulva undergo programmed cell death. In males, LIN-39 and MAB-5 act in their respective domains to allow CA and CP precursors (P3–P11.aap) to survive and undergo a final round of cell division during L3 (Clark *et al.*, 1993). In *lin-39*; *mab-5* mutants in both sexes, all Pn.aap cells die. It is not yet known how Hox proteins interact with cell cycle and cell death regulators to regulate development of these sex-specific motor neurons.

4.4. The male tail

Even a glance at the *C. elegans* male and hermaphrodite posterior provides dramatic evidence of sexual specialization in *C. elegans*. Whereas in the hermaphrodite the tail is tapered and the hindgut and epidermis relatively unspecialized, the posterior of the male includes elaborate structures that mediate mating behavior and sperm transfer (Fig. 1.7). The cell lineage of the male tail is an elaboration on a comparatively simple hermaphrodite program of development; most of the posterior mating structures of the male arise from blast cells whose counterparts do not divide postembryonically in hermaphrodites (Sulston et al., 1980).

4.4.1. The hindgut: Proctodeum and copulatory spicules

In contrast to the simple hindgut of the hermaphrodite, the male proctodeum is highly complex. The proctodeum is composed of a cloaca into which both the vas deferens and intestine open and which houses the copulatory spicules. The spicules are sclerotized organs composed of the dendrites of two sensory neurons (SPV, SPD) surrounded by socket and sheath syncytia. The spicules are inserted into the hermaphrodite vulva during mating to facilitate sperm transfer (Liu and Sternberg, 1995). The roof of the cloaca is hardened to form the gubernaculum, which diverts the spicules ventrally toward the cloaca.

4.4.2. Rectal epithelial cells are specified by transcription factors active in both sexes

Despite their dramatic structural and functional differences, the hermaphrodite rectum and the male proctodeum are derived from the same rectal epithelial cells (Fig. 1.8) (Sulston and Horvitz, 1977; Sulston et al., 1980). These cells, called B, F, U, and Y, arise from sex nonspecific embryonic lineages (Sulston et al., 1983). The fates of these cells require several transcription factors, including the T box protein MAB-9, the PAX protein EGL-38, the Zinc finger protein EGL-48, and the Hox protein EGL-5, and

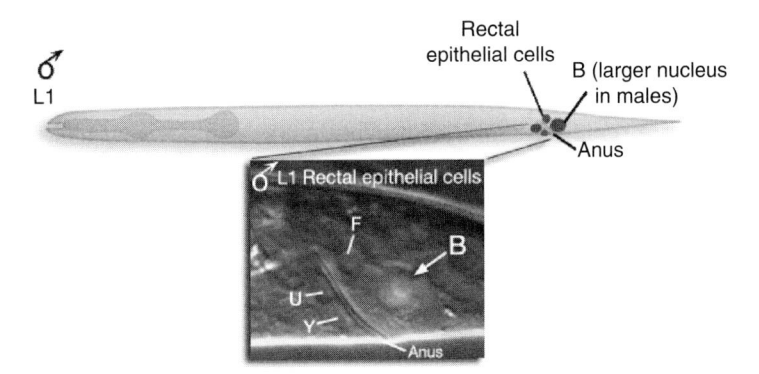

Figure 1.8 Rectal epithelial blast cells. Nomarski micrograph of the rectal epithelial blast cells B, F, U, and Y, in a recently hatched L1 male. Signaling among these cells specifies the fates of reproductive structures in the hindgut. The largest cell, B, produces 42 male-specific progeny, including the copulatory spicules. B, F, U, and Y contribute to the epithelial lining of the cloaca. (Lints and Hall, 2005. Adapted with permission from WormAtlas, http://wormatlas.org/).

loss of these proteins therefore dramatically disrupts development of the proctodeum in males, but causes more subtle rectal defects in hermaphrodites (Chamberlin et al., 1997, 1999; Chisholm, 1991).

4.4.3. Male-specific signals pattern the proctodeum

At hatching, the effects of the sex-determination pathway on the rectal epithelial blast cells become apparent. At this point, divisions of B, F, U, and Y cease in hermaphrodites, but continue in males. The large size of the male B cell portends its elaborate lineage, which will contribute a total of 42 male-specific progeny, notably including all cells of the copulatory spicules (Sulston and Horvitz, 1977; Sulston et al., 1980). Y, F, and U produce fewer cells, contributing with B to the epithelial lining of the cloaca and neurons of the post-cloacal sensilla (Sulston and Horvitz, 1977; Sulston et al., 1980).

The male-specific divisions and differentiation that lead to production of spicules by the B cell depends on intercellular signaling. The first division of B is asymmetric, and requires a Wnt signal (LIN-44) originating in tail hypodermal cells and acting through a planar cell polarity pathway (Herman and Horvitz, 1994; Herman et al., 1995; Wu and Herman, 2006, 2007). The spicules derive from the larger anterior daughter of B (B.a), whose fate requires the PAX transcription factor VAB-3 (Chamberlin and Sternberg, 1995). The posterior daughter (B.p) gives rise to dorsal proctodeum and neurons. The first eight descendants of B.a migrate to form a ring surrounding the rectum, next assorting into anterior and posterior groups of four cells. Cell ablation studies have determined that the fates of these cells depend on a complex network of signals originating from neighboring rectal epithelial cells (Chamberlin and Sternberg, 1993). For example, an EGF signal from F and U involving the LIN-3 ligand and LET-23 receptor helps to define anterior fate (Chamberlin and Sternberg, 1994). The fates of posterior cells are influenced by a signal originating from a descendant of Y (Y.p) (Chamberlin and Sternberg, 1993). Differentiation and morphogenesis of the spicules during the late L4 larval stage requires further cell migration and signaling mediated by components of a TGFβ signaling pathway (Baird and Ellazar, 1999; Jiang and Sternberg, 1999).

4.4.4. Sensory rays

The most distinctive features of the male tail are the nine bilateral pairs of sensory rays, which are projections of the posterior hypodermis that extend inside a thin cuticular covering known as the fan (Fig. 1.9). Each ray contains two sensory neurons and a structural cell, and most open to the exterior environment. Rays are required during mating to establish and maintain contact with the hermaphrodite (Liu and Sternberg, 1995).

Rays are descendants of the hypodermal seam cells, a row of lateral blast cells that remain unfused with the hypodermal syncytium. In both sexes, the anterior seam cells V1–V4 undergo postembryonic self-renewing divisions,

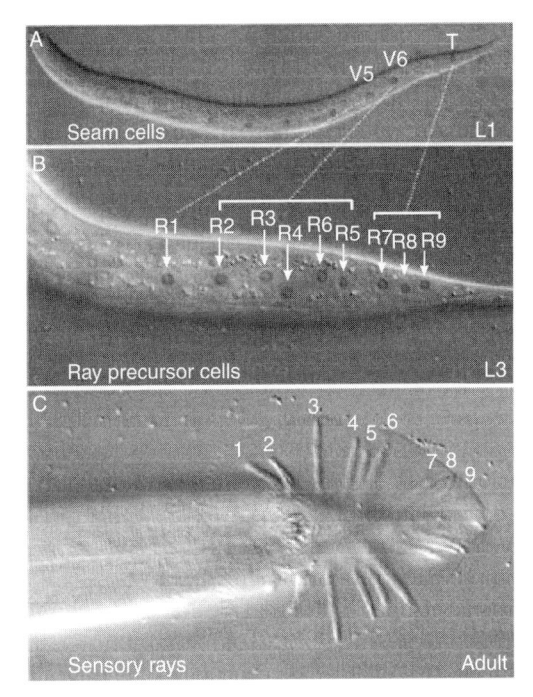

Somatic Sexual Differentiation in *C. elegans* 21

Figure 1.9 Sensory rays are descendants of hypodermal seam cells. (A) L1 worm showing positions of hypodermal seam cell nuclei (highlighted in blue). The posterior seam cells V5, V6, and T give rise to the V and T rays. (B) L3 male tail, showing positions of the nuclei of sensory ray precursor cells, R1–R9 (highlighted in blue). (C) Adult V rays (1–6) and T rays (7–9) are descendants of the sensory ray precursor cells.

differentiating into cuticular ridges called alae at the final molt, and V5 additionally produces a neuroblast known as the postdeirid (Sulston and Horvitz, 1977). In hermaphrodites, V5 and V6 also produce alae, while the T cell generates cells with neuronal and hypodermal fates.

In males, V5, V6, and T are neurogenic, generating the V (rays 1–6) and T (rays 7–9) sensory rays (Sulston and Horvitz, 1977; Sulston *et al.*, 1980). V5, V6, and T divide during the first three larval stages to produce nine ray precursor cells, R1–R9. During the L3 and L4 larval stages, R1–R9 undergo three divisions known as the ray sublineage, ultimately producing the RnA and RnB neurons, and the structural cell Rnst. The rays extend from the body at the end of the L4 larval stage.

4.4.5. LIN-32 promotes neurogenesis in the V5 and V6 seam cells

The production of V rays in males, as well as the postdeirid neuroblast in both sexes, is mediated by the proneural bHLH transcription factors LIN-32, a homolog of *Drosophila* Atonal, and its E/Daughterless binding partner

HLH-2 (Portman and Emmons, 2000; Zhao and Emmons, 1995). LIN-32 is both necessary and sufficient for sensory ray production; *lin-32* mutants lack V rays, while expression of *lin-32* in V1–V4 in either sex causes these seam cells to produce V rays inappropriate (Zhao and Emmons, 1995). The spatial and temporal regulation of *lin-32* activity is therefore key to V ray development.

Genetic screens for males lacking sensory rays have identified components of a regulatory hierarchy that controls *lin-32* expression in the V ray lineages. V ray development is initiated by the Hox protein MAB-5, which acts with a second Hox protein, EGL-5, to promote *lin-32* expression (Emmons, 1999; Ferreira et al., 1999; Yi et al., 2000). MAB-5 expression is initiated in V6 by the homeodomain transcription factor PAL-1 (Cowing and Kenyon, 1992; Hunter et al., 1999). Expression of *mab-5* in V5 is under PAL-1-independent control; however, regulators of MAB-5 in V5 have not yet been identified.

The Hox-mediated specification of V rays is potentiated by MAB-3, a transcription factor homologous to *Drosophila* Doublesex (Dsx) (Raymond et al., 1998). MAB-3 promotes V ray formation by repressing expression of *ref-1*, which encodes an antineural bHLH protein (Alper and Kenyon, 2001; Ross et al., 2005). In *mab-3* mutants, REF-1 is inappropriately expressed in the V ray lineage, where it prevents *lin-32* expression (Ross et al., 2005). As a result, *mab-3* mutant males lack V rays. Thus, the regulation of proneural and antineural bHLH proteins by a conserved sexual regulator promotes male-specific neurogenesis. A second antineural bHLH protein, LIN-22, limits neurogenesis to V5 and V6 by preventing anterior expression of MAB-5 and LIN-32 (Wrischnik and Kenyon, 1997).

4.4.6. Regulators of V ray identity

After ray precursor fate has been specified, a presumptive V ray makes a number of developmental choices before assuming its unique identity. Each ray must undergo a distinct program of morphogenesis that determines its characteristic shape and anteroposterior position, and may express different neurotransmitters than its neighbors (reviewed in Emmons, 2005). Within each ray sublineage, RnA, RnB, and Rnst also adopt distinct characteristics. For example, RnA and RnB in a single ray may express different neurotransmitters and synapse with distinct targets (Lints et al., 2004; Troemel et al., 1995).

In addition to their role in initiating V ray specification, the Hox genes are key regulators of V ray identity. MAB-5 and EGL-5 are each expressed in distinct subsets of Rn descendants (Ferreira et al., 1999; Salser and Kenyon, 1996), and genetic manipulations that alter the dose of Hox genes in the V ray lineage cause transformations of ray identity (Chisholm, 1991; Chow and Emmons, 1994). Based on these observations, it has been proposed that different combinations of MAB-5 and EGL-5 expression in different rays comprise a Hox code that contributes to specification of ray identity (Emmons, 1999;

Ferreira et al., 1999). However, Hox expression patterns alone cannot account for V ray identity; a TGFβ pathway, the Pax-6 homolog MAB-18, and the Dsx-related protein MAB-23 collaborate with the Hox genes to specify the characteristics of individual rays, such as neurotransmitter profiles (Baird et al., 1991; Lints, 2002, 2004; Lints and Emmons, 1999, 2002; Lints et al., 2004; Savage et al., 1996; Zhang and Emmons, 1995).

4.5. The nervous system and sex-specific behavior

The more elaborate copulatory specializations of the male reflect, in part, its much more active role in mating. The hermaphrodite, by contrast, is a passive partner whose primary sex-specific behavior is the expulsion of embryos (reviewed in Schafer, 2005). Analyses of mate searching and copulation in males and egg laying in hermaphrodites have shed light on the control of behavior at the cellular and genetic level. Further, we focus on the sex-specific behaviors of the male.

4.5.1. Male mating behavior

Male mating behavior consists of a series of sub-behaviors beginning with mate searching and culminating in sperm transfer (Fig. 1.10). These sub-behaviors and the cells required for their execution have been extensively analyzed using laser ablation (Liu and Sternberg, 1995). This work broke ground for subsequent genetic analysis aimed at identifying molecules controlling male-specific behaviors (for a detailed review see Barr and Garcia, 2006).

The male's search for a mate is influenced by hermaphrodite-derived cues that attract and retain males (Lipton et al., 2004; Simon and Sternberg, 2002). A recent study demonstrated that males chemotax to the peak of a gradient of hermaphrodite-conditioned medium (White et al., 2007). This attraction to hermaphrodite pheromone requires the male-specific CEM neurons as well as the nonsex-specific AWA and AWC olfactory neurons. Ablation of the CEMs, AWAs, or AWCs in sexually mature males reduces their attraction to hermaphrodite-conditioned medium. Absence of the OSM-9 TRPV (transient receptor protein vanilloid) channel subunit from any of these classes of neurons phenocopies this effect, indicating that an OSM-9-containing channel is a molecular mediator of hermaphrodite detection by these cells. Intriguingly, ablation of CEMs, AWCs, or AWAs prior to sexual maturity does not impair attraction behavior, suggesting that the remaining neurons compensate for those that have been destroyed. In addition to this long-range chemotaxis, a postulated short-range cue retains males in the vicinity of hermaphrodites (Lipton et al., 2004). Retention behavior is regulated by a number of factors including serotonin levels, feeding status, and the presence of the germ line.

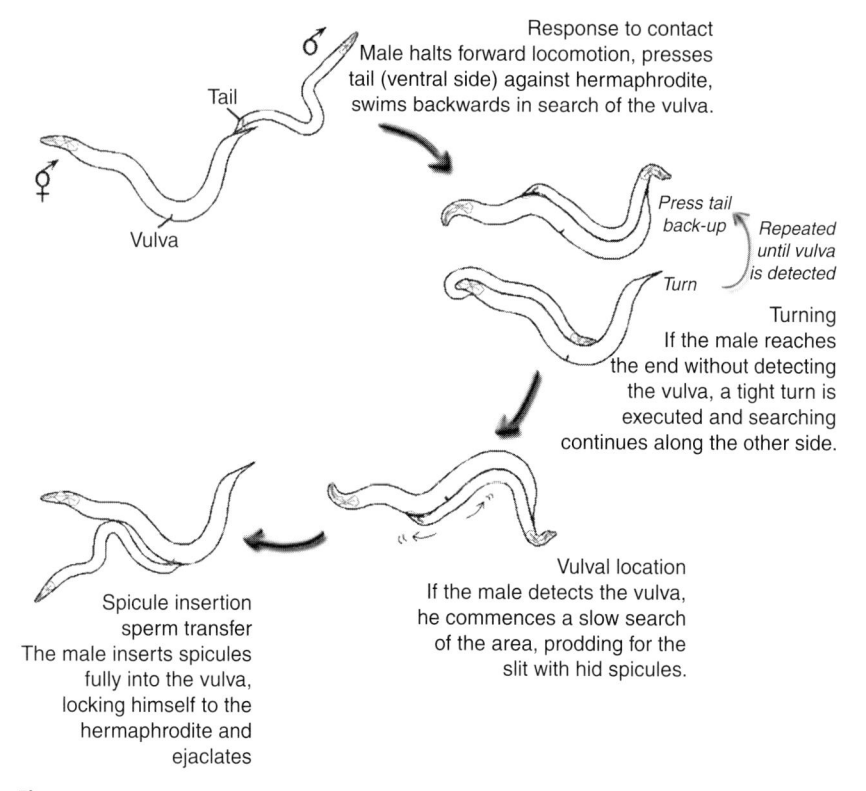

Figure 1.10 Steps of male mating behavior. Based on Loer and Kenyon (1993) and Liu and Sternberg (1995). Adapted with permission from Handbook of *C. elegans* male anatomy, WormAtlas, http://wormatlas.org/.

The next step in mating behavior, response, is initiated when the male's sensory rays contact a hermaphrodite. The male moves backward, scanning the hermaphrodite's body with the ventral side of his tail. Response requires the sensory ray neurons, most of whose dendrites are exposed on the dorsal (rays 1, 5, and 7) or ventral side (rays 2, 4, and 8) of the tail, as well as the hook, post-coacal sensilla, and spicules (Liu and Sternberg, 1995). The polycystin transmembrane proteins LOV-1 and PKD-2, expressed on the sensory cilia of B-type sensory neurons of rays 1–5 and 7–9 and the hook neuron HOB, are necessary for response (Barr and Sternberg, 1999; Barr *et al.*, 2001; Peden and Barr, 2005).

Upon reaching the head or tail of the hermaphrodite, the male makes a tight ventral turn around the end of his partner to continue scanning along the opposite side of the body. Turning requires the T rays (7–9), as well as the CP motor neurons and their targets, the diagonal sex muscles (Liu and Sternberg, 1995; Loer and Kenyon, 1993). The neurotransmitters dopamine and serotonin are key regulators of turning behavior. Serotonin, which is produced by the CP neurons and the B-type neurons of rays 1,3, and 9, is required for normal turning, and induces turning when applied exogenously (Carnell et al., 2005; Lints et al., 2004; Loer and Kenyon, 1993). Males lacking dopamine or the dopaminergic A-type neurons of rays 5, 7, and 9 are deficient in turning (Lints and Emmons, 1999; Liu and Sternberg, 1995). Turning is modulated by male-specific mechanosensory input from the touch receptor neurons (TRNs), which are present in both sexes (Liu et al., 2007). Males lacking normal TRNs or the Deg/EnaC mechanosensory channel encoded by *mec-4* and *mec-10* make multiple attempts at turning, suggesting that the TRNs inhibit repetitive turning. This function of the TRNs requires FMRF-like neuropeptides.

Once the male locates the vulva, he stops and begins to move in synchrony with the hermaphrodite to maintain the precise position of his cloaca over the vulva. Vulval location is mediated by the hook and its associated neurons, HOA and HOB, and requires the activities of PKD-2 and LOV-1 (Barr and Sternberg, 1999; Barr et al., 2001). Contact between the cloaca and the vulva triggers the male to insert his spicules with a rapid prodding movement, mediated by the ryanodine receptor calcium channel UNC-68 (Garcia et al., 2001). This is followed by a prolonged contraction that requires the voltage-gated calcium channel EGL-19. The two stages of contraction are controlled by the spicule protractor muscles and coordinated by the activity of two distinct cholinergic motor inputs (Garcia et al., 2001). Timing of spicule insertion is regulated by the voltage-gated potassium channel UNC-103 (Garcia and Sternberg, 2003). Ejaculation occurs during the prolonged second stage of spicule muscle contraction. The SPV spicule neurons coordinate these activities, as males lacking SPV neurons ejaculate prematurely (Liu and Sternberg, 1995). Both spicule insertion and sperm transfer are also coordinated with general behaviors controlled by the core nervous system. Input from the pharyngeal NSM secretory motor neurons links feeding to spicule protraction and input from the ventral nerve cord motor neurons is required to initiate sperm transfer (Gruninger et al., 2006; Schindelman et al., 2006).

4.6. Sex-specific cell death

In the case of two groups of neurons, sexual dimorphism arises, not by sex-specific cell division, but by sex-specific cell death. The two hermaphrodite-specific neurons (HSNs) and four male-specific CEM neurons are born in both sexes, but survive in only one. Death of the HSNs in males is mediated by the proapoptotic BH3 (Bcl-2 homology) protein EGL-1 (Conradt and Horvitz, 1998, 1999). *egl-1* transcription is repressed in hermaphrodites by TRA-1, allowing the HSNs to survive. The sexually dimorphic survival of the CEMs is controlled by the homeobox gene *ceh-30*, which, like *egl-1*, is likely a direct target of transcriptional regulation by TRA-1 (Schwartz and Horvitz, 2007).

4.7. Covert sexual specialization

While it is clear that a significant proportion of *C. elegans* somatic cells show overt sexual dimorphism, the proportion of cells that are covertly sexually specialized may be quite significant. These are cells that appear identical (or nearly identical) in the two sexes, but express sex-specific genes and perform sex-specific functions. For example, the intestines of the two sexes are superficially alike. However, the hermaphrodite intestine produces abundant yolk proteins (vitellogenins) that are taken up by the gonad and accumulate in oocytes, while the male intestine does not produce yolk (Kimble and Sharrock, 1983). Vitellogenin (*vit*) gene transcription is directly repressed in males by MAB-3 (Shen and Hodgkin, 1988; Yi and Zarkower, 1999). *mab-3* in turn is a direct target of the sex determination pathway in the intestine; TRA-1 represses *mab-3* transcription in hermaphrodites to allow yolk production (Yi and Zarkower, 1999).

Covert sexual specialization also exists in the nervous system, where neurons that are present in both sexes, but have sex-specific connectivity or gene expression, have been identified. The neurons that coordinate general behaviors, such as locomotion and detection of food, with sex-specific behaviors such as egg laying and mating are, by virtue of their synapses with HSNs or male tail neurons, sexually dimorphic (White *et al.*, 1986). There also is evidence that chemosensory neurons express odorant receptors sex-specifically. One such receptor, *srd-1*, is expressed in the ADF sensory neurons of males, but not hermaphrodites (Troemel *et al.*, 1995). Consistent with this sex-specific expression of neuronal gene products, males and hermaphrodites show sexually distinct preferences for volatile attractants (Lee and Portman, 2007). Strikingly, hermaphrodites whose nervous system is genetically masculinized do not develop a significant number of male-specific neurons, but do exhibit male-typical olfactory preferences, suggesting that this sexually dimorphic behavior is mediated by neurons common to both sexes. As further analyses of sex-specific behaviors, gene expression, and nervous system structure are completed, more examples are likely to emerge (Wormatlas male wiring project, http://worms.aecom.yu.edu/).

5. THE INTERFACE BETWEEN SEX DETERMINATION AND SEX DIFFERENTIATION

TRA-1 activity is required for all aspects of somatic sexual differentiation. It is the terminal regulatory gene in the global sex determination pathway, and thus it serves as the molecular interface between the X/A ratio and the sexual regulators that play more specialized roles in specific tissues and lineages.

Because sexual dimorphism is so extensive, TRA-1 may control more cell fates than any other *C. elegans* regulatory factor. Directly or indirectly it must control the activity of many genes, including some of those described in the preceding section. In this section we consider two issues. The first is the mechanism by which TRA-1 regulates the transcription of its direct targets. The second is the identity and nature of the genes functioning downstream of TRA-1.

5.1. Transcriptional regulation by TRA-1

Despite the pervasive role of TRA-1 in sexual regulation, only three direct somatic targets have so far been identified: *mab-3*, whose transcription is regulated by TRA-1 in the intestine; *egl-1*, whose transcription is regulated by TRA-1 in the HSN neurons; and *ceh-30*, whose transcription is regulated by TRA-1 in the CEM neurons. *In vitro* and so far *in vivo*, TRA-1 binds the same DNA recognition element as Ci/GLI proteins (Conradt and Horvitz, 1999; Yi et al., 2000; Zarkower and Hodgkin, 1993). These sites can be in diverse locations: the site in *egl-1* is almost 6 kb downstream of the transcription unit, whereas the site in *mab-3* is 1.5 kb upstream. On its three known targets, TRA-1 acts as a transcriptional repressor, blocking expression in hermaphrodites. As described earlier, TRA-1A is proteolytically cleaved, analogously to CI/GLI proteins. Uncleaved CI and GLI3 have been shown to act as transcriptional activators (Koebernick and Pieler, 2002). It remains an open question whether uncleaved TRA-1A can activate transcription.

Repression by GLI and GLI3 relies on corepressors and recruitment of histone deacetylases (HDACs). No corepressors have been identified for TRA-1, but a cooperating protein complex including the PLZF-like transcription factor TRA-4 has been shown to regulate some aspects of sexual differentiation, particularly in the embryo and early larva (Grote and Conradt, 2006). Genetic epistasis and other tests suggest a model in which a complex containing TRA-4, the histone chaperone NASP, and the HDAC HDA-1 acts in parallel with TRA-1 to repress target gene transcription. At least in *egl-1*, TRA-1 and TRA-4 exert their repression via the same region of DNA. Since TRA-4 also is a sequence-specific C2H2 zinc finger protein, it is possible that TRA-1 and TRA-4 sites will be found in close proximity in a number of target genes.

TRA-1 appears to repress transcription by interfering with nearby enhancer elements (Conradt and Horvitz, 1999; Yi et al., 2000). Such short-range repression is a mechanism by which a global regulator like TRA-1 can act with precision, imposing sex-specific transcription on a gene in some cells or tissues without affecting enhancer elements that control expression elsewhere. It is harder to envision how a widely expressed activator could achieve such selective control in regulating many target genes. Thus short-range repression may provide superior flexibility during evolution by allowing a regulator to "sample" new targets and to fine-tune the expression of others (Cai et al., 1996; Gray et al., 1994; Zarkower, 2001).

5.2. Translating global decisions into local action

The genes acting downstream of TRA-1 lie at the boundary between sex determination and sexual differentiation. A functional distinction between these processes is that failure of sex determination causes the adoption of the inappropriate sexual fate in a cell or tissue, whereas failure in sexual differentiation causes the incomplete adoption of the appropriate fate. The downstream sexual regulators illustrate that the border between these processes can be indistinct. In *mab-3* mutant males the intestine produces yolk and can be regarded as feminized, but the lateral hypodermis in the tail undergoes incomplete male development with no sign of feminization. Similarly, *fkh-6* acts to determine sex in the early XO gonad, but promotes later differentiation in the XX gonad.

Several themes are apparent in downstream regulation of sexual differentiation. A prevalent feature in development of sexually dimorphic structures is the involvement of conserved regulatory modules. For example, Notch signaling and bHLH-mediated neurogenesis are deployed sex-specifically at key points to promote sexual dimorphism in the worm. Another recurring example is that of the Hox proteins, which promote sex-specific gonad and sex muscle migration, competency of hook and vulval precursors in the ventral hypodermis, cell death in neuronal lineages, and specification and patterning of sensory rays. The Hox proteins can be expressed sex-specifically, or exhibit sex-specific activity (Chisholm, 1991; Clark *et al.*, 1993; Ferreira *et al.*, 1999; Kenyon, 1986; Salser and Kenyon, 1996; Salser *et al.*, 1993; Wang *et al.*, 1993). In either case, sex-specificity must depend on an interface between the Hox proteins and the sex-determination pathway. Such is the case in *Drosophila*, where interactions between the *Hox* genes and DSX control sexually dimorphic abdominal pigmentation, proliferation in the genital discs, and embryonic gonad development (DeFalco *et al.*, 2004; Keisman *et al.*, 2001; Kopp *et al.*, 2000; Sanchez *et al.*, 2001). With the exception of REF-1 and EGL-27 in the ventral hypodermal cells, mediators linking the *C. elegans* sex determination pathway to Hox activity are as yet unidentified. Given the importance of other conserved regulators in promoting sexual differentiation in *C. elegans*, it seems likely that their homologs will be found to perform similar functions in the divergence between male and female developmental programs in other animals.

6. SOMETHING OLD, SOMETHING NEW: CONSERVATION OF NEMATODE SEXUAL REGULATORS

Sex is ancient and exceedingly popular as a mode of reproduction, and meiotic regulators are highly conserved from yeast to human. Somatic sexual dimorphism presumably evolved to enhance the efficiency with which gametes are produced and deployed to produce progeny. It may seem puzzling, therefore, given the deep conservation of many meiotic regulators, that the pathways controlling somatic sex in worms, flies, and mammals are highly dissimilar.

At least part of the explanation is that sex determination is a labile process at both the macro and the micro levels. This can be inferred from the diversity of primary sex determining signals employed (Bull, 1983), and the rapid evolution of many sex-determining genes (de Bono and Hodgkin, 1996; Kuwabara, 1996b; O'Neil and Belote, 1992; Whitfield *et al.*, 1993). One likely reason for the fluidity of sex determining mechanisms is that the minimal requirement for a sex determination system is merely that some individuals in a population adopt each sex, and this might be accomplished by a variety of means. This has been experimentally demonstrated by manipulation of the *C. elegans* pathway. In a genetic tour de force, Hodgkin showed that straightforward manipulations, including point mutations in regulatory genes and chromosomal rearrangements, can allow any of seven autosomal sex-determining genes to act as primary sexual determinants and each of the five *C. elegans* autosomes to serve as a sex chromosome (Hodgkin, 2002). There is every reason to expect that similarly dramatic and rapid changes occur outside the laboratory.

Despite the impressive lack of similarity in sex determination pathways between worms, flies, and mammals, there is at least one indication of deep conservation in sex determination, involving the *Doublesex/mab-3* (DM) gene family. These genes encode transcriptional regulators that share a distinctive zinc finger-like DNA binding motif (Raymond *et al.*, 1998; Zhu *et al.*, 2000). The founding member of the gene family, *Doublesex*, controls most aspects of somatic sex determination in flies. In *C. elegans*, *mab-3* and *mab-23* are dedicated regulators of male sexual differentiation, as described earlier, although they play more limited roles than *Dsx*. DM genes regulate sexual development in vertebrates as well, controlling testicular differentiation in mammals and acting as male-determining genes in some fish (Matsuda *et al.*, 2002; Nanda *et al.*, 2002; Raymond *et al.*, 2000). The association of a DM gene with sexual reproduction in a coral (Miller *et al.*, 2003) indicates a pre-Cambrian origin for these genes and possibly for their involvement in sexual development.

Wilkins has proposed that the nematode sex determination pathway formed "bottom up" by successive recruitment of upstream repressors (Wilkins, 1995). This model predicts that genes with older roles in sexual regulation should be found at the bottom of the pathway and newcomers at the top. The presence of ancient DM genes downstream of *tra-1* supports this idea, but in worms these genes are not components of the global pathway. Among the global regulators, evidence is very incomplete, but some of the genes near the bottom of the pathway may have regulated sex for some time. In particular, a functional *tra-1* homolog has been found in a nematode species separated from *C. elegans* by 200–300 MY (Pires-daSilva and Sommer, 2004). A *her-1* homolog is present in the filarial parasite *Brugia malayi*, diverged by as much as 400 MY, but it has not been shown to have a sex determining function (Streit *et al.*, 1999). Intriguingly, human FEM1B is part of a CBC complex and interacts with human FEM-2, though the function of these proteins is unknown (Kamura *et al.*, 2004; Tan *et al.*, 2001). FEM-3 appears to be nematode-specific (Haag *et al.*, 2002).

Similarity of TRA-2 with Patched and TRA-1 with Ci/GLI suggests that a hedgehog signaling module was recruited to control sexual development during nematode evolution and then underwent rapid change, including loss of components such as hedgehog and smoothened, eventually coming under the control of *her-1*, *sdc-2*, and *xol-1*. The ancestral hedgehog pathway has been lost, possibly replaced by other signaling pathways. Studies of sex determination in other nematode species are approaching maturity and should provide clearer glimpses of the origin of the *C. elegans* sex determination pathway. For a recent review of evolution of nematode sex determination, see Haag (2005).

7. The Future

Sex determination was among the first developmental processes for which genetic pathways were defined, in *C. elegans* and in other species. Subsequent molecular and evolutionary analyses have revealed much additional detail and highlighted questions for the future. How do the X-linked and autosomal signal elements interact to regulate *xol-1* with such exquisite dose sensitivity? What is the molecular mechanism by which XOL-1 regulates *sdc-2*? What are the other direct targets of TRA-1? How does sex regulation intersect with spatial/temporal regulation and, in particular, how is Hox protein activity made sex-specific? What was the evolutionary history of the current *C. elegans* sex-determining pathway?

ACKNOWLEDGMENTS

We are grateful to Drs. Patricia Kuwabara and Barbara Meyer and the WormAtlas Project (David Hall, Robyn Lints, Zeynep Altun) for generously providing illustrations, and to Hillel Schwartz and Dr. H. Robert Horvitz for sharing data prior to publication.

REFERENCES

Ahringer, J., Rosenquist, T. A., Lawson, D. N., and Kimble, J. (1992). The *Caenorhabditis elegans* sex determining gene fem-3 is regulated post-transcriptionally. *EMBO J.* **11**, 2303–2310.
Akerib, C. C., and Meyer, B. J. (1994). Identification of X chromosome regions in *Caenorhabditis elegans* that contain sex-determination signal elements. *Genetics* **138**, 1105–1125.
Alper, S., and Kenyon, C. (2001). REF-1, a protein with two bhlh domains, alters the pattern of cell fusion in *C. elegans* by regulating Hox protein activity. *Development* **128**, 1793–1804.
Baird, S. E., and Ellazar, S. A. (1999). Tgfbeta-like signaling and spicule development in *Caenorhabditis elegans*. *Dev. Biol.* **212**, 93–100.
Baird, S. E., Fitch, D. H., Kassem, I. A., and Emmons, S. W. (1991). Pattern formation in the nematode epidermis: Determination of the arrangement of peripheral sense organs in the *C. elegans* male tail. *Development* **113**, 515–526.
Barnes, T. M., and Hodgkin, J. (1996). The tra-3 sex determination gene of *Caenorhabditis elegans* encodes a member of the calpain regulatory protease family. *EMBO J.* **15**, 4477–4484.
Barr, M. M., and Garcia, L. R. (2006). Male mating behavior. "WormBook".
Barr, M. M., and Sternberg, P. W. (1999). A polycystic kidney-disease gene homologue required for male mating behaviour in *C. elegans*. *Nature* **401**, 386–389.
Barr, M. M., DeModena, J., Braun, D., Nguyen, C. Q., Hall, D. H., and Sternberg, P. W. (2001). The *Caenorhabditis elegans* autosomal dominant polycystic kidney disease gene homologs lov-1 and pkd-2 act in the same pathway. *Curr. Biol.* **11**, 1341–1346.
Blelloch, R., Anna-Arriola, S. S., Gao, D., Li, Y., Hodgkin, J., and Kimble, J. (1999). The gon-1 gene is required for gonadal morphogenesis in *Caenorhabditis elegans*. *Dev. Biol.* **216**, 382–393.
Branda, C. S., and Stern, M. J. (2000). Mechanisms controlling sex myoblast migration in *Caenorhabditis elegans* hermaphrodites. *Dev. Biol.* **226**, 137–151.
Bull, J. J. (1983). "The Evolution of Sex-Determining Mechanisms." Benjamin/Cummings, Menlo Park, CA.
Burdine, R. D., Chen, E. B., Kwok, S. F., and Stern, M. J. (1997). Egl-17 encodes an invertebrate fibroblast growth factor family member required specifically for sex myoblast migration in *Caenorhabditis elegans*. *Proc. Natl. Acad. Sci. USA* **94**, 2433–2437.
Burdine, R. D., Branda, C. S., and Stern, M. J. (1998). EGL-17(FGF) expression coordinates the attraction of the migrating sex myoblasts with vulval induction in *C. elegans*. *Development* **125**, 1083–1093.
Cai, H. N., Arnosti, D. N., and Levine, M. (1996). Long-range repression in the *Drosophila* embryo. *Proc. Natl. Acad. Sci. USA* **93**, 9309–9314.
Carmi, I., and Meyer, B. J. (1999). The primary sex determination signal of *Caenorhabditis elegans*. *Genetics* **152**, 999–1015.
Carmi, I., Kopczynski, J. B., and Meyer, B. J. (1998). The nuclear hormone receptor SEX-1 is an X-chromosome signal that determines nematode sex. *Nature* **396**, 168–173.
Carnell, L., Illi, J., Hong, S. W., and McIntire, S. L. (2005). The G-protein-coupled serotonin receptor SER-1 regulates egg laying and male mating behaviors in *Caenorhabditis elegans*. *J. Neurosci.* **25**, 10671–10681.
Ch'ng, Q., and Kenyon, C. (1999). Egl-27 generates anteroposterior patterns of cell fusion in *C. elegans* by regulating Hox gene expression and Hox protein function. *Development* **126**, 3303–3312.
Chamberlin, H. M., and Sternberg, P. W. (1993). Multiple cell interactions are required for fate specification during male spicule development in *Caenorhabditis elegans*. *Development* **118**, 297–324.

Chamberlin, H. M., and Sternberg, P. W. (1994). The lin-3/let-23 pathway mediates inductive signalling during male spicule development in *Caenorhabditis elegans*. *Development* **120**, 2713–2721.

Chamberlin, H. M., and Sternberg, P. W. (1995). Mutations in the *Caenorhabditis elegans* gene vab-3 reveal distinct roles in fate specification and unequal cytokinesis in an asymmetric cell division. *Dev. Biol.* **170**, 679–689.

Chamberlin, H. M., Palmer, R. E., Newman, A. P., Sternberg, P. W., Baillie, D. L., and Thomas, J. H. (1997). The PAX gene egl-38 mediates developmental patterning in *Caenorhabditis elegans*. *Development* **124**, 3919–3928.

Chamberlin, H. M., Brown, K. B., Sternberg, P. W., and Thomas, J. H. (1999). Characterization of seven genes affecting *Caenorhabditis elegans* hindgut development. *Genetics* **153**, 731–742.

Chang, W., Tilmann, C., Thoemke, K., Markussen, F. H., Mathies, L. D., Kimble, J., and Zarkower, D. (2004). A forkhead protein controls sexual identity of the *C. elegans* male somatic gonad. *Development* **131**, 1425–1436.

Chasnov, J. R., So, W. K., Chan, C. M., and Chow, K. L. (2007). The species, sex, and stage specificity of a *Caenorhabditis* sex pheromone. *Proc. Natl. Acad. Sci. USA* **104**, 6730–6735.

Chen, H., Thiagalingam, A., Chopra, H., Borges, M. W., Feder, J. N., Nelkin, B. D., Baylin, S. B., and Ball, D. W. (1997). Conservation of the *Drosophila* lateral inhibition pathway in human lung cancer: A hairy-related protein (HES-1) directly represses achaete-scute homolog-1 expression. *Proc. Natl. Acad. Sci. USA* **94**, 5355–5360.

Chin-Sang, I. D., and Spence, A. M. (1996). *Caenorhabditis elegans* sex-determining protein FEM-2 is a protein phosphatase that promotes male development and interacts directly with FEM-3. *Genes Dev.* **10**, 2314–2325.

Chisholm, A. (1991). Control of cell fate in the tail region of *C. elegans* by the gene egl-5. *Development* **111**, 921–932.

Chow, K. L., and Emmons, S. W. (1994). HOM-C/Hox genes and four interacting loci determine the morphogenetic properties of single cells in the nematode male tail. *Development* **120**, 2579–2592.

Chu, D. S., Dawes, H. E., Lieb, J. D., Chan, R. C., Kuo, A. F., and Meyer, B. J. (2002). A molecular link between gene-specific and chromosome-wide transcriptional repression. *Genes Dev.* **16**, 796–805.

Clark, S. G., Chisholm, A. D., and Horvitz, H. R. (1993). Control of cell fates in the central body region of *C. elegans* by the homeobox gene lin-39. *Cell* **74**, 43–55.

Colavita, A., Krishna, S., Zheng, H., Padgett, R. W., and Culotti, J. G. (1998). Pioneer axon guidance by UNC-129, a *C. elegans* TGF-beta. *Science* **281**, 706–709.

Conradt, B., and Horvitz, H. R. (1998). The *C. elegans* protein EGL-1 is required for programmed cell death and interacts with the Bcl-2-like protein CED-9. *Cell* **93**, 519–529.

Conradt, B., and Horvitz, H. R. (1999). The TRA-1A sex determination protein of *C. elegans* regulates sexually dimorphic cell deaths by repressing the egl-1 cell death activator gene. *Cell* **98**, 317–327.

Cowing, D. W., and Kenyon, C. (1992). Expression of the homeotic gene mab-5 during *Caenorhabditis elegans* embryogenesis. *Development* **116**, 481–490.

Cram, E. J., Shang, H., and Schwarzbauer, J. E. (2006). A systematic RNA interference screen reveals a cell migration gene network in *C. elegans*. *J. Cell Sci.* **119**, 4811–4818.

Csankovszki, G., McDonel, P., and Meyer, B. J. (2004). Recruitment and spreading of the *C. elegans* dosage compensation complex along X chromosomes. *Science* **303**, 1182–1185.

Dawes, H. E., Berlin, D. S., Lapidus, D. M., Nusbaum, C., Davis, T. L., and Meyer, B. J. (1999). Dosage compensation proteins targeted to X chromosomes by a determinant of hermaphrodite fate. *Science* **284**, 1800–1804.

de Bono, M., and Hodgkin, J. (1996). Evolution of sex determination in *Caenorhabditis*: Unusually high divergence of tra-1 and its functional consequences. *Genetics* **144**, 587–595.

de Bono, M., Zarkower, D., and Hodgkin, J. (1995). Dominant feminizing mutations implicate protein–protein interactions as the main mode of regulation of the nematode sex-determining gene tra-1. *Genes Dev.* **9**, 155–167.

DeFalco, T., Le Bras, S., and Van Doren, M. (2004). Abdominal-B is essential for proper sexually dimorphic development of the *Drosophila* gonad. *Mech. Dev.* **121**, 1323–1333.

DeVore, D. L., Horvitz, H. R., and Stern, M. J. (1995). An FGF receptor signaling pathway is required for the normal cell migrations of the sex myoblasts in *C. elegans* hermaphrodites. *Cell* **83**, 611–620.

Doniach, T. (1986). Activity of the sex-determining gene tra-2 is modulated to allow spermatogenesis in the *C. elegans* hermaphrodite. *Genetics* **114**, 53–76.

Emmons, S. W. (1999). Cell fate determination in *Caenorhabditis elegans* ray development. "Cell Lineage and Fate Determination" (S. A. Moody, Ed.), Vol. 1, pp. 139–155. Academic Press, San Diego.

Emmons, S. W. (2005). Male development. "WormBook" (T. C. e. R. Community, Ed.)

Ferreira, H. B., Zhang, Y., Zhao, C., and Emmons, S. W. (1999). Patterning of *Caenorhabditis elegans* posterior structures by the abdominal-B homolog, egl-5. *Dev. Biol.* **207**, 215–228.

Garcia, L. R., and Sternberg, P. W. (2003). *Caenorhabditis elegans* UNC-103 ERG-like potassium channel regulates contractile behaviors of sex muscles in males before and during mating. *J. Neurosci.* **23**, 2696–2705.

Garcia, L. R., Mehta, P., and Sternberg, P. W. (2001). Regulation of distinct muscle behaviors controls the *C. elegans* male's copulatory spicules during mating. *Cell* **107**, 777–788.

Gaudet, J., VanderElst, I., and Spence, A. M. (1996). Post-transcriptional regulation of sex determination in *Caenorhabditis elegans*: Widespread expression of the sex-determining gene fem-1 in both sexes. *Mol. Biol. Cell* **7**, 1107–1121.

Goodwin, E. B., Okkema, P. G., Evans, T. C., and Kimble, J. (1993). Translational regulation of tra-2 by its 3′ untranslated region controls sexual identity in *C. elegans*. *Cell* **75**, 329–339.

Gray, S., Szymanski, P., and Levine, M. (1994). Short-range repression permits multiple enhancers to function autonomously within a complex promoter. *Genes Dev.* **8**, 1829–1838.

Greenwald, I. S., Sternberg, P. W., and Horvitz, H. R. (1983). The lin-12 locus specifies cell fates in *Caenorhabditis elegans*. *Cell* **34**, 435–444.

Grote, P., and Conradt, B. (2006). The PLZF-like protein TRA-4 cooperates with the Gli-like transcription factor TRA-1 to promote female development in *C. elegans*. *Dev. Cell* **11**, 561–573.

Gruninger, T. R., Gualberto, D. G., LeBoeuf, B., and Garcia, L. R. (2006). Integration of male mating and feeding behaviors in *Caenorhabditis elegans*. *J. Neurosci.* **26**, 169–179.

Haag, E. S. (2005). The evolution of nematode sex determination: *C. elegans* as a reference point for comparative biology. *The C. elegans Research Community*.

Haag, E. S., Wang, S., and Kimble, J. (2002). Rapid coevolution of the nematode sex-determining genes fem-3 and tra-2. *Curr. Biol.* **12**, 2035–2041.

Hamaoka, B. Y., Dann, C. E., III, Geisbrecht, B. V., and Leahy, D. J. (2004). Crystal structure of *Caenorhabditis elegans* HER-1 and characterization of the interaction between HER-1 and TRA-2A. *Proc. Natl. Acad. Sci. USA* **101**, 11673–11678.

Harfe, B. D., Vaz Gomes, A., Kenyon, C., Liu, J., Krause, M., and Fire, A. (1998). Analysis of a *Caenorhabditis elegans* twist homolog identifies conserved and divergent aspects of mesodermal patterning. *Genes Dev.* **12**, 2623–2635.

Hedgecock, E. M., Culotti, J. G., and Hall, D. H. (1990). The unc-5, unc-6, and unc-40 genes guide circumferential migrations of pioneer axons and mesodermal cells on the epidermis in *C. elegans*. *Neuron* **4,** 61–85.

Herman, M. A. (2006). Hermaphrodite cell-fate specification. "WormBook" (T. C. e. R. Community, Ed.).

Herman, M. A., and Horvitz, H. R. (1994). The *Caenorhabditis elegans* gene lin-44 controls the polarity of asymmetric cell divisions. *Development* **120,** 1035–1047.

Herman, M. A., Vassilieva, L. L., Horvitz, H. R., Shaw, J. E., and Herman, R. K. (1995). The *C. elegans* gene lin-44, which controls the polarity of certain asymmetric cell divisions, encodes a Wnt protein and acts cell nonautonomously. *Cell* **83,** 101–110.

Hodgkin, J. (1980). More sex-determination mutants of *Caenorhabditis elegans*. *Genetics* **96,** 649–664.

Hodgkin, J. (1987a). A genetic analysis of the sex-determining gene, tra-1, in the nematode *Caenorhabditis elegans*. *Genes Dev.* **1,** 731–745.

Hodgkin, J. (1987b). Sex determination and dosage compensation in *Caenorhabditis elegans*. *Annu. Rev. Genet.* **21,** 133–154.

Hodgkin, J. (2002). Exploring the envelope. Systematic alteration in the sex-determination system of the nematode *caenorhabditis elegans*. *Genetics* **162,** 767–780.

Hodgkin, J. A., and Brenner, S. (1977). Mutations causing transformation of sexual phenotype in the nematode *Caenorhabditis elegans*. *Genetics* **86,** 275–287.

Hodgkin, J., Zellan, J. D., and Albertson, D. G. (1994). Identification of a candidate primary sex determination locus, fox-1, on the X chromosome of *Caenorhabditis elegans*. *Development* **120,** 3681–3689.

Huang, X., Huang, P., Robinson, M. K., Stern, M. J., and Jin, Y. (2003). UNC-71, a disintegrin and metalloprotease (ADAM) protein, regulates motor axon guidance and sex myoblast migration in *C. elegans*. *Development* **130,** 3147–3161.

Hunter, C. P., and Wood, W. B. (1992). Evidence from mosaic analysis of the masculinizing gene her-1 for cell interactions in *C. elegans* sex determination. *Nature* **355,** 551–555.

Hunter, C. P., Harris, J. M., Maloof, J. N., and Kenyon, C. (1999). Hox gene expression in a single *Caenorhabditis elegans* cell is regulated by a caudal homolog and intercellular signals that inhibit wnt signaling. *Development* **126,** 805–814.

Jager, S., Schwartz, H. T., Horvitz, H. R., and Conradt, B. (2004). The *Caenorhabditis elegans* F-box protein SEL-10 promotes female development and may target FEM-1 and FEM-3 for degradation by the proteasome. *Proc. Natl. Acad. Sci. USA* **101,** 12549–12554.

Jiang, L. I., and Sternberg, P. W. (1999). Socket cells mediate spicule morphogenesis in *Caenorhabditis elegans* males. *Dev. Biol.* **211,** 88–99.

Kamura, T., Maenaka, K., Kotoshiba, S., Matsumoto, M., Kohda, D., Conaway, R. C., Conaway, J. W., and Nakayama, K. I. (2004). VHL-box and SOCS-box domains determine binding specificity for Cul2-Rbx1 and Cul5-Rbx2 modules of ubiquitin ligases. *Genes Dev.* **18,** 3055–3065.

Keisman, E. L., Christiansen, A. E., and Baker, B. S. (2001). The sex determination gene doublesex regulates the A/P organizer to direct sex-specific patterns of growth in the *Drosophila* genital imaginal disc. *Dev. Cell* **1,** 215–225.

Kenyon, C. (1986). A gene involved in the development of the posterior body region of *C. elegans*. *Cell* **46,** 477–487.

Kimble, J. (1981). Alterations in cell lineage following laser ablation of cells in the somatic gonad of *Caenorhabditis elegans*. *Dev. Biol.* **87,** 286–300.

Kimble, J., and Hirsh, D. (1979). The postembryonic cell lineages of the hermaphrodite and male gonads in *Caenorhabditis elegans*. *Dev. Biol.* **70,** 396–417.

Kimble, J., and Sharrock, W. J. (1983). Tissue-specific synthesis of yolk proteins in *Caenorhabditis elegans*. *Dev. Biol.* **96,** 189–196.

Klein, R. D., and Meyer, B. J. (1993). Independent domains of the Sdc-3 protein control sex determination and dosage compensation in *C. elegans*. *Cell* **72**, 349–364.

Koebernick, K., and Pieler, T. (2002). Gli-type zinc finger proteins as bipotential transducers of Hedgehog signaling. *Differentiation* **70**, 69–76.

Kopp, A., Duncan, I., and Carroll, S. B. (2000). Genetic control and evolution of sexually dimorphic characters in *Drosophila*. *Nature* **408**, 553–559.

Kuwabara, P. E. (1996a). A novel regulatory mutation in the *C. elegans* sex determination gene tra-2 defines a candidate ligand/receptor interaction site. *Development* **122**, 2089–2098.

Kuwabara, P. E. (1996b). Interspecies comparison reveals evolution of control regions in the nematode sex-determining gene tra-2. *Genetics* **144**, 597–607.

Kuwabara, P. E. (2007). A complex solution to a sexual dilemma. *Dev. Cell* **13**, 6–8.

Kuwabara, P. E., Lee, M. H., Schedl, T., and Jefferis, G. S. (2000). A *C. elegans* patched gene, ptc-1, functions in germ-line cytokinesis. *Genes Dev.* **14**, 1933–1944.

Lee, K., and Portman, D. S. (2007). Neural sex modifies the function of a *C. elegans* sensory circuit. *Curr. Biol.* **17**, 1858–1863.

Lints, R., and Emmons, S. W. (1999). Patterning of dopaminergic neurotransmitter identity among *Caenorhabditis elegans* ray sensory neurons by a tgfbeta family signaling pathway and a Hox gene. *Development* **126**, 5819–5831.

Lints, R., and Emmons, S. W. (2002). Regulation of sex-specific differentiation and mating behavior in *C. elegans* by a new member of the DM domain transcription factor family. *Genes Dev.* **16**, 2390–2402.

Lints, R. and Hall, D. H. (2005). Handbook of *C. elegans* Male Anatomy. In WormAtlas. http://www.wormatlas.org/handbook/contents.htm.

Lints, R., Jia, L., Kim, K., Li, C., and Emmons, S. W. (2004). Axial patterning of *C. elegans* male sensilla identities by selector genes. *Dev. Biol.* **269**, 137–151.

Lipton, J., Kleemann, G., Ghosh, R., Lints, R., and Emmons, S. W. (2004). Mate searching in *Caenorhabditis elegans*: A genetic model for sex drive in a simple invertebrate. *J. Neurosci.* **24**, 7427–7434.

Liu, J., and Fire, A. (2000). Overlapping roles of two Hox genes and the exd ortholog ceh-20 in diversification of the *C. elegans* postembryonic mesoderm. *Development* **127**, 5179–5190.

Liu, K. S., and Sternberg, P. W. (1995). Sensory regulation of male mating behavior in *Caenorhabditis elegans*. *Neuron* **14**, 79–89.

Liu, T., Kim, K., Li, C., and Barr, M. M. (2007). Fmrfamide-like neuropeptides and mechanosensory touch receptor neurons regulate male sexual turning behavior in *Caenorhabditis elegans*. *J. Neurosci.* **27**, 7174–7182.

Loer, C. M., and Kenyon, C. J. (1993). Serotonin-deficient mutants and male mating behavior in the nematode *Caenorhabditis elegans*. *J. Neurosci.* **13**, 5407–5417.

Lum, D. H., Kuwabara, P. E., Zarkower, D., and Spence, A. M. (2000). Direct protein-protein interaction between the intracellular domain of TRA-2 and the transcription factor TRA-1A modulates feminizing activity in *C. elegans*. *Genes Dev.* **14**, 3153–3165.

Luz, J. G., Hassig, C. A., Pickle, C., Godzik, A., Meyer, B. J., and Wilson, I. A. (2003). XOL-1, primary determinant of sexual fate in *C. elegans*, is a GHMP kinase family member and a structural prototype for a class of developmental regulators. *Genes Dev.* **17**, 977–990.

Madl, J. E., and Herman, R. K. (1979). Polyploids and sex determination in *Caenorhabditis elegans*. *Genetics* **93**, 393–402.

Maloof, J. N., and Kenyon, C. (1998). The Hox gene lin-39 is required during *C. elegans* vulval induction to select the outcome of Ras signaling. *Development* **125**, 181–190.

Mathies, L. D., Henderson, S. T., and Kimble, J. (2003). The *C. elegans* Hand gene controls embryogenesis and early gonadogenesis. *Development* **130**, 2881–2892.

Mathies, L. D., Schvarzstein, M., Morphy, K. M., Blelloch, R., Spence, A. M., and Kimble, J. (2004). TRA-1/GLI controls development of somatic gonadal precursors in *C. elegans*. *Development* **131**, 4333–4343.

Matsuda, M., Nagahama, Y., Shinomiya, A., Sato, T., Matsuda, C., Kobayashi, T., Morrey, C. E., Shibata, N., Asakawa, S., Shimizu, N., Hori, H., Hamaguchi, S., et al. (2002). DMY is a Y-specific DM-domain gene required for male development in the medaka fish. *Nature* **417**, 559–563.

McDonel, P., Jans, J., Peterson, B. K., and Meyer, B. J. (2006). Clustered DNA motifs mark X chromosomes for repression by a dosage compensation complex. *Nature* **444**, 614–618.

Mehra, A., Gaudet, J., Heck, L., Kuwabara, P. E., and Spence, A. M. (1999). Negative regulation of male development in *Caenorhabditis elegans* by a protein–protein interaction between TRA-2A and FEM-3. *Genes Dev.* **13**, 1453–1463.

Meighan, C. M., and Schwarzbauer, J. E. (2007). Control of *C. elegans* hermaphrodite gonad size and shape by vab-3/Pax6-mediated regulation of integrin receptors. *Genes Dev.* **21**, 1615–1620.

Meyer, B. J. (2005). X-chromosome dosage compensation. "Wormbook".

Meyer, B. J., and Casson, L. P. (1986). *Caenorhabditis elegans* compensates for the difference in X chromosome dosage between the sexes by regulating transcript levels. *Cell* **47**, 871–881.

Miller, S. W., Hayward, D. C., Bunch, T. A., Miller, D. J., Ball, E. E., Bardwell, V. J., Zarkower, D., and Brower, D. L. (2003). A DM domain protein from a coral, Acropora millepora, homologous to proteins important for sex determination. *Evol. Dev.* **5**, 251–258.

Miskowski, J., Li, Y., and Kimble, J. (2001). The *sys-1* gene and sexual dimorphism during gonadogenesis in *Caenorhabditis elegans*. *Dev. Biol.* **230**, 61–73.

Nanda, I., Kondo, M., Hornung, U., Asakawa, S., Winkler, C., Shimizu, A., Shan, Z., Haaf, T., Shimizu, N., Shima, A., Schmid, M., and Schartl, M. (2002). A duplicated copy of DMRT1 in the sex-determining region of the Y chromosome of the medaka, Oryzias latipes. *Proc. Natl. Acad. Sci. USA* **99**, 11778–11783.

Nicoll, M., Akerib, C. C., and Meyer, B. J. (1997). X-chromosome-counting mechanisms that determine nematode sex. *Nature* **388**, 200–204.

Nigon, V. (1949). Les modalities de la reproduction et le determinisme de sexe chez quelques Nematodes libres. *Ann. Sci. Nat. Zool. Ser.* **11**, 1–132.

Nigon, V. (1951). Polyploidie experimentale chez un Nematode libre, *Rhabditis elegans* Maupas. *Bull. Biol. Fr. Belg.* **85**, 187–225.

O'Neil, M. T., and Belote, J. M. (1992). Interspecific comparison of the transformer gene of *Drosophila* reveals an unusually high degree of evolutionary divergence. *Genetics* **131**, 113–128.

Peden, E. M., and Barr, M. M. (2005). The KLP-6 kinesin is required for male mating behaviors and polycystin localization in *Caenorhabditis elegans*. *Curr. Biol.* **15**, 394–404.

Perry, M. D., Li, W., Trent, C., Robertson, B., Fire, A., Hageman, J. M., and Wood, W. B. (1993). Molecular characterization of the her-1 gene suggests a direct role in cell signaling during *Caenorhabditis elegans* sex determination. *Genes Dev.* **7**, 216–228.

Perry, M. D., Trent, C., Robertson, B., Chamblin, C., and Wood, W. B. (1994). Sequenced alleles of the *Caenorhabditis elegans* sex-determining gene her-1 include a novel class of conditional promoter mutations. *Genetics* **138**, 317–327.

Pilgrim, D., McGregor, A., Jackle, P., Johnson, T., and Hansen, D. (1995). The *C. elegans* sex-determining gene fem-2 encodes a putative protein phosphatase. *Mol. Biol. Cell* **6**, 1159–1171.

Pires-daSilva, A., and Sommer, R. J. (2004). Conservation of the global sex determination gene tra-1 in distantly related nematodes. *Genes Dev.* **18**, 1198–1208.

Portman, D. S., and Emmons, S. W. (2000). The basic helix-loop-helix transcription factors LIN-32 and HLH-2 function together in multiple steps of a *C. elegans* neuronal sublineage. *Development* **127,** 5415–5426.

Powell, J. R., Jow, M. M., and Meyer, B. J. (2005). The T-box transcription factor SEA-1 is an autosomal element of the X:A signal that determines *C. elegans* sex. *Dev. Cell* **9,** 339–349.

Raymond, C. S., Shamu, C. E., Shen, M. M., Seifert, K. J., Hirsch, B., Hodgkin, J., and Zarkower, D. (1998). Evidence for evolutionary conservation of sex-determining genes. *Nature* **391,** 691–695.

Raymond, C. S., Murphy, M. W., O'Sullivan, M. G., Bardwell, V. J., and Zarkower, D. (2000). Dmrt1, a gene related to worm and fly sexual regulators, is required for mammalian testis differentiation. *Genes Dev.* **14,** 2587–2595.

Ross, J. M., Kalis, A. K., Murphy, M. W., and Zarkower, D. (2005). The DM domain protein MAB-3 promotes sex-specific neurogenesis in *C. elegans* by regulating bhlh proteins. *Dev. Cell* **8,** 881–892.

Salser, S. J., and Kenyon, C. (1996). A *C. elegans* hox gene switches on, off, on and off again to regulate proliferation, differentiation and morphogenesis. *Development* **122,** 1651–1661.

Salser, S. J., Loer, C. M., and Kenyon, C. (1993). Multiple HOM-C gene interactions specify cell fates in the nematode central nervous system. *Genes Dev.* **7,** 1714–1724.

Sanchez, L., Gorfinkiel, N., and Guerrero, I. (2001). Sex determination genes control the development of the *Drosophila* genital disc, modulating the response to Hedgehog, Wingless and Decapentaplegic signals. *Development* **128,** 1033–1043.

Savage, C., Das, P., Finelli, A. L., Townsend, S. R., Sun, C. Y., Baird, S. E., and Padgett, R. W. (1996). *Caenorhabditis elegans* genes sma-2, sma-3, and sma-4 define a conserved family of transforming growth factor beta pathway components. *Proc. Natl. Acad. Sci. USA* **93,** 790–794.

Schafer, W. R. (2005). Egg-laying. "WormBook".

Schedl, T., Graham, P. L., Barton, M. K., and Kimble, J. (1989). Analysis of the role of tra-1 in germline sex determination in the nematode *Caenorhabditis elegans*. *Genetics* **123,** 755–769.

Schindelman, G., Whittaker, A. J., Thum, J. Y., Gharib, S., and Sternberg, P. W. (2006). Initiation of male sperm-transfer behavior in *Caenorhabditis elegans* requires input from the ventral nerve cord. *BMC Biol.* **4,** 26.

Schvarzstein, M., and Spence, A. M. (2006). The *C. elegans* sex-determining GLI protein TRA-1A is regulated by sex-specific proteolysis. *Dev. Cell* **11,** 733–740.

Schwartz, H. T., and Horvitz, H. R. (2007). The *C. elegans* protein CEH-30 protects male-specific neurons from apoptosis independently of the Bcl-2 homolog CED-9. *Genes Dev.* **21,** 3181–3194.

Segal, S. P., Graves, L. E., Verheyden, J., and Goodwin, E. B. (2001). RNA-regulated TRA-1 nuclear export controls sexual fate. *Dev. Cell* **1,** 539–551.

Shen, M. M., and Hodgkin, J. (1988). Mab-3, a gene required for sex-specific yolk protein expression and a male-specific lineage in *C. elegans*. *Cell* **54,** 1019–1031.

Siegfried, K. R., and Kimble, J. (2002). POP-1 controls axis formation during early gonadogenesis in *C. elegans*. *Development* **129,** 443–453.

Siegfried, K. R., Kidd, A. R., III, Chesney, M. A., and Kimble, J. (2004). The sys-1 and sys-3 genes cooperate with Wnt signaling to establish the proximal–distal axis of the *Caenorhabditis elegans* gonad. *Genetics* **166,** 171–186.

Simon, J. M., and Sternberg, P. W. (2002). Evidence of a mate-finding cue in the hermaphrodite nematode *Caenorhabditis elegans*. *Proc. Natl. Acad. Sci. USA* **99,** 1598–1603.

Skipper, M., Milne, C. A., and Hodgkin, J. (1999). Genetic and molecular analysis of fox-1, a numerator element involved in *Caenorhabditis elegans* primary sex determination. *Genetics* **151**, 617–631.

Sokol, S. B., and Kuwabara, P. E. (2000). Proteolysis in *Caenorhabditis elegans* sex determination: Cleavage of TRA-2A by TRA-3. *Genes Dev.* **14**, 901–906.

Spence, A. M., Coulson, A., and Hodgkin, J. (1990). The product of fem-1, a nematode sex-determining gene, contains a motif found in cell cycle control proteins and receptors for cell–cell interactions. *Cell* **60**, 981–990.

Starostina, N. G., Lim, J. M., Schvarzstein, M., Wells, L., Spence, A. M., and Kipreos, E. T. (2007). A CUL-2 ubiquitin ligase containing three FEM proteins degrades TRA-1 to regulate *C. elegans* sex determination. *Dev. Cell* **13**, 127–139.

Stern, M. J., and Horvitz, H. R. (1991). A normally attractive cell interaction is repulsive in two *C. elegans* mesodermal cell migration mutants. *Development* **113**, 797–803.

Sternberg, P. W. (2005). Vulval development. "WormBook".

Sternberg, P. W., and Horvitz, H. R. (1986). Pattern formation during vulval development in *C. elegans*. *Cell* **44**, 761–772.

Steven, R., Kubiseski, T. J., Zheng, H., Kulkarni, S., Mancillas, J., Ruiz Morales, A., Hogue, C. W., Pawson, T., and Culotti, J. (1998). UNC-73 activates the Rac gtpase and is required for cell and growth cone migrations in *C. elegans*. *Cell* **92**, 785–795.

Streit, A., Li, W., Robertson, B., Schein, J., Kamal, I. H., Marra, M., and Wood, W. B. (1999). Homologs of the *Caenorhabditis elegans* masculinizing gene her-1 in *C. briggsae* and the filarial parasite *Brugia malayi*. *Genetics* **152**, 1573–1584.

Stringham, E., Pujol, N., Vandekerckhove, J., and Bogaert, T. (2002). Unc-53 controls longitudinal migration in *C. elegans*. *Development* **129**, 3367–3379.

Sulston, J. E., and Horvitz, H. R. (1977). Post-embryonic cell lineages of the nematode, *Caenorhabditis elegans*. *Dev. Biol.* **56**, 110–156.

Sulston, J. E., Albertson, D. G., and Thomson, J. N. (1980). The *Caenorhabditis elegans* male: Postembryonic development of nongonadal structures. *Dev. Biol.* **78**, 542–576.

Sulston, J. E., Schierenberg, E., White, J. G., and Thomson, J. N. (1983). The embryonic cell lineage of the nematode *Caenorhabditis elegans*. *Dev. Biol.* **100**, 64–119.

Tan, K. M., Chan, S. L., Tan, K. O., and Yu, V. C. (2001). The *Caenorhabditis elegans* sex-determining protein FEM-2 and its human homologue, hfem-2, are Ca2+/calmodulin-dependent protein kinase phosphatases that promote apoptosis. *J. Biol. Chem.* **276**, 44193–44202.

Thomas, J. H., Stern, M. J., and Horvitz, H. R. (1990). Cell interactions coordinate the development of the *C. elegans* egg-laying system. *Cell* **62**, 1041–1052.

Tilmann, C., and Kimble, J. (2005). Cyclin D regulation of a sexually dimorphic asymmetric cell division. *Dev. Cell* **9**, 489–499.

Troemel, E. R., Chou, J. H., Dwyer, N. D., Colbert, H. A., and Bargmann, C. I. (1995). Divergent seven transmembrane receptors are candidate chemosensory receptors in *C. elegans*. *Cell* **83**, 207–218.

Villeneuve, A. M., and Meyer, B. J. (1990). The regulatory hierarchy controlling sex determination and dosage compensation in *Caenorhabditis elegans*. *Adv. Genet.* **27**, 117–188.

Wang, S., and Kimble, J. (2001). The TRA-1 transcription factor binds TRA-2 to regulate sexual fates in *Caenorhabditis elegans*. *EMBO J.* **20**, 1363–1372.

Wang, B. B., Muller-Immergluck, M. M., Austin, J., Robinson, N. T., Chisholm, A., and Kenyon, C. (1993). A homeotic gene cluster patterns the anteroposterior body axis of *C. elegans*. *Cell* **74**, 29–42.

White, J. G. (1988). The anatomy. "The Nematode *Caenorhabditis elegans*" (W. B. Wood, Ed.), pp. 81–121. Cold Spring Harbor Laboratory, Cold Spring Harbor, NY.

White, J. G., Southgate, J. N., Thomson, J. N., and Brenner, S. (1986). The structure of the nervous system of the nematode *Caenorhabditis elegans*. *Philos. Trans. R. Soc.* **314**, 1–340.

White, J. Q., Nicholas, T. J., Gritton, J., Truong, L., Davidson, E. R., and Jorgensen, E. M. (2007). The sensory circuitry for sexual attraction in *C. elegans* males. *Curr. Biol.* **17,** 1847–1857.

Whitfield, L. S., Lovell-Badge, R., and Goodfellow, P. N. (1993). Rapid sequence evolution of the mammalian sex-determining gene SRY. *Nature* **364,** 713–715.

Wilkins, A. S. (1995). Moving up the hierarchy: A hypothesis on the evolution of a genetic sex determination pathway. *Bioessays* **17,** 71–77.

Wrischnik, L. A., and Kenyon, C. J. (1997). The role of lin-22, a hairy/enhancer of split homolog, in patterning the peripheral nervous system of *C. elegans*. *Development* **124,** 2875–2888.

Wu, M., and Herman, M. A. (2006). A novel noncanonical Wnt pathway is involved in the regulation of the asymmetric B cell division in *C. elegans*. *Dev. Biol.* **293,** 316–329.

Wu, M., and Herman, M. A. (2007). Asymmetric localizations of LIN-17/Fz and MIG-5/Dsh are involved in the asymmetric B cell division in *C. elegans*. *Dev. Biol.* **303,** 650–662.

Yi, W., and Zarkower, D. (1999). Similarity of DNA binding and transcriptional regulation by *Caenorhabditis elegans* MAB-3 and *Drosophila melanogaster* DSX suggests conservation of sex determining mechanisms. *Development* **126,** 873–881.

Yi, W., Ross, J. M., and Zarkower, D. (2000). Mab-3 is a direct tra-1 target gene regulating diverse aspects of *C. elegans* male sexual development and behavior. *Development* **127,** 4469–4480.

Yonker, S. A., and Meyer, B. J. (2003). Recruitment of *C. elegans* dosage compensation proteins for gene-specific versus chromosome-wide repression. *Development* **130,** 6519–6532.

Zarkower, D. (2001). Establishing sexual dimorphism: Conservation amidst diversity? *Nat. Rev. Genet.* **2,** 175–185.

Zarkower, D., and Hodgkin, J. (1992). Molecular analysis of the *C. elegans* sex-determining gene tra-1: A gene encoding two zinc finger proteins. *Cell* **70,** 237–249.

Zarkower, D., and Hodgkin, J. (1993). Zinc fingers in sex determination: Only one of the two *C. elegans* Tra-1 proteins binds DNA *in vitro*. *Nucleic Acids Res.* **21,** 3691–3698.

Zhang, Y., and Emmons, S. W. (1995). Specification of sense-organ identity by a *Caenorhabditis elegans* Pax-6 homologue. *Nature* **377,** 55–59.

Zhao, C., and Emmons, S. W. (1995). A transcription factor controlling development of peripheral sense organs in *C. elegans*. *Nature* **373,** 74–78.

Zhu, L., Wilken, J., Phillips, N. B., Narendra, U., Chan, G., Stratton, S. M., Kent, S. B., and Weiss, M. A. (2000). Sexual dimorphism in diverse metazoans is regulated by a novel class of intertwined zinc fingers. *Genes Dev.* **14,** 1750–1764.

CHAPTER TWO

SEX DETERMINATION IN THE *CAENORHABDITIS ELEGANS* GERM LINE

Ronald E. Ellis

Contents

1. Nematode Gender: Hermaphrodites, Females, and Males	42
1.1. Hermaphrodites	42
1.2. Females	44
1.3. Males	44
2. The Core Sex-Determination Pathway	44
2.1. The ratio of X-chromosomes to autosomes determines sex	45
2.2. The secreted protein HER-1 promotes male development and spermatogenesis	45
2.3. HER-1 inhibits the TRA-2 receptor	46
2.4. The FEM proteins promote TRA-1 degradation	46
2.5. The FEM proteins are also regulated by proteolysis	47
3. Transcriptional Control of Germ Cell Fates by TRA-1	47
3.1. A cleavage product of TRA-1 promotes oogenesis	48
3.2. Full length TRA-1 is found in germ cells	48
3.3. TRA-1 regulates *fog-1* and *fog-3* to control germ cell fates	48
3.4. FOG-1 and FOG-3 promote spermatogenesis	49
3.5. In *C. elegans*, the FEM proteins play an additional role in spermatogenesis	50
4. Modulating the Sex-Determination Pathway to Create Hermaphrodites	50
4.1. Translational repression of *tra-2* allows hermaphrodites to make sperm	50
4.2. TRA-2 acts through parallel pathways to regulate TRA-1	52
4.3. RPN-10 regulates the degradation of TRA-2	53
5. Maintaining a Balance Between TRA-2 and FEM-3 Levels	53
5.1. Additional regulators of *fem-3* translation	54
6. Do FOG-1 Levels Control Germ Cell Fates?	55

Department of Molecular Biology, School of Osteopathic Medicine, B303 Science Center, The University of Medicine and Dentistry of New Jersey, 2 Medical Center Drive, Stratford, New Jersey 08084

7. Mysteries	55
7.1. What causes the sperm/oocyte switch?	55
7.2. How does the control of mitosis influence sexual development?	57
7.3. What are the architectural rules for translational regulation?	57
7.4. How has this pathway evolved?	57
7.5. What other somatic signals influence germ cell fates?	58
8. Conclusion	58
Acknowledgments	59
References	59

1. Nematode Gender: Hermaphrodites, Females, and Males

The vast majority of animals determine their sexual identity when young, and produce oocytes if female, or sperm if male. In most species, this decision is not easily modified. For example, in fruit flies, mutations that alter sexual identity in germ cells usually result in tumors or cell death, rather than in females making sperm or males making oocytes (Chapter 4). By contrast, *XX* nematodes from some species are self-fertile hermaphrodites, which produce both sperm and oocytes in the same gonad from a common pool of germ cells (Fig. 2.1).

1.1. Hermaphrodites

In *Caenorhabditis elegans*, *XX* animals are hermaphrodites. The hermaphrodite gonad is composed of a large tube with two arms that meet in a central uterus (Fig. 2.1). At the distal end of each arm, a somatic cell causes adjacent germ cells to remain in mitosis and proliferate (reviewed by Kimble and Crittenden, 2007). Once germ cells move away from the distal tip cell, they enter a transition zone and begin meiosis. Germ cells quickly progress through early prophase I to the pachytene stage, where they arrest for many hours. These pachytene cells actively produce messenger RNAs and proteins that diffuse throughout the germ line, and which help form mature oocytes. As each germ cell approaches the bend in the gonad (Fig. 2.1), it resumes meiosis, and either develops into a mature oocyte or undergoes apoptosis.

Throughout these developmental stages, the germ cells remain part of a syncytium. Each nucleus is located on the surface of the tube that forms the germ line and is partially enclosed. Although membranes separate it from the surrounding somatic gonad and from its immediate neighbors, they do

Sex Determination in the *C. elegans* Germ Line

Virgin female

Hermaphrodite

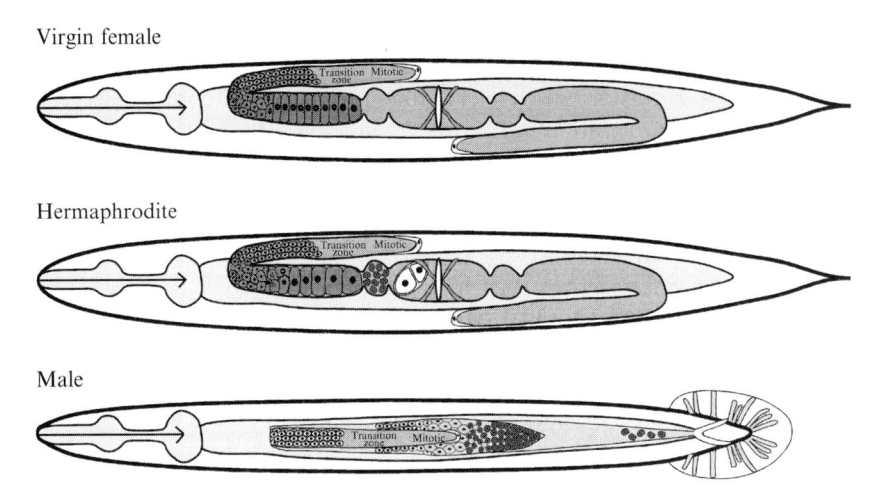

Male

Figure 2.1 Nematodes have three sexes. Diagrams of an adult *XX* female, *XX* hermaphrodite, and *XO* male. Anterior is to the left and ventral is up. In the *XX* animals, cells in pachytene are pink, oocytes are red, dying cells are gray, and sperm are dark blue. Since the females lack sperm, their oocytes are "stacking" in the ovary, and the uterus is empty. By contrast, the hermaphrodite oocytes are being fertilized, released into the uterus, and laid. In the male, cells in pachytene and primary spermatocytes are light blue, residual bodies are light blue circles surrounded by spermatids, and spermatids and sperm are dark blue. Active sperm are shown with pseudopods in both the hermaphrodite spermatheca, and in the vas deferens of the male, just prior to ejaculation. In all animals, the somatic gonad is gray, except for the distal tip cells, which are yellow. The vulva is an oval centered over the uterus, and connected to the sex muscles, which are colored orange. The pharynx and intestine are green.

not extend very far toward the interior, so each "cell" retains an opening to a large core of cytoplasm that runs through the center of the germ line tube. Thus, molecules can be shared among developing germ cells until late in oogenesis. In many other animals, germ cells also have cytoplasmic connections during early stages of development.

Hermaphrodites are self-fertile because the first germ cells to differentiate during larval development form sperm. Late in the fourth larval stage, the germ line switches permanently to oogenesis and the sperm that were formed crawl into the spermatheca, a storage organ that can also hold sperm from males, and wait there for the chance to fertilize their own oocytes. Thus, *XX* nematodes are self-fertile hermaphrodites. However, they lack the equipment needed to mate with each other. These characteristics led Sydney Brenner to select *C. elegans* as a model animal because self-fertilization simplifies genetic crosses (Brenner, 1974). However, they also raise unique developmental problems—in particular, how do otherwise female animals successfully produce sperm?

1.2. Females

Many closely related nematodes, like *Caenorhabditis remanei*, produce XX females. These animals look like hermaphrodites in almost every respect, but do not make sperm (Fig. 2.1). Among nematodes, male/female species predominate, and phylogenetic studies imply that this is the ancestral state (Fitch, 2002). Self-fertile hermaphrodites are also common and appear to have evolved independently in many nematode families. Even within the genus *Caenorhabditis*, hermaphrodites are likely to have evolved on at least two different occasions, producing *C. elegans* and *Caenorhabditis briggsae* (Cho et al., 2004; Kiontke et al., 2004). This observation raises critical questions, such as why the core sex-determination pathway in nematodes is so easily altered to create XX hermaphrodites?

1.3. Males

Most species of nematodes (with the exception of some parthenogenetic ones) also produce males. These animals have a single *X* chromosome, and their bodies are adapted to find and mate with females or hermaphrodites (Wolff and Zarkower, 2008).

The male gonad forms a long tube that ends at the cloaca in the tail (Fig. 2.1). As with females or hermaphrodites, the distal tip cells at one end of the gonad create a stem cell niche. As germ cells move away from the distal tip, they enter meiosis, progress to the pachytene stage, and arrest. However, once they resume differentiation they develop into primary spermatocytes, each of which goes through meiosis to produce four spermatids and a residual body.

Nematode sperm are amoeboid (reviewed by L'Hernault, 2006). They remain inactive until mating, at which point they are ejaculated into the female, activated by an unknown signal, and begin to crawl towards the spermatheca. Since a protease inhibitor helps prevent premature activation of male sperm (Stanfield and Villeneuve, 2006), the activation process is probably controlled by an extracellular signal, perhaps even by a secreted protease.

In hermaphrodites, both male sperm and self sperm crawl reside in the spermatheca, and compete to fertilize oocytes. In general, the male sperm are larger than those made by hermaphrodites, and compete better for fertilization (LaMunyon and Ward, 1998; Singson et al., 1999).

2. THE CORE SEX-DETERMINATION PATHWAY

Unlike the case with many species, the core genes of the sex-determination pathway play similar roles in the nematode soma (Wolff and Zarkower, 2008) and germ line (Fig. 2.2).

Sex Determination in the *C. elegans* Germ Line

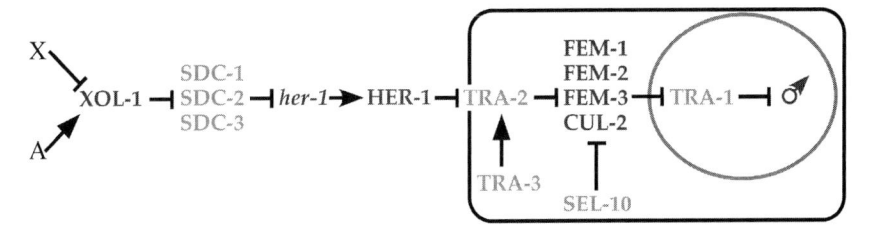

```
X                              FEM-1
 \      SDC-1                  FEM-2
  XOL-1 ⊣ SDC-2 ⊣ her-1 ➤ HER-1 ⊣ TRA-2 ⊣ FEM-3 ⊣ TRA-1 ⊣ ♂
 ↗      SDC-3                  CUL-2
A                          ↑
                         TRA-3  ⊥
                                SEL-10
```

Figure 2.2 A signal transduction pathway controls somatic sex in *C. elegans*. Proteins are uppercase, and the *her-1* gene (which is regulated at the transcriptional level by the SDC proteins) is lowercase. The black rectangle groups factors that act in target cells; since TRA-2 is the HER-1 receptor, it is depicted spanning the cell membrane. The gray oval indicates the cell nucleus, where the transcription factor TRA-1 regulates the expression of sex-specific genes. Factors that promote male fates are blue, and those that promote female fates are red. Arrows indicate positive interactions, and —indicates negative interactions.

2.1. The ratio of *X*-chromosomes to autosomes determines sex

In nematodes, the sex-determination process begins with the ability of cells to sense the ratio of *X*-chromosomes to autosomes, and either express *xol-1* in males, or repress it in hermaphrodites (reviewed by Wolff and Zarkower, 2008). Thus, XOL-1 levels integrate information about the $X:A$ ratio and determine how the animal will develop. In males, XOL-1 represses three *sdc* genes, which allows the transcription of *her-1* and subsequent male development. Since *XX* animals do not express XOL-1, the SCD proteins are free to block transcription of *her-1*, allowing hermaphrodite development. In addition, the SDC proteins establish normal dosage compensation.

2.2. The secreted protein HER-1 promotes male development and spermatogenesis

HER-1 is a small, secreted protein that causes cells to adopt male fates (Perry *et al.*, 1993). In this sense, it is like male sex hormones in mammals. However, HER-1 is produced throughout the body, rather than only in the somatic gonad (Hunter and Wood, 1992). In effect, HER-1 coordinates sexual fates among different tissues—high levels (which diffuse quickly in worms) cause male development, and low levels allow female or hermaphrodite development. In mutants where *her-1* is inactive, *XO* animals develop as hermaphrodites and eventually make oocytes, so *her-1* is required for continuous spermatogenesis in males (Hodgkin, 1980). Furthermore, when *XX* animals express HER-1 from a transgene that is active only in muscle, the rest of the body is masculinized, and many animals make only sperm, which shows that HER-1 causes spermatogenesis, as well as male somatic fates (Perry *et al.*, 1993).

Mosaic analyses suggest that HER-1 expression from two tissues plays a critical role in controlling germ cell fates; these tissues are the surrounding somatic gonad, and the intestine (Hunter and Wood, 1992). The somatic gonad is, of course, in extensive contact with the developing germ cells (Fig. 2.1). The intestine is located near the gonad in the pseudo-cœlom, and is the major site in worms for the production and secretion of proteins. For example, yolk proteins are made by the intestine and absorbed by germ cells (Kimble and Sharrock, 1983).

2.3. HER-1 inhibits the TRA-2 receptor

In both the germ line and soma, HER-1 inactivates its target, the TRA-2 receptor (Kuwabara and Kimble, 1995; Kuwabara et al., 1992; Okkema and Kimble, 1991). The putative binding site for HER 1 was identified by a mutation that creates a constitutively active TRA-2 receptor; this mutation causes XO animals to become hermaphrodites (Hodgkin and Albertson, 1995; Kuwabara, 1996a). Potential sites of interaction on HER-1 were identified by X-ray crystallography and the analysis of HER-1 mutants (Hamaoka et al., 2004). However, it is not yet known how either the structure or function of TRA-2 is changed by interacting with HER-1.

TRA-3 is a calpain protease (Barnes and Hodgkin, 1996) that acts like a positive regulator of TRA-2 *in vivo* (Hodgkin, 1980) and that cleaves TRA-2 *in vitro* (Sokol and Kuwabara, 2000). Thus, one simple model is that TRA-3 normally cleaves TRA-2 to yield an active, intracellular fragment. If so, perhaps the cleavage reaction does not occur when HER-1 is binding TRA-2.

What does activated TRA-2 do? *In vitro* assays and the yeast two-hybrid system show that TRA-2 can bind FEM-3 (Mehra et al., 1999), which acts at the next step in the sex-determination pathway. Thus, TRA-2 might directly control its activity.

2.4. The FEM proteins promote TRA-1 degradation

The three *fem* genes were identified years ago by mutations that cause all animals to develop as females, and analyses of double mutants show that all three genes act downstream of *tra-2* (Doniach and Hodgkin, 1984; Hodgkin, 1986; Kimble et al., 1984). Cloning revealed that FEM-1 has ankyrin repeats (Spence et al., 1990), FEM-2 is a type 2C protein phosphatase (Pilgrim et al., 1995), and FEM-3 is a novel protein (Ahringer et al., 1992). Unfortunately, this information did not show how they controlled TRA-1. Recently, mutations in a fourth gene, *cul-2*, were shown to have a similar phenotype (Starostina et al., 2007). Although animals lacking all *cul-2* activity are dead, homozygous *cul-2* XO mutants from heterozygous mothers show extensive feminization in both the soma and germ line.

In fact, they strongly resemble *fem-3 XO* mutants produced by heterozygous mothers.

CUL-2 binds FEM-1, and is a component of the E3 ubiquitin–ligase complex. When TRA-1 is expressed in human cells, CUL-2 and the FEM proteins promote its ubiquitinylation and degradation (Starostina *et al.*, 2007). Thus, the FEM proteins appear to work with CUL-2 to catalyze the degradation of TRA-1. These results explain why Schvarzstein and Spence (2006) found that TRA-1 was present at high levels in hermaphrodites, but at low levels in males.

2.5. The FEM proteins are also regulated by proteolysis

The F-box protein SEL-10 binds FEM-1 and FEM-3, and targets them for ubiquitinylation and degradation when expressed in mammalian cells (Jager *et al.*, 2004). Mutations that inactivate *sel-10* partially masculinize the body, but do not affect the germ line. However, they suppress *tra-2(mx)* mutations, which normally cause all germ cells to become oocytes (see following sections). Thus, SEL-10 appears to act upstream of the *fem* genes to maintain low levels of FEM protein activity.

3. TRANSCRIPTIONAL CONTROL OF GERM CELL FATES BY TRA-1

Although null mutations in *tra-1* can transform XX animals into fertile males, they do not affect all tissues equally. Most somatic tissues are completely transformed to male fates (Hodgkin and Brenner, 1977). The somatic gonad is often transformed to male fates, but sometimes develops abnormally (Hodgkin, 1987; Mathies *et al.*, 2004). But although larval germ cells develop as sperm, later ones become oocytes, even in XO animals (Hodgkin, 1987; Schedl *et al.*, 1989). Thus, the role of *tra-1* in germ cells is complex. Early in life it promotes oogenesis, but later on it promotes spermatogenesis. How does this happen?

TRA-1A, the major product of *tra-1*, is a zinc-finger protein related to Cubitus interruptus in flies and the Gli proteins in mammals (Zarkower and Hodgkin, 1992). Both Ci and the Gli proteins can act as activators or repressors (Alexandre *et al.*, 1996; Ruiz i Altaba, 1999), much like TRA-1 seems to do in the *C. elegans* germ line. Sequences near the C-terminus of these proteins are needed for them to activate transcription, and cleavage produces an N-terminal zinc-finger protein that represses transcription (reviewed by Jiang, 2002).

3.1. A cleavage product of TRA-1 promotes oogenesis

Recent studies showed that in *C. elegans*, TRA-1A is cleaved to produce a feminizing product, TRA-1^{100} (Schvarzstein and Spence, 2006). Furthermore, nonsense mutants that produce only truncated forms of TRA-1 cause female development in the soma, and oogenesis in the germ line. Thus, TRA-1^{100} is likely to promote oogenesis. Since the only known targets of TRA-1 in germ cells are required for spermatogenesis (see the following sections), TRA-1^{100} is probably a repressor. Although the FEM proteins and CUL-2 cause full-length TRA-1 to be degraded, they do not seem to affect TRA-1^{100} (Starostina *et al.*, 2007).

3.2. Full length TRA-1 is found in germ cells

In flies, full-length Ci activates the transcription of target genes. If TRA-1^{100} represses male genes, does full-length TRA-1 activate them? Surprisingly, animals that lack a germ line do not accumulate full-length TRA-1 (Schvarzstein and Spence, 2006). However, hermaphrodites that produce only sperm do accumulate it. One model that could explain these data is that full-length TRA-1 promotes spermatogenesis, and the cleavage product TRA-1^{100} promotes oogenesis (Fig. 2.3). Since null mutants still make both types of gametes, neither TRA-1 nor TRA-1^{100} is required for either fate.

3.3. TRA-1 regulates *fog-1* and *fog-3* to control germ cell fates

In the soma, *tra-1* represses genes needed for male fates, like *mab-3* in the intestine or *egl-1* in the HSN neurons (reviewed by Wolff and Zarkower, 2008). However, the two promoters with the greatest number of TRA-1

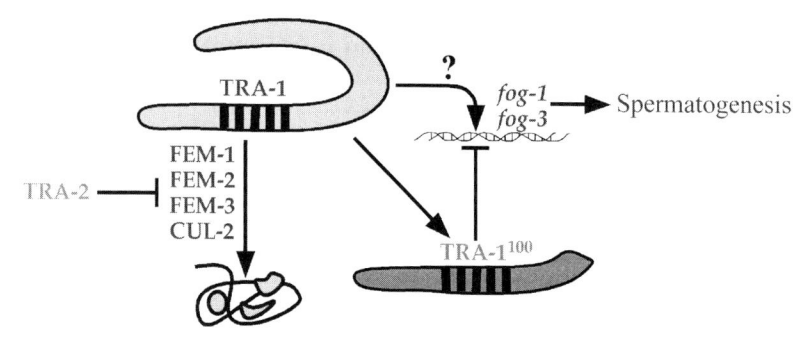

Figure 2.3 The Gli homolog TRA-1 both promotes and inhibits spermatogenesis. Proteins are uppercase, and the *fog-1* and *fog-3* genes (which are regulated at the transcriptional level by TRA-1) are lowercase. Factors that promote male fates are blue, and those that promote female fates are red. The proposal that full-length TRA-1 promotes spermatogenesis is consistent with several observations (see text), but has not been proven. Arrows indicate positive interactions and — indicates negative interactions.

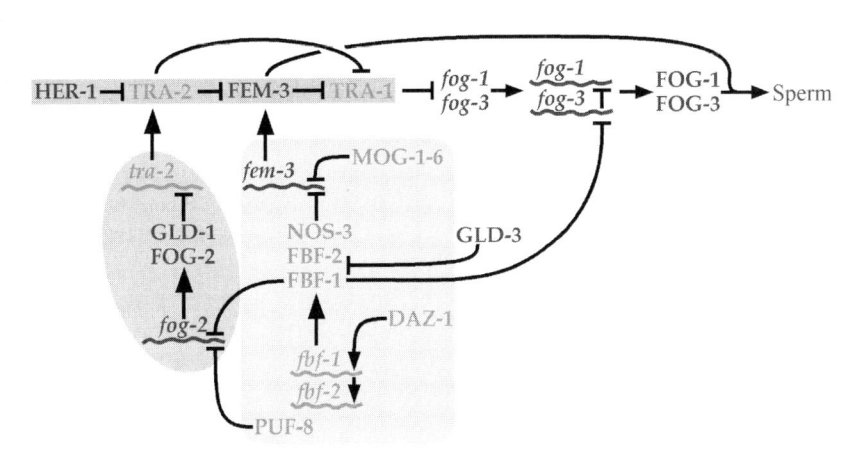

Figure 2.4 The core pathway is modulated in *C. elegans* to create *XX* hermaphrodites. Core members of the pathway are indicated by a gray box. Of the genes shown in Fig. 2.2, only those downstream of *her-1* are described here. For simplicity, FEM-1, FEM-2, CUL-2, SEL-10, and TRA-3 are also omitted, since their activities, although important, do not appear to be modulated in the germ line. Proteins are uppercase, genes are lowercase, and messenger RNAs are lowercase with a squiggly line. Factors that promote male fates are blue, and ones that promote female fates are red. The regulatory network that represses translation of *tra-2* messages is grouped by a blue oval, and the one that regulates *fem-3* translation is grouped by a pink box. Arrows indicate positive interactions, and — indicates negative interactions.

binding sites belong to *fog-1* and *fog-3*, which regulate germ cell fates (Chen and Ellis, 2000; Jin *et al.*, 2001b). Since the analysis of double mutants shows that they act downstream of *tra-1*, both *fog-1* and *fog-3* act at the end of the sex-determination pathway in the germ line (Figs. 2.3 and 2.4).

The *fog-1* and *fog-3* promoters have three to six TRA-1 binding sites, but no other conserved features. These sites have been tested for *fog-3*: the wild type promoter binds TRA-1 in gel shift assays, but mutant promoters with altered TRA-1 sites do not (Chen and Ellis, 2000). In *XX* larvae, mutations in *tra-1* cause a dramatic increase in the level of *fog-3* transcripts, which suggests that TRA-1 represses *fog-3*. However, in *fog-3* transgenes, point mutations in some of the TRA-1 binding sites have the surprising effect of inactivating the promoter, rather than increasing its efficiency, so TRA-1 might also promote *fog-3* expression (Chen and Ellis, 2000). Perhaps full-length TRA-1 activates transcription, and the cleaved TRA-1[100] represses it (Fig. 2.3).

3.4. FOG-1 and FOG-3 promote spermatogenesis

Mutations in *fog-1* or *fog-3* cause all germ cells to differentiate as oocytes, even in males, but do not affect the soma (Barton and Kimble, 1990; Ellis and Kimble, 1995). Thus, both genes normally promote spermatogenesis.

The *fog-1* gene produces two major transcripts, but only the larger one has a known function. It encodes a cytoplasmic polyadenylation element binding protein (Jin et al., 2001b; Luitjens et al., 2000). CPEB proteins contain two RNA Recognition Motifs and a zinc-finger, which cooperate to bind RNA (reviewed by de Moor et al., 2005). In *Xenopus*, CPEB proteins bind the 3'-untranslated region of target mRNAs, blocking the translation of some targets and promoting the polyadenylation and translation of others. An analysis of more than 30 mutations confirmed that the RNA-binding domains of FOG-1 are essential (Jin et al., 2001a). Furthermore, FOG-1 binds its own message *in vitro* (Jin et al., 2001a). Thus, FOG-1 is likely to specify germ cell fates by regulating translation. Its key targets are not yet known.

FOG-3 is related to the Tob and BTG proteins of mammals (Chen et al., 2000). These proteins are found in all animals, and share an amino-terminal domain of about 115 residues. Although Tob proteins are implicated in interactions with many developmental regulatory proteins, their molecular functions are not known (Jia and Meng, 2007).

3.5. In *C. elegans*, the FEM proteins play an additional role in spermatogenesis

Because the FEM proteins regulate TRA-1 stability, one might expect that *tra-1; fem* double mutants would make sperm as well as oocytes, just like *tra-1* mutants. However, these animals only make oocytes (Doniach and Hodgkin, 1984; Hodgkin, 1986). Thus, the three FEM proteins play an additional role in spermatogenesis. Since mutations in the *fem* genes prevent spermatogenesis even when *fog-3* transcript levels are high, this second activity works at the end of the regulatory pathway (Chen and Ellis, 2000). Whatever the FEM proteins are doing is specific to *C. elegans*, since in *C. briggsae* mutations in the *fem* genes do not block spermatogenesis (Hill et al., 2006).

4. Modulating the Sex-Determination Pathway to Create Hermaphrodites

C. elegans hermaphrodites are female in almost all respects, except that they produce sperm. Since hermaphrodites do not express HER-1, how do they initiate spermatogenesis?

4.1. Translational repression of *tra-2* allows hermaphrodites to make sperm

Mutant screens identified two genes that are required for XX animals to make sperm, but not for males to do so—*fog-2* (Schedl and Kimble, 1988) and *gld-1* (Francis et al., 1995b). Mutations in *fog-2* only prevent spermatogenesis in *XX* animals, so *fog-2* strains produce males and females just like diœcious species. Mutations in *gld-1* affect other aspects of germ cell development in *XX* animals, making them sterile (Francis *et al.*, 1995a,b).

Three lines of evidence suggest that FOG-2 and GLD-1 repress translation of *tra-2* messages (Fig. 2.4). First, mutations in the 3'-untranslated region of *tra-2* that prevent interactions with GLD-1 (Jan *et al.*, 1999) also prevent spermatogenesis (Doniach, 1986; Goodwin *et al.*, 1993). Second, in the soma, GLD-1 can repress a reporter gene that is fused to the wild-type *tra-2* 3'-UTR, but not one fused to a mutant 3'-UTR (Jan *et al.*, 1999). Third, TRA-2 protein levels increase in *gld-1* mutants (Jan *et al.*, 1999).

GLD-1 contains a maxi-KH RNA-binding domain (Jones and Schedl, 1995), and is likely to act as a dimer (Ryder *et al.*, 2004). Its first identified target was *tra-2* mRNA (Jan *et al.*, 1999), but it also binds dozens of other messages (e.g., Lee and Schedl, 2001; Marin and Evans, 2003; Mootz *et al.*, 2004; Schumacher *et al.*, 2005), which explains its pleiotropic effects on germ cells. One of these additional targets is *tra-1* (Lakiza *et al.*, 2005). Many of these targets share a hexanucleotide sequence that appears to form part of the GLD-1 binding site (Ryder *et al.*, 2004). In hermaphrodites, GLD-1 protein is present at low levels in mitotic cells and in maturing oocytes, but at high levels in between (Jones *et al.*, 1996), which fits well with models in which it represses target messages during early stages of meiosis.

Although GLD-1 can bind *tra-2* messages on its own, it cannot regulate translation without FOG-2. These two proteins interact in the yeast two-hybrid system and *in vitro*, and can form a ternary complex with *tra-2* mRNA (Clifford *et al.*, 2000). The carboxyl-terminus of FOG-2 mediates the interaction with GLD-1, and has been evolving rapidly (Nayak *et al.*, 2005), which implies that this system might be a recent innovation. The amino-terminus of FOG-2 binds the Skp-1-related protein SKR-1, presumably through its F-box.

How FOG-2 and GLD-1 repress translation remains a mystery. Most F-box proteins recruit targets to a ubiquitin–ligase complex, which ubiquitinylates the targets, leading to their degradation (reviewed by Kipreos, 2005). However, FOG-2 cannot cause GLD-1 to be degraded, since they act together to repress *tra-2*. Perhaps FOG-2 recruits other regulatory proteins to the *tra-2* 3'-UTR. One possible candidate is LAF-1, an essential protein that has been implicated in the regulation of *tra-2* translation (Goodwin *et al.*, 1997). In addition, ATX-2, which interacts with Poly-A binding protein, might regulate translation of *tra-2*, since *atx-2(RNAi)* causes constitutive spermatogenesis in many animals, but is completely suppressed by the *tra-2(q122gf)* mutation (Ciosk *et al.*, 2004; Maine *et al.*, 2004).

The expression of FOG-2 itself is also under the control of translational regulators. Two proteins related to *Drosophila* Pumilio, PUF-8 and FBF-1, act redundantly to promote oogenesis and lower FOG-2 protein levels

(Bachorik and Kimble, 2005). FBF-1 also acts at a second point in the sex-determination pathway, and will be discussed in greater detail in the following sections.

4.2. TRA-2 acts through parallel pathways to regulate TRA-1

Dominant *tra-2* mutations have played a major role in elucidating how sexual fate is controlled in the germ line. As described previously, a mutation that prevents TRA-2 from binding HER-1 transforms *XO* animals from males to hermaphrodites, highlighting the importance of this signal-transduction pathway in the germ line. Mutations that prevent *tra-2* messenger RNAs from binding GLD-1 transform hermaphrodites into females, showing that translational control allows *XX* animals to make sperm. A third class of dominant mutations is just as intriguing, but is not yet fully understood.

Several dominant mutations in TRA-2 prevent hermaphrodites from making sperm, but do not affect males (Doniach, 1986; Kuwabara *et al.*, 1998). Since these mutations feminize germ cells but masculinize the soma when homozygous, they have been called *mixomorphic* alleles. Each of them is a missense mutation that alters part of the intracellular domain of TRA-2. The affected region of the protein normally binds TRA-1, but the mixomorphic mutations disrupt this interaction (Lum *et al.*, 2000; Wang and Kimble, 2001). One explanation of these results is that TRA-2 normally interacts with TRA-1 to promote hermaphrodite spermatogenesis. On the other hand, the mixomorphic alleles behave as if they increase TRA-2 activity in some assays (Doniach, 1986; Schedl and Kimble, 1988), so it remains possible that the TRA-1/TRA-2 complex promotes oogenesis.

Since TRA-2 regulates TRA-1 directly through this interaction, and indirectly by binding to FEM-3 (Mehra *et al.*, 1999), it controls germ cell fates through parallel pathways (Fig. 2.4). The TRA-2/TRA-1 interaction also occurs in *C. briggsae* (Wang and Kimble, 2001), so these parallel pathways have been conserved during evolution, which suggests that they form a critical part of the regulatory circuit. However, the existence of both TRA-2/FEM-3 and TRA-2/TRA-1 interactions makes it difficult to interpret some double mutant phenotypes.

How does a membrane protein like TRA-2 find a transcription factor like TRA-1? If TRA-3 indeed cleaves TRA-2 *in vivo*, the fact that *tra-3* mutants develop as males suggests that this cleavage activates TRA-2. Perhaps the intracellular portion of TRA-2, often called TRA-2$_{ic}$, is freed to find TRA-1 in the cytoplasm or nucleus. Since TRA-2$_{ic}$ binds the carboxyl-terminus of TRA-1, which is removed during processing, one possibility is that TRA-2$_{ic}$ protects TRA-1 from processing or degradation, thus favoring spermatogenesis. Some germ cells express a second *tra-2* transcript, *tra-2B*, which encodes an intracellular protein similar to TRA-2$_{ic}$

(Kuwabara *et al.*, 1998); however, this transcript is only made during oogenesis and its function is not known.

4.3. RPN-10 regulates the degradation of TRA-2

Knocking down two components of the 26s proteosome, RPN-10 and UFD-2, also prevents hermaphrodite spermatogenesis (Shimada *et al.*, 2006). By contrast, the males appear normal, although sensitive assays reveal small changes in their expression of some sex-related genes. Since this effect on hermaphrodite spermatogenesis is suppressed by mutations in *tra-2*, both *rpn-10* and *ufd-2* probably act upstream of TRA-2 to promote its degradation. This model is supported by the fact that TRA-2 protein levels increase in the intestinal nuclei of *rpn-10* mutants. Since RPN-10 and UFD-2 can regulate TRA-2 in the soma, perhaps the specificity of the mutant phenotype is caused by the heightened sensitivity of hermaphrodite germ cells to small changes in TRA-2 activity.

5. Maintaining a Balance Between TRA-2 and FEM-3 Levels

Several gain-of-function mutations in *fem-3* also have a dramatic effect on germ cells fates, causing animals to produce sperm constitutively (Barton *et al.*, 1987). Each of these *fem-3(gf)* mutations affects a point mutation element (or PME) located in the 3'-UTR (Ahringer and Kimble, 1991; Ahringer *et al.*, 1992), causing an increase in the length of the poly(A) tail. Taken together, these data suggest that increased translation of *fem-3* in these mutants leads to spermatogenesis. How does this system work? Two proteins, FBF-1 and FBF-2, bind the PME sequence and repress translation of *fem-3* (Zhang *et al.*, 1997). These FBF proteins are similar in sequence, and partially but not completely redundant in function (Lamont *et al.*, 2004). If the activities of both FBF proteins are knocked down by RNA interference, most hermaphrodites make more sperm than normal, as one would expect if FEM-3 activity were increased (Zhang *et al.*, 1997). Eliminating FBF activity altogether also blocks the proliferation of germ cells.

The FBF proteins are homologous to *Drosophila* Pumilio, which interacts with Nanos to block translation of hunchback mRNA. Yeast two-hybrid studies and *in vitro* assays show that the FBF proteins bind the Nanos homolog NOS-3 (Kraemer *et al.*, 1999). Although they do not bind NOS-1 or NOS-2 under the same conditions, RNA interference shows that all three NOS proteins act redundantly to prevent spermatogenesis. One possibility is that NOS-3 forms a complex with the FBF proteins that represses *fem-3* translation, and that NOS-1 or NOS-2 can replace NOS-3, but require *fem-3* mRNA in order to form the complex.

The activity of the FBF proteins appears to be regulated in at least three ways. First, GLD-3 is a homolog of bicaudal-C that binds to FBF-1 and FBF-2 (Eckmann *et al.*, 2002). Since mutations in *gld-3* promote oogenesis in both sexes, GLD-3 normally inhibits FBF activity. Analysis of GLD-3 mutants that cannot bind FBF, and of *fbf-1 fbf-2 gld-3* triple mutants, shows that GLD-3 acts through the FBF proteins to control sexual fate in germ cells (Eckmann *et al.*, 2002, 2004). Second, the expression of the FBF proteins also appears to be positively influenced by the translational regulator DAZ-1 (Karashima *et al.*, 2000; Otori *et al.*, 2006). Third, the Notch pathway regulates FBF activity (Lamont *et al.*, 2004).

As was true with GLD-1, the FBF proteins and GLD-3 regulate other aspects of germ cell development. In particular, GLD-3 is required for germ cells to complete spermatogenesis (Eckmann *et al.*, 2002), and to begin meiosis (Eckmann *et al.*, 2004). The FBF proteins are needed for the production of high quality sperm (Luitjens *et al.*, 2000), and to prevent premature entry into meiosis (Crittenden *et al.*, 2002).

5.1. Additional regulators of *fem-3* translation

General screens for mutations that cause hermaphrodites to produce only sperm, like the *fem-3(gf)* alleles described above, identified six genes, *mog-1* through *mog-6* (Graham and Kimble, 1993; Graham *et al.*, 1993). Their effect on sex-determination is suppressed by mutations in the *fem* genes, but not by mutations in *fog-2*, which suggests that they repress one of the *fem* genes. However, most embryos produced by these double mutants die. A mutation in any of the *mog* genes activates a reporter construct fused to the *fem-3* 3'-UTR (Gallegos *et al.*, 1998). Since this reporter is also upregulated by mutations in the PME, perhaps the *mog* genes cooperate with FBF-1 and FBF-2 to repress translation of *fem-3*. However, these reporter experiments used a transgene that was only expressed in the soma, and the *fbf* genes are not expressed in the soma (Zhang *et al.*, 1997). Thus, the *mog* genes might also cooperate with other members of the FBF family, some of which are expressed in somatic tissues (Walser *et al.*, 2006).

MOG-1, MOG-4, and MOG-5 are DEAH helicases that are likely to bind RNA (Puoti and Kimble, 1999, 2000), and MOG-6 is a cyclophilin (Belfiore *et al.*, 2004). Each of these four proteins interacts with MEP-1, a zinc-finger protein that is also required for the repression of the *fem-3* 3'-UTR, but which has additional functions too.

In *C. elegans*, MAG-1 (the homolog of Mago Nashi from *Drosophila*) and RNP-4 (the homolog of yeast Y14) also influence the switch from spermatogenesis to oogenesis (Kawano *et al.*, 2004; Li *et al.*, 2000). Knocking down either gene with RNA interference causes animals to produce sperm throughout their lives, and results in the death of any embryos that they manage to produce. MAG-1 binds RNP-4 (Kawano *et al.*, 2004), and both proteins are thought to act as part of the Exon Junction Complex, which marks splice sites in mature messenger RNAs (reviewed by Singh and Lykke-Andersen, 2003). Since mammalian Magoh and Y14 promote increased levels of translation while remaining associated with mRNAs (Nott *et al.*, 2004), perhaps MAG-1 and RNP-4 promote translation of a message needed for oogenesis.

6. Do FOG-1 Levels Control Germ Cell Fates?

Males that lack *tra-1* often switch from spermatogenesis to oogenesis (Hodgkin, 1987; Schedl *et al.*, 1989). In these animals, spermatogenesis still requires *fog-1* and *fog-3*, so some factor besides TRA-1 also regulates *fog-1* and *fog-3* activity. This factor might be the FBF proteins or an unknown regulator. Males with a single copy of *fog-1* also switch from spermatogenesis to oogenesis, but males with a single copy of *fog-3* are normal, so perhaps *fog-1* is the major target of this regulation (Barton and Kimble, 1990; Ellis and Kimble, 1995). The idea that FOG-1 activity determines germ cell fates is supported by the observations that (1) FOG-1 is expressed in germ cells when sexual identity is being determined, and that (2) the level of FOG-1 expression correlates with the number of sperm that will be made (Lamont and Kimble, 2007).

7. Mysteries

Although we know a great deal about how sexual fate is determined in nematode germ cells, many important questions remain unanswered.

7.1. What causes the sperm/oocyte switch?

As seen above, many kinds of mutations cause hermaphrodites to produce only sperm or only oocytes. However, it does not follow that the affected genes normally control when germ cells switch from spermatogenesis to oogenesis. These genes might instead set the stage for the decision, or help implement it.

For example, several studies imply that the relative activities of TRA-2 and FEM-3 play a critical role in determining whether germ cells become sperm or eggs. In particular, the *tra-2(gf)* mutants make higher levels of TRA-2 protein, which causes oogenesis. By contrast, gain-of-function mutations that increase the levels of FEM-3 have the opposite effect, causing animals to make sperm throughout their lives (Barton et al., 1987). But *tra-2(gf); fem-3(gf)* double mutants are normal hermaphrodites, which shows that the critical factor is the relative level of these two proteins (Schedl and Kimble, 1988).

Four models could explain how the relative levels of TRA-2 and FEM-3 activity change to cause hermaphrodites to make oocytes (Fig. 2.5). First, a developmental change could lead to a dramatic increase in TRA-2 levels as animals become adults, resulting in oogenesis (Fig. 2.5A). One such change would be a decline in the levels of FOG-2, which represses translation of *tra-2* messages. However, FOG-2 levels do not appear to change significantly at this time (Clifford et al., 2000). Alternatively, the level of FEM-3 could decrease dramatically (Fig. 2.5B). One potential cause would be an increase in the activities of repressors like the FBF proteins. However, the levels of *fbf* transcripts do not increase in adult hermaphrodites; instead, they decline slightly (Zhang et al., 1997). Third, both TRA-2 and FEM-3 levels might change (Fig. 2.5C). These changes could be caused by unknown regulators of TRA-2 or FEM-3 activity that become active as animals mature to adults. Final, TRA-2 and FEM-3 might accumulate at different rates, which would gradually tip the balance in favor of TRA-2 (Fig. 2.5D).

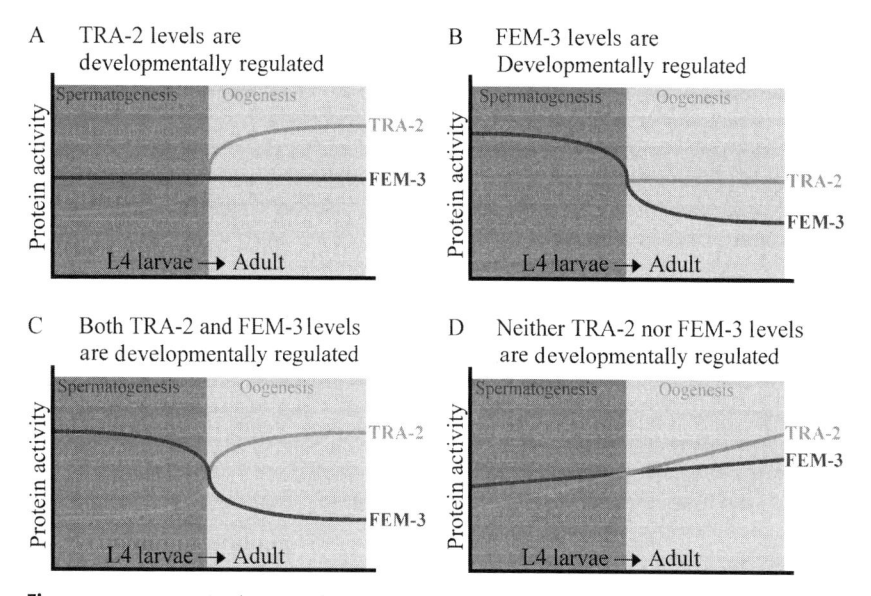

Figure 2.5 How is the switch from spermatogenesis to oogenesis controlled? The relative levels of TRA-2 and FEM-3 appear to determine germ cell fates. How the activities of these proteins change as hermaphrodites become adults is not known. Four different models could explain the switch these animals make from spermatogenesis to oogenesis.

7.2. How does the control of mitosis influence sexual development?

Many of the genes that regulate sexual fate in germ cells also help control when these cells enter meiosis (reviewed by Kimble and Crittenden, 2007; Kimble and Page, 2007). Examples include *fog-1* (Thompson *et al.*, 2005), *fog-3* (Thompson *et al.*, 2005), *fbf-1* (Lamont *et al.*, 2004), *fbf-2* (Lamont *et al.*, 2004), *puf-8* (Subramaniam and Seydoux, 2003), *gld-1* (Francis *et al.*, 1995a), and *gld-3* (Eckmann *et al.*, 2004). This correlation could be explained by (1) each of these genes playing redundant roles in two different regulatory pathways or (2) feedback between the pathways controlling sex and meiosis, so that mutations that influenced one process would by necessity influence the other. Which model is correct remains to be seen.

7.3. What are the architectural rules for translational regulation?

As seen in Fig. 2.4, the translational regulatory networks active in the germ line involve many pleiotropic proteins, like *fbf-1* or *gld-1*. What are the rules that guide the functions of these translational regulators? One possibility is that this pathway is complex because it was shaped by several contingent events during evolution, and was particularly sensitive to small changes since sex-determination in hermaphrodites operates close to the margin that separates spermatogenesis from oogenesis. However, it is also possible that there is a spatial logic to the organization of these networks that is not captured in genetic diagrams. For example, overlapping patterns of translational regulatory proteins might define precise regions along the distal/proximal axis of the germ line, so as to activate or repress mRNAs in each of these regions. If so, translational controls would create the kind of complex spatial patterning associated with transcriptional regulation (reviewed by Rivera-Pomar and Jackle, 1996) or mRNA localization (reviewed by Lecuyer *et al.*, 2007) in the *Drosophila* embryo.

7.4. How has this pathway evolved?

The elegans group of nematodes includes two male/hermaphrodite species and several male/female ones. Since female species appear to be the ancestral state (Cho *et al.*, 2004; Kiontke *et al.*, 2004), one major question is how hermaphroditism itself evolved in *C. elegans* and *C. briggsae*. In addition, by studying how sexual fate is controlled in each of these species, we could learn (1) what elements of the pathway have been stable and (2) what type of modifications to the core pathway are allowed.

So far, the core pathway appears to be conserved between *C. elegans*, *C. briggsae*, and *C. remanei*. For example, *fog-3* is similar in all three species

(Chen et al., 2001), *tra-2* and *fem-3* function similarly and interact in all three species (Haag et al., 2002; Kuwabara, 1996b), the MX region of TRA-2 is conserved in all three species, its interaction TRA-1 has been shown to be conserved in *C. briggsae* (Wang and Kimble, 2001), and *tra-1* itself is conserved in sequence and function in *C. briggsae* (de Bono and Hodgkin, 1996) and *C. remanei* (S. Cho and R. E. Ellis, unpublished results).

However, very interesting differences arise when one examines how this core pathway has been modified. For example, *C. briggsae* lacks a FOG-2 homolog and inactivation of GLD-1 cause spermatogenesis rather than preventing it, as in *C. elegans* (Nayak et al., 2005). Instead, *C. briggsae* used a novel F-box protein to promote hermaphrodite spermatogenesis (Y. Guo and R. E. Ellis, manuscript in preparation). Similarly, the FEM proteins play two roles in spermatogenesis in *C. elegans*, but only one in *C. briggsae* (Hill et al., 2006). Thus, modifications to the core pathway might be very fluid during evolution.

7.5. What other somatic signals influence germ cell fates?

Two independent lines of evidence suggest that HER-1 is not the only somatic signal to control germ cell fates. First, in *XX* animals, the ablation of a single sheath/spermathecal precurser cell sometimes causes all germ cells to differentiate as oocytes (McCarter et al., 1997). Since these animals do not express HER-1, a different signal might be involved.

Second, the *fshr-1* gene acts in the somatic gonad to control several aspects of germ cell development (Cho et al., 2007). Surprisingly, *fshr-1; fbf-1; fbf-2* triple mutants bypass the requirement for *fog-1* and *fog-3* in spermatogenesis. Even more surprising, these effects are enhanced by mutations that block cell death. Although some results suggest that FSHR-1 acts through a heterotrimeric G-protein, how it controls downstream portions of the sex determination pathway remains unknown.

Since ablation of germ cells sometimes causes *XX* animals to make only oocytes (McCarter et al., 1997), it is possible that both of these pathways act indirectly, by regulating proliferation in the developing germ line. As discussed above, numerous studies have shown that many genes that regulate sexual fate in the germ line also control proliferation and vice versa (Kimble and Crittenden, 2007; Thompson et al., 2005).

8. Conclusion

This is a particularly exciting time to be studying sex determination in the germ line. As shown above, many of the genes that control this decision have been identified in *C. elegans* and we are in a position to address deep questions about how this regulatory circuit works and how it evolved. In addition, the next 5 years should see a flowering of comparative biology in which the roles of each of these genes are elucidated in a wide variety of species.

ACKNOWLEDGMENTS

I thank Judith Kimble and David Zarkower for comments and advice, and the American Cancer Society and National Science Foundation for grants that supported this work.

REFERENCES

Ahringer, J., and Kimble, J. (1991). Control of the sperm-oocyte switch in *Caenorhabditis elegans* hermaphrodites by the *fem-3* 3' untranslated region. *Nature* **349**(6307), 346–348.

Ahringer, J., Rosenquist, T. A., Lawson, D. N., and Kimble, J. (1992). The *Caenorhabditis elegans* sex determining gene *fem-3* is regulated post-transcriptionally. *EMBO J.* **11**(6), 2303–2310.

Alexandre, C., Jacinto, A., and Ingham, P. W. (1996). Transcriptional activation of hedgehog target genes in *Drosophila* is mediated directly by the cubitus interruptus protein, a member of the GLI family of zinc finger DNA-binding proteins. *Genes Dev.* **10**(16), 2003–2013.

Bachorik, J. L., and Kimble, J. (2005). Redundant control of the *Caenorhabditis elegans* sperm/oocyte switch by PUF-8 and FBF-1, two distinct PUF RNA-binding proteins. *Proc. Natl. Acad. Sci. USA* **102**(31), 10893–10897.

Barnes, T. M., and Hodgkin, J. (1996). The *tra-3* sex determination gene of *Caenorhabditis elegans* encodes a member of the calpain regulatory protease family. *EMBO J.* **15**(17), 4477–4484.

Barton, M. K., and Kimble, J. (1990). *Fog-1*, a regulatory gene required for specification of spermatogenesis in the germ line of *Caenorhabditis elegans*. *Genetics* **125**(1), 29–39.

Barton, M. K., Schedl, T. B., and Kimble, J. (1987). Gain-of-function mutations of *fem-3*, a sex-determination gene in *Caenorhabditis elegans*. *Genetics* **115**(1), 107–119.

Belfiore, M., Pugnale, P., Saudan, Z., and Puoti, A. (2004). Roles of the *C. elegans* cyclophilin-like protein MOG-6 in MEP-1 binding and germline fates. *Development* **131**(12), 2935–2945.

Brenner, S. (1974). The genetics of *Caenorhabditis elegans*. *Genetics* **77**(1), 71–94.

Chen, P., and Ellis, R. E. (2000). TRA-1A regulates transcription of fog-3, which controls germ cell fate in *C. elegans*. *Development* **127**(14), 3119–3129.

Chen, P. J., Singal, A., Kimble, J., and Ellis, R. E. (2000). A novel member of the tob family of proteins controls sexual fate in *Caenorhabditis elegans* germ cells. *Dev. Biol.* **217**(1), 77–90.

Chen, P. J., Cho, S., Jin, S. W., and Ellis, R. E. (2001). Specification of germ cell fates by FOG-3 has been conserved during nematode evolution. *Genetics* **158**(4), 1513–1525.

Cho, S., Jin, S. W., Cohen, A., and Ellis, R. E. (2004). A phylogeny of *Caenorhabditis* reveals frequent loss of introns during nematode evolution. *Genome Res.* **14**(7), 1207–1220.

Cho, S., Rogers, K. W., and Fay, D. S. (2007). The *C. elegans* glycopeptide hormone receptor ortholog, FSHR-1, regulates germline differentiation and survival. *Curr. Biol.* **17**(3), 203–212.

Ciosk, R., DePalma, M., and Priess, J. R. (2004). ATX-2, the *C. elegans* ortholog of ataxin 2, functions in translational regulation in the germline. *Development* **131**(19), 4831–4841.

Clifford, R., Lee, M. H., Nayak, S., Ohmachi, M., Giorgini, F., and Schedl, T. (2000). FOG-2, a novel F-box containing protein, associates with the GLD-1 RNA binding protein and directs male sex determination in the *C. elegans* hermaphrodite germline. *Development* **127**(24), 5265–5276.

Crittenden, S. L., Bernstein, D. S., Bachorik, J. L., Thompson, B. E., Gallegos, M., Petcherski, A. G., Moulder, G., Barstead, R., Wickens, M., and Kimble, J. (2002). A conserved RNA-binding protein controls germline stem cells in *Caenorhabditis elegans*. *Nature* **417**(6889), 660–663.

de Bono, M., and Hodgkin, J. (1996). Evolution of sex determination in *Caenorhabditis*: Unusually high divergence of tra-1 and its functional consequences. *Genetics* **144**(2), 587–595.

de Moor, C. H., Meijer, H., and Lissenden, S. (2005). Mechanisms of translational control by the 3' UTR in development and differentiation. *Semin. Cell Dev. Biol.* **16**(1), 49–58.

Doniach, T. (1986). Activity of the sex-determining gene tra-2 is modulated to allow spermatogenesis in the *C. elegans* hermaphrodite. *Genetics* **114**(1), 53–76.

Doniach, T., and Hodgkin, J. (1984). A sex-determining gene, fem-1, required for both male and hermaphrodite development in *Caenorhabditis elegans*. *Dev. Biol.* **106**(1), 223–235.

Eckmann, C. R., Kraemer, B., Wickens, M., and Kimble, J. (2002). GLD-3, a bicaudal-C homolog that inhibits FBF to control germline sex determination in *C. elegans*. *Dev. Cell* **3**(5), 697–710.

Eckmann, C. R., Crittenden, S. L., Suh, N., and Kimble, J. (2004). GLD-3 and control of the mitosis/meiosis decision in the germline of *Caenorhabditis elegans*. *Genetics* **168**(1), 147–160.

Ellis, R. E., and Kimble, J. (1995). The fog-3 gene and regulation of cell fate in the germ line of *Caenorhabditis elegans*. *Genetics* **139**(2), 561–577.

Fitch, D. H. A. (2002). Phylogeny. In "Wormatlas," wormatlas.org.

Francis, R., Barton, M. K., Kimble, J., and Schedl, T. (1995a). Gld-1, a tumor suppressor gene required for oocyte development in *Caenorhabditis elegans*. *Genetics* **139**(2), 579–606.

Francis, R., Maine, E., and Schedl, T. (1995b). Analysis of the multiple roles of gld-1 in germline development: Interactions with the sex determination cascade and the glp-1 signaling pathway. *Genetics* **139**(2), 607–630.

Gallegos, M., Ahringer, J., Crittenden, S., and Kimble, J. (1998). Repression by the 3' UTR of fem-3, a sex-determining gene, relies on a ubiquitous mog-dependent control in *Caenorhabditis elegans*. *EMBO J.* **17**(21), 6337–6347.

Goodwin, E. B., Okkema, P. G., Evans, T. C., and Kimble, J. (1993). Translational regulation of tra-2 by its 3' untranslated region controls sexual identity in *C. elegans*. *Cell* **75**(2), 329–339.

Goodwin, E. B., Hofstra, K., Hurney, C. A., Mango, S., and Kimble, J. (1997). A genetic pathway for regulation of tra-2 translation. *Development* **124**(3), 749–758.

Graham, P. L., and Kimble, J. (1993). The mog-1 gene is required for the switch from spermatogenesis to oogenesis in *Caenorhabditis elegans*. *Genetics* **133**(4), 919–931.

Graham, P. L., Schedl, T., and Kimble, J. (1993). More mog genes that influence the switch from spermatogenesis to oogenesis in the hermaphrodite germ line of *Caenorhabditis elegans*. *Dev. Genet.* **14**(6), 471–484.

Haag, E. S., Wang, S., and Kimble, J. (2002). Rapid coevolution of the nematode sex-determining genes fem-3 and tra-2. *Curr. Biol.* **12**(23), 2035–2041.

Hamaoka, B. Y., Dann, C. E., III, Geisbrecht, B. V., and Leahy, D. J. (2004). Crystal structure of *Caenorhabditis elegans* HER-1 and characterization of the interaction between HER-1 and TRA-2A. *Proc. Natl. Acad. Sci. USA* **101**(32), 11673–11678.

Hill, R. C., de Carvalho, C. E., Salogiannis, J., Schlager, B., Pilgrim, D., and Haag, E. S. (2006). Genetic flexibility in the convergent evolution of hermaphroditism in *Caenorhabditis* nematodes. *Dev. Cell* **10**(4), 531–538.

Hodgkin, J. (1980). More sex-determination mutants of *Caenorhabditis elegans*. *Genetics* **96**(3), 649–664.

Hodgkin, J. (1986). Sex determination in the nematode *C. elegans*: Analysis of tra-3 suppressors and characterization of fem genes. *Genetics* **114**(1), 15–52.

Hodgkin, J. (1987). A genetic analysis of the sex-determining gene, tra-1, in the nematode *Caenorhabditis elegans*. *Genes Dev.* **1**(7), 731–745.

Hodgkin, J., and Albertson, D. G. (1995). Isolation of dominant XO-feminizing mutations in *Caenorhabditis elegans*: New regulatory tra alleles and an X chromosome duplication with implications for primary sex determination. *Genetics* **141**(2), 527–542.

Hodgkin, J. A., and Brenner, S. (1977). Mutations causing transformation of sexual phenotype in the nematode *Caenorhabditis elegans*. *Genetics* **86**(2): Pt. 1, 275–287.

Hunter, C. P., and Wood, W. B. (1992). Evidence from mosaic analysis of the masculinizing gene her-1 for cell interactions in *C. elegans* sex determination. *Nature* **355**(6360), 551–555.

Jager, S., Schwartz, H. T., Horvitz, H. R., and Conradt, B. (2004). The *Caenorhabditis elegans* F-box protein SEL-10 promotes female development and may target FEM-1 and FEM-3 for degradation by the proteasome. *Proc. Natl. Acad. Sci. USA* **101**(34), 12549–12554.

Jan, E., Motzny, C. K., Graves, L. E., and Goodwin, E. B. (1999). The STAR protein, GLD-1, is a translational regulator of sexual identity in *Caenorhabditis elegans*. *EMBO J.* **18**(1), 258–269.

Jia, S., and Meng, A. (2007). Tob genes in development and homeostasis. *Dev. Dyn.* **236**(4), 913–921.

Jiang, J. (2002). Degrading Ci: Who is Cul-pable? *Genes Dev.* **16**(18), 2315–2321.

Jin, S. W., Arno, N., Cohen, A., Shah, A., Xu, Q., Chen, N., and Ellis, R. E. (2001a). In *Caenorhabditis elegans*, the RNA-binding domains of the cytoplasmic polyadenylation element binding protein FOG-1 are needed to regulate germ cell fates. *Genetics* **159**(4), 1617–1630.

Jin, S. W., Kimble, J., and Ellis, R. E. (2001b). Regulation of cell fate in *Caenorhabditis elegans* by a novel cytoplasmic polyadenylation element binding protein. *Dev. Biol.* **229**(2), 537–553.

Jones, A. R., and Schedl, T. (1995). Mutations in gld-1, a female germ cell-specific tumor suppressor gene in *Caenorhabditis elegans*, affect a conserved domain also found in Src-associated protein Sam68. *Genes Dev.* **9**(12), 1491–1504.

Jones, A. R., Francis, R., and Schedl, T. (1996). GLD-1, a cytoplasmic protein essential for oocyte differentiation, shows stage- and sex-specific expression during *Caenorhabditis elegans* germline development. *Dev. Biol.* **180**(1), 165–183.

Karashima, T., Sugimoto, A., and Yamamoto, M. (2000). *Caenorhabditis elegans* homologue of the human azoospermia factor DAZ is required for oogenesis but not for spermatogenesis. *Development* **127**(5), 1069–1079.

Kawano, T., Kataoka, N., Dreyfuss, G., and Sakamoto, H. (2004). Ce-Y14 and MAG-1, components of the exon-**exon junction complex, are required for embryogenesis and germline sexual switching in *Caenorhabditis elegans*. *Mech. Dev.* **121**(1), 27–35.

Kimble, J., and Crittenden, S. L. (2007). Control of germline stem cells, entry into meiosis, and the sperm/oocyte decision in *C. elegans*. *Annu. Rev. Cell Dev. Biol.* **23**, 405–433.

Kimble, J., and Page, D. C. (2007). The mysteries of sexual identity. The germ cell's perspective. *Science* **316**(5823), 400–401.

Kimble, J., and Sharrock, W. J. (1983). Tissue-specific synthesis of yolk proteins in *Caenorhabditis elegans*. *Dev. Biol.* **96**(1), 189–196.

Kimble, J., Edgar, L., and Hirsh, D. (1984). Specification of male development in *Caenorhabditis elegans*: The fem genes. *Dev. Biol.* **105**(1), 234–239.

Kiontke, K., Gavin, N. P., Raynes, Y., Roehrig, C., Piano, F., and Fitch, D. H. (2004). *Caenorhabditis* phylogeny predicts convergence of hermaphroditism and extensive intron loss. *Proc. Natl. Acad. Sci. USA* **101**(24), 9003–9008.

Kipreos, E. T. (2005). Ubiquitin-mediated pathways in *C. elegans*. *In* "Wormbook," ed. The *C. elegans* Research Community.

Kraemer, B., Crittenden, S., Gallegos, M., Moulder, G., Barstead, R., Kimble, J., and Wickens, M. (1999). NANOS-3 and FBF proteins physically interact to control the sperm–oocyte switch in *Caenorhabditis elegans*. *Curr. Biol.* **9**(18), 1009–1018.

Kuwabara, P. E. (1996a). A novel regulatory mutation in the *C. elegans* sex determination gene tra-2 defines a candidate ligand/receptor interaction site. *Development* **122**(7), 2089–2098.

Kuwabara, P. E. (1996b). Interspecies comparison reveals evolution of control regions in the nematode sex-determining gene tra-2. *Genetics* **144**(2), 597–607.

Kuwabara, P. E., and Kimble, J. (1995). A predicted membrane protein, TRA-2A, directs hermaphrodite development in *Caenorhabditis elegans*. *Development* **121**(9), 2995–3004.

Kuwabara, P. E., Okkema, P. G., and Kimble, J. (1992). Tra-2 encodes a membrane protein and may mediate cell communication in the *Caenorhabditis elegans* sex determination pathway. *Mol. Biol. Cell* **3**(4), 461–473.

Kuwabara, P. E., Okkema, P. G., and Kimble, J. (1998). Germ-line regulation of the *Caenorhabditis elegans* sex-determining gene tra-2. *Dev. Biol.* **204**(1), 251–262.

Lakiza, O., Frater, L., Yoo, Y., Villavicencio, E., Walterhouse, D., Goodwin, E. B., and Iannaccone, P. (2005). STAR proteins quaking-6 and GLD-1 regulate translation of the homologues GLI1 and tra-1 through a conserved RNA 3' UTR-based mechanism. *Dev. Biol.* **287**(1), 98–110.

Lamont, L. B., and Kimble, J. (2007). Developmental expression of FOG-1/CPEB protein and its control in the *Caenorhabditis elegans* hermaphrodite germ line. *Dev. Dyn.* **236**(3), 871–879.

Lamont, L. B., Crittenden, S. L., Bernstein, D., Wickens, M., and Kimble, J. (2004). FBF-1 and FBF-2 regulate the size of the mitotic region in the *C. elegans* germline. *Dev. Cell* **7**(5), 697–707.

LaMunyon, C. W., and Ward, S. (1998). Larger sperm outcompete smaller sperm in the nematode *Caenorhabditis elegans*. *Proc. Biol. Sci.* **265**(1409), 1997–2002.

Lecuyer, E., Yoshida, H., Parthasarathy, N., Alm, C., Babak, T., Cerovina, T., Hughes, T. R., Tomancak, P., and Krause, H. M. (2007). Global analysis of mRNA localization reveals a prominent role in organizing cellular architecture and function. *Cell* **131**(1), 174–187.

Lee, M. H., and Schedl, T. (2001). Identification of *in vivo* mRNA targets of GLD-1, a maxi-KH motif containing protein required for *C. elegans* germ cell development. *Genes Dev.* **15**(18), 2408–2420.

L'Hernault, S. W. (2006). Spermatogenesis. *In* "Wormbook," ed. The *C. elegans* Research Community, www.wormbook.org.

Li, W., Boswell, R., and Wood, W. B. (2000). Mag-1, a homolog of *Drosophila* mago nashi, regulates hermaphrodite germ-line sex determination in *Caenorhabditis elegans*. *Dev. Biol.* **218**(2), 172–182.

Luitjens, C., Gallegos, M., Kraemer, B., Kimble, J., and Wickens, M. (2000). CPEB proteins control two key steps in spermatogenesis in *C. elegans*. *Genes Dev.* **14**(20), 2596–2609.

Lum, D. H., Kuwabara, P. E., Zarkower, D., and Spence, A. M. (2000). Direct protein–protein interaction between the intracellular domain of TRA-2 and the transcription factor TRA-1A modulates feminizing activity in *C. elegans*. *Genes Dev.* **14**(24), 3153–3165.

Maine, E. M., Hansen, D., Springer, D., and Vought, V. E. (2004). *Caenorhabditis elegans* atx-2 promotes germline proliferation and the oocyte fate. *Genetics* **168**(2), 817–830.

Marin, V. A., and Evans, T. C. (2003). Translational repression of a *C. elegans* Notch mRNA by the STAR/KH domain protein GLD-1. *Development* **130**(12), 2623–2632.

Mathies, L. D., Schvarzstein, M., Morphy, K. M., Blelloch, R., Spence, A. M., and Kimble, J. (2004). TRA-1/GLI controls development of somatic gonadal precursors in *C. elegans*. *Development* **131**(17), 4333–4343.

McCarter, J., Bartlett, B., Dang, T., and Schedl, T. (1997). Soma-germ cell interactions in *Caenorhabditis elegans*: Multiple events of hermaphrodite germline development require the somatic sheath and spermathecal lineages. *Dev. Biol.* **181**(2), 121–143.

Mehra, A., Gaudet, J., Heck, L., Kuwabara, P. E., and Spence, A. M. (1999). Negative regulation of male development in *Caenorhabditis elegans* by a protein–protein interaction between TRA-2A and FEM-3. *Genes Dev.* **13**(11), 1453–1463.

Mootz, D., Ho, D. M., and Hunter, C. P. (2004). The STAR/Maxi-KH domain protein GLD-1 mediates a developmental switch in the translational control of *C. elegans* PAL-1. *Development* **131**(14), 3263–3272.

Nayak, S., Goree, J., and Schedl, T. (2005). Fog-2 and the evolution of self-fertile hermaphroditism in *Caenorhabditis*. *PLoS Biol.* **3**(1), e6.

Nott, A., Le Hir, H., and Moore, M. J. (2004). Splicing enhances translation in mammalian cells: An additional function of the exon junction complex. *Genes Dev.* **18**(2), 210–222.

Okkema, P. G., and Kimble, J. (1991). Molecular analysis of tra-2, a sex determining gene in *C. elegans*. *EMBO J.* **10**(1), 171–176.

Otori, M., Karashima, T., and Yamamoto, M. (2006). The *Caenorhabditis elegans* homologue of deleted in azoospermia is involved in the sperm/oocyte switch. *Mol. Biol. Cell* **17**(7), 3147–3155.

Perry, M. D., Li, W., Trent, C., Robertson, B., Fire, A., Hageman, J. M., and Wood, W. B. (1993). Molecular characterization of the her-1 gene suggests a direct role in cell signaling during *Caenorhabditis elegans* sex determination. *Genes Dev.* **7**(2), 216–228.

Pilgrim, D., McGregor, A., Jackle, P., Johnson, T., and Hansen, D. (1995). The *C. elegans* sex-determining gene fem-2 encodes a putative protein phosphatase. *Mol. Biol. Cell* **6**(9), 1159–1171.

Puoti, A., and Kimble, J. (1999). The *Caenorhabditis elegans* sex determination gene mog-1 encodes a member of the DEAH-Box protein family. *Mol. Cell Biol.* **19**(3), 2189–2197.

Puoti, A., and Kimble, J. (2000). The hermaphrodite sperm/oocyte switch requires the *Caenorhabditis elegans* homologs of PRP2 and PRP22. *Proc. Natl. Acad. Sci. USA* **97**(7), 3276–3281.

Rivera-Pomar, R., and Jackle, H. (1996). From gradients to stripes in *Drosophila* embryogenesis: Filling in the gaps. *Trends Genet.* **12**(11), 478–483.

Ruiz i Altaba, A. (1999). Gli proteins encode context-dependent positive and negative functions: Implications for development and disease. *Development* **126**(14), 3205–3216.

Ryder, S. P., Frater, L. A., Abramovitz, D. L., Goodwin, E. B., and Williamson, J. R. (2004). RNA target specificity of the STAR/GSG domain post-transcriptional regulatory protein GLD-1. *Nat. Struct. Mol. Biol.* **11**(1), 20–28.

Schedl, T., and Kimble, J. (1988). Fog-2, a germ-line-specific sex determination gene required for hermaphrodite spermatogenesis in *Caenorhabditis elegans*. *Genetics* **119**(1), 43–61.

Schedl, T., Graham, P. L., Barton, M. K., and Kimble, J. (1989). Analysis of the role of tra-1 in germline sex determination in the nematode *Caenorhabditis elegans*. *Genetics* **123**(4), 755–769.

Schumacher, B., Hanazawa, M., Lee, M. H., Nayak, S., Volkmann, K., Hofmann, E. R., Hengartner, M., Schedl, T., and Gartner, A. (2005). Translational repression of *C. elegans* p53 by GLD-1 regulates DNA damage-induced apoptosis. *Cell* **120**(3), 357–368.

Schvarzstein, M., and Spence, A. M. (2006). The *C. elegans* sex-determining GLI protein TRA-1A is regulated by sex-specific proteolysis. *Dev. Cell* **11**(5), 733–740.

Shimada, M., Kanematsu, K., Tanaka, K., Yokosawa, H., and Kawahara, H. (2006). Proteasomal ubiquitin receptor RPN-10 controls sex determination in *Caenorhabditis elegans*. *Mol. Biol. Cell* **17**(12), 5356–5371.

Singh, G., and Lykke-Andersen, J. (2003). New insights into the formation of active nonsense-mediated decay complexes. *Trends Biochem. Sci.* **28**(9), 464–466.

Singson, A., Hill, K. L., and L'Hernault, S. W. (1999). Sperm competition in the absence of fertilization in *Caenorhabditis elegans*. *Genetics* **152**(1), 201–208.

Sokol, S. B., and Kuwabara, P. E. (2000). Proteolysis in *Caenorhabditis elegans* sex determination: Cleavage of TRA-2A by TRA-3. *Genes Dev.* **14**(8), 901–906.

Spence, A. M., Coulson, A., and Hodgkin, J. (1990). The product of fem-1, a nematode sex-determining gene, contains a motif found in cell cycle control proteins and receptors for cell–cell interactions. *Cell* **60**(6), 981–990.

Stanfield, G. M., and Villeneuve, A. M. (2006). Regulation of sperm activation by SWM-1 is required for reproductive success of *C. elegans* males. *Curr. Biol.* **16**(3), 252–263.

Starostina, N. G., Lim, J. M., Schvarzstein, M., Wells, L., Spence, A. M., and Kipreos, E. T. (2007). A CUL-2 ubiquitin ligase containing three FEM proteins degrades TRA-1 to regulate *C. elegans* sex determination. *Dev. Cell* **13**(1), 127–139.

Subramaniam, K., and Seydoux, G. (2003). Dedifferentiation of primary spermatocytes into germ cell tumors in *C. elegans* lacking the pumilio-like protein PUF-8. *Curr. Biol.* **13**(2), 134–139.

Thompson, B. E., Bernstein, D. S., Bachorik, J. L., Petcherski, A. G., Wickens, M., and Kimble, J. (2005). Dose-dependent control of proliferation and sperm specification by FOG-1/CPEB. *Development* **132**(15), 3471–3481.

Walser, C. B., Battu, G., Hoier, E. F., and Hajnal, A. (2006). Distinct roles of the Pumilio and FBF translational repressors during *C. elegans* vulval development. *Development* **133**(17), 3461–3471.

Wang, S., and Kimble, J. (2001). The TRA-1 transcription factor binds TRA-2 to regulate sexual fates in *Caenorhabditis elegans*. *EMBO J.* **20**(6), 1363–1372.

Wolff, J. R., and Zarkower, D. (2008). Somatic sexual differentiation in *C. elegans*. *In* "Sex Determination & Differentiation," (M. Van Doren, ed.). Academic Press.★★

Zarkower, D., and Hodgkin, J. (1992). Molecular analysis of the *C. elegans* sex-determining gene tra-1: A gene encoding two zinc finger proteins. *Cell* **70**(2), 237–249.

Zhang, B., Gallegos, M., Puoti, A., Durkin, E., Fields, S., Kimble, J., and Wickens, M. P. (1997). A conserved RNA-binding protein that regulates sexual fates in the *C. elegans* hermaphrodite germ line. *Nature* **390**(6659), 477–484.

CHAPTER THREE

The Creation of Sexual Dimorphism in the *Drosophila* Soma

Nicole Camara, Cale Whitworth, *and* Mark Van Doren

Contents

1. Sex Determination	66
1.1. X chromosome dosage controls sex determination	66
1.2. Sex lethal	67
1.3. Transformer	69
1.4. Doublesex	70
2. The Creation of Sexual Dimorphism	74
2.1. Sexual dimorphism in the gonad: Ovary vs. Testis	74
2.2. The genital disc	85
2.3. Sex combs and sexual dimorphism in the leg	88
2.4. Sexual dimorphism in the abdomen	90
2.5. Sexual dimorphism in the fat body	92
2.6. Muscle of Lawrence	92
3. A New Look at an Old Problem	93
3.1. Not all cells decide their own sex	93
3.2. Not all cells even know their own sex	94
3.3. Integrating sex and patterning	95
3.4. Of flies and men	95
References	96

Abstract

Animals have evolved a fascinating array of mechanisms for conducting sexual reproduction. These include producing the sex-specific gametes, as well as mechanisms for attracting a mate, courting a mate, and getting the gametes together. These processes require that males and females take on dramatically different forms (sexual dimorphism). Here, we will explore the problem of how sex is determined in *Drosophila*, and pay particular attention to how information about sexual identity is used to instruct males and females to develop differently. Along the way, we will highlight new work that challenges some of the traditional views about sex determination. In

Department of Biology, Johns Hopkins University, Baltimore, MD 21218

Drosophila, it is commonly thought that every cell decides its own sex based on its sex chromosome constitution (XX vs. XY). However, we now know that many cell types undergo nonautonomous sex determination, where they are told what sex to be through signals from surrounding cells, independent of their own chromosomal content. Further, it now appears that not all cells even "know" their sex, since key members of the sex determination pathway are not expressed in all cells. Thus, our understanding of how sex is determined, and how sexual identity is used to create sexual dimorphism, has changed considerably.

1. Sex Determination

1.1. X chromosome dosage controls sex determination

Sex determination is usually controlled by a "switch" which initiates the male or female mode of development. Sex determination switches vary widely in different animal species, and can be controlled by either genetic (i.e., sex chromosome constitution) or environmental mechanisms. In flies, XX individuals are female and XY individuals are male (Bridges, 1916; Stevens, 1908). Sex is regulated by the number of X chromosomes (Bridges, 1913, 1916), since XXY individuals are fertile females, while XO individuals are males, but are sterile because genes on the Y chromosome are required for spermatogenesis. Multiple regions of the X chromosome contribute to its "female-promoting" character (Dobzahnsky, 1931; Patterson *et al.*, 1937). It has long been thought that X chromosomes are counted relative to the number of autosomes (X:A ratio), since individuals with two X chromosomes but three sets of autosomes (2X:3A) are intersexual, while haploid individuals (1X:1A) and individuals with three sets of each (3X:3A) are female (Bridges, 1921, 1930). Recently, however, the role of the autosomes has been attributed to effects on the embryo independent of X chromosome counting, suggesting that X chromosome number alone is the primary sex determination signal (Erickson and Quintero, 2007). Regardless, it is clear that in normal diploid animals, the number of X chromosomes determines sex. Eventually, the process of X chromosome dosage compensation increases gene expression from the single X chromosome in males, so that it matches expression from the two X chromosomes in females (Lucchesi *et al.*, 2005). Thus, X chromosomes must be "counted" while there is still a difference in expression of X chromosome genes.

1.2. Sex lethal

The key downstream target that is regulated by X chromosome number is the gene *Sex lethal* (*Sxl*), which controls both sex determination and X chromosome dosage compensation (Fig. 3.1; reviewed in Cline and Meyer, 1996).

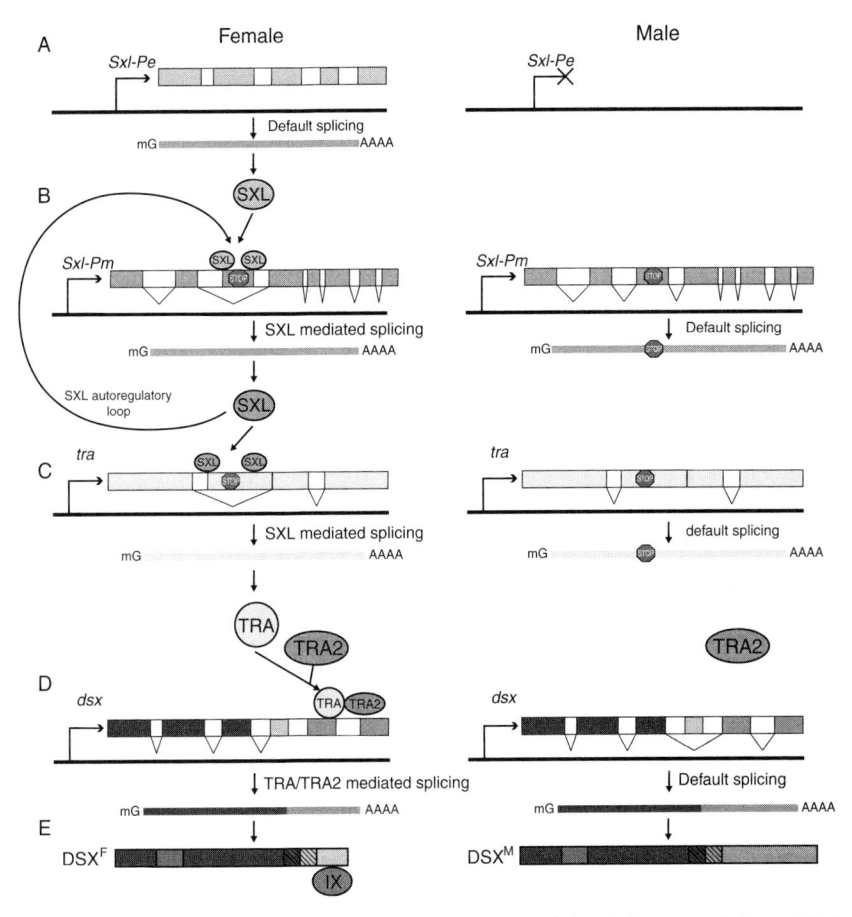

Figure 3.1 The *D. melanogaster* sex determination heirarchy. (A) Transcription of *Sxl* from the early promoter (Pe) is activated in females, but remains inactive in males. (B) SXL protein, found only in females, directs alternative splicing of the *Sxl Pm* transcript to an active form via removal of an early stop codon. This auto-regulatory loop maintains SXL expression in females. In males, default splicing retains the early stop codon. (C) SXL regulates splicing of *tra* in females, resulting in removal of an early stop codon and production of TRA protein. In males, default splicing of *tra* retains the early stop condon. (D) In females, TRA and TRA2 direct splicing of the *doublesex* transcript into the female-specific isoform (dsx^F). The lack of functional TRA protein in males results in default splicing of the *dsx* transcript into the male-specific isoform (DSX^M). (E) Schematic of the two DSX isoforms. The DNA binding domain (DM domain) and the N-terminal dimerization domain are indicated in red. The C-terminal dimerization domain is indicated by stripes and consists of a non sex-specific region and a sex-specific region. Also shown is IX, which interacts specifically with the N-terminal tail of DSX^F. (See Color Insert.)

Sxl encodes an RNA binding protein that can regulate both alternative RNA splicing and mRNA translation (reviewed in Penalva and Sanchez, 2003). *Sxl* has two promoters and is initially expressed from the "establishment promoter" (*Pe*), which is activated in XX (female) embryos (Keyes *et al.*, 1992; Salz *et al.*, 1989). The transcript produced from *Pe* does not require regulated splicing to produce an mRNA that encodes a functional SXL protein. Subsequently, *Pe* becomes inactive and *Sxl* is transcribed from the "maintenance promoter" (*Pm*), which is active in both males and females. The transcript from *Pm* requires alternative splicing, regulated by SXL, in order to produce an mRNA encoding a functional SXL protein (Bell *et al.*, 1988). Since SXL protein is initially made only in females (from the *Pe* transcript), only females can splice the *Pm* transcript into a form encoding SXL. This auto-regulatory loop is critical for maintaining female-specific SXL expression throughout the life of the animal (Bell *et al.*, 1991; Cline, 1984; Sanchez and Nothiger, 1983). The complicated switch from *Pe* to *Pm* is necessary so that X chromosome number can be "read out" (by *Pe*) during the short time window before dosage compensation equalizes X chromosome gene expression in males and females; afterwards the maintenance of sex-specific SXL expression (by *Pm*) is dependent on SXL auto-regulation rather than X chromosome number.

Activation of Sxl   Sex-specific expression of *Sxl* is regulated by several X chromosome loci, *sisterlessA* (*sisA*), *sisB* (or *scute*), *sisC*, and *runt*, which promote activation of *Pe* (reviewed in Cline and Meyer, 1996; Sanchez *et al.*, 1994). *sisA*, *sisB*, and *runt* encode transcription factors of the bZIP, bHLH, and RUNT domain families (Erickson and Cline, 1993; Kania *et al.*, 1990; Villares and Cabrera, 1987), respectively, while *sisC* (also known as *outstretched* or *unpaired*) encodes a secreted ligand that can lead to activation of the STAT92E transcription factor (Jinks *et al.*, 2000; Sefton *et al.*, 2000). These genes are expressed in the early embryo, just after activation of the zygotic genome and before dosage compensation affects X chromosome gene expression. Thus, the two-fold increase in expression of these loci in XX animals is able to initiate sex-specific expression of *Sxl Pe*, and some of these factors have been shown to directly regulate *Sxl* (Avila and Erickson, 2007; Kramer *et al.*, 1999; Yang *et al.*, 2001). Activation of *Sxl* by the X chromosome elements also requires factors not on the X chromosome, such as *daughterless*, a partner for *sisB/scute* (Cabrera and Alonso, 1991; Caudy *et al.*, 1988; Cline, 1980; Cronmiller *et al.*, 1988; Parkhurst *et al.*, 1990; Van Doren *et al.*, 1991), and *STAT92E* (Jinks *et al.*, 2000; Sefton *et al.*, 2000). In addition, the activators of *Sxl Pe* must overcome the action of negative regulators such as *extramacrochaete*, which blocks the activity of *sisB/scute*, and *deadpan*, which works with *groucho* as a repressor of *Sxl Pe* (Barbash and Cline, 1995; Ellis *et al.*, 1990; Garrell and Modolell, 1990; Paroush *et al.*, 1994; Van Doren *et al.*, 1991; Younger-Shepherd *et al.*, 1992). The negative regulators ensure that *Sxl Pe* will be properly repressed in males.

Targets of SXL In addition to regulating the splicing of its own RNA, there are two other known targets of SXL: *male-specific lethal 2* (*msl-2*), which is a key regulator of X chromosome dosage compensation, and *transformer* (*tra*), which regulates the differences in phenotype between males and females (sexual dimorphism). SXL acts as a repressor of *msl-2* at the level of both alternative splicing and translational regulation, so that MSL2 protein is only produced in males, leading to the upregulation of the single X chromosome (reviewed in Cline and Meyer, 1996; Lucchesi *et al.*, 2005; Meller and Kuroda, 2002; Penalva and Sanchez, 2003). SXL regulates *tra* by alternative splicing (Boggs *et al.*, 1987; Butler *et al.*, 1986), such that the female *tra* mRNA produces a functional TRA protein, while the male *tra* mRNA does not (Fig. 3.1). The activity of SXL in RNA splicing is regulated by several additional factors, such as *sans-fille, virilizer*, and *fl(2)d* (Flickinger and Salz, 1994; Granadino *et al.*, 1992b; Hilfiker-Kleiner *et al.*, 1994; Penn *et al.*, 2008), which themselves are not regulated sex-specifically. There is also evidence that additional targets for SXL exist. The MSL2 complex alone does not account for all of *Sxl*'s effects on dosage compensation (Meller and Kuroda, 2002), and SXL must also have at least one other target in the germline, where *Sxl* is required but *tra* and *msl-2* are not (see Hempel *et al.*, this volume).

1.3. Transformer

TRA is also an alternative splicing factor (Burtis and Baker, 1989; Nagoshi *et al.*, 1988) and controls all known aspects of sexual dimorphism in the *Drosophila* soma, apart from the difference in body size between males and females (McKeown *et al.*, 1988; Sturtevant, 1945). *tra* also regulates the nonautonomous somatic control over germline sex determination (Evans and Cline, 2007; Steinmann-Zwicky, 1994), but not the germline autonomous control over this process (see Hempel *et al.*, this volume). Although it has been speculated that there may be additional targets of *Sxl* besides *tra* that regulate aspects of sex determination in the soma (e.g., Evans and Cline, 2007), no examples of this have been found which can be clearly separated from *Sxl*'s effects on dosage compensation. Thus, *tra* remains the only known factor that is regulated by SXL to control sex-specific development of the soma.

TRA's function in alternative splicing requires the obligate partner Transformer 2 (TRA2; Amrein *et al.*, 1988; Fujihara *et al.*, 1978; Goralski *et al.*, 1989). Since TRA2 is active in both sexes (Mattox *et al.*, 1990), the sex-specific nature of this alternative splicing cascade is dependent on TRA. There are only two known targets for sex-specific alternative splicing by TRA/TRA2, *doublesex* (*dsx*) and *fruitless* (*fru*), which each encode sex-specific transcription factors. *dsx* primarily controls male versus female morphology, while *fru* controls male versus female behavior (along with *dsx* and other factors; Sanders and Arbeitman, 2008; Shirangi *et al.*, 2006; reviewed in Shirangi and McKeown, 2007). Since the topic of this review is the creation of sexual dimorphism, we will focus our attention on *dsx*.

1.4. Doublesex

dsx was first reported by Hildreth (1965) as a spontaneous mutation which creates intersexes in both male and female flies. Adults mutant for *dsx* have both male and female features present in their external genitalia and reproductive tracts, giving the gene its name (Hildreth, 1965). Interestingly, the role of *dsx* in sex determination and sexual differentiation is much more conserved than that of upstream genes such as *Sxl* (reviewed by Graham *et al.*, 2003; Pomiankowski *et al.*, 2004; Schutt and Nothiger, 2000). *Sxl* is only thought to act in sex determination in related *Drosophila* species, and not in other insects studied. In contrast, the role for *dsx* is highly conserved in different insects, and homologs of *dsx* (*dsx, mab-3* related transcription factors, DMRTs) have been implicated in sex-specific development in worms, frogs, fish, reptiles, birds, platypus, mice, and man (reviewed in Ferguson-Smith, 2007; Hong *et al.*, 2007; Zarkower, 2002). In mouse, several DMRTs are expressed in the developing gonad (Kim *et al.*, 2003), and DMRT1 is required for proper testis development (Kim *et al.*, 2007; Raymond *et al.*, 2000). In humans, deletions of chromosomal region 9p.24 are associated with male to female sex reversal. Multiple *dsx* homologs are found in this region (Ottolenghi *et al.*, 2000; Raymond *et al.*, 1999), at least one of which is expressed in the developing human testis (Moniot *et al.*, 2000). Despite the central role that *dsx* and its homologs play in sex determination, we know surprisingly little about *dsx* target genes and how *dsx* regulates sexual dimorphism.

Sex-specific control of DSX action   TRA and TRA2 regulate the splicing of the *dsx* transcript to produce male- and female-specific mRNAs (Fig. 3.1; Baker and Wolfner, 1988; Burtis and Baker, 1989; Nagoshi *et al.*, 1988). These each encode functional but distinct forms of DSX protein, DSX Female (DSX^F) and DSX Male (DSX^M), which have a common N-terminus but different C-termini (Burtis and Baker, 1989). The common N-terminus contains the DNA binding domain (the *dsx/mab-3* or DM domain) consisting of an atypical Zn-binding module and an adjacent DNA recognition helix (Erdman and Burtis, 1993; Narendra *et al.*, 2002). DSX binds DNA as a dimer, and there are two distinct domains that mediate dimerization (An *et al.*, 1996; Cho and Wensink, 1996; Erdman *et al.*, 1996). When DSX^F and DSX^M are both present at the same time ($dsx^D/+$), the resulting phenotype is similar to a *dsx* null mutant, indicating that the two proteins can interfere with each other's activity (Nagoshi and Baker, 1990). Whether this is due to the formation of inactive heterodimers or because the two proteins generally have opposite effects on the same target genes, is not known.

Since DSX^F and DSX^M have the same DNA binding domain, and the full-length proteins have similar DNA binding properties *in vitro* (Burtis *et al.*, 1991; Cho and Wensink, 1997), it appears that the sex-specific functions of DSX are not due to differences in their DNA binding specificity. Instead, their sex-specific functions are likely mediated by how their different C-terminal domains interact with other proteins (DSX cofactors) to influence gene expression. One such cofactor, *intersex* (*ix*), is required for *dsx* function in females, but not males (Garrett-Engele *et al.*, 2002; Waterbury *et al.*, 1999). *ix* mutant females appear intersexual, like *dsx* mutants, but *ix* mutant males are normal (Baker and Ridge, 1980; Chase and Baker, 1995). *ix* interacts specifically with the C-terminal region of DSX^F (Garrett-Engele *et al.*, 2002) and is homologous to a member of the mammalian mediator co-activator complex (Sato *et al.*, 2003), suggesting a role in transcriptional regulation. Since *ix* cannot account for all the differences in function between DSX^F and DSX^M, it is likely that other DSX cofactors exist. In addition to allowing DSX^F and DSX^M to have different effects on the same target genes, it is possible that some cofactors have their own DNA binding specificity, allowing DSX^F and DSX^M to regulate distinct targets *in vivo*.

DSX target genes Currently, we have little data with which to determine how DSX^F and DSX^M act to control sex-specific gene regulation. The only known direct target of DSX is the fat body enhancer (FBE), which controls expression of two yolk protein genes (*Yp1* and *Yp2*), that are expressed in females but not males (reviewed by Bownes, 1994) (See note added in proof). This enhancer is regulated by both DSX^M and DSX^F, and the two proteins have opposite effects on gene expression: DSX^F activates these genes in females while DSX^M represses them in males (Coschigano and Wensink, 1993). In *dsx* mutants (both XX and XY), the activity of the FBE is intermediate between normal males and normal females. Both DSX^F or DSX^M can bind to the same sites within the FBE, and these sites are essential for *dsx*-dependent regulation by this element (Burtis *et al.*, 1991; Coschigano and Wensink, 1993). The DSX^F-specific cofactor *ix* is essential for activation of the FBE by DSX^F, and FBE activity is intermediate in *ix* mutant females, similar to *dsx* mutants (Garrett-Engele *et al.*, 2002; Waterbury *et al.*, 1999). This indicates that DSX^F cannot function as a regulator of the FBE in the absence of *ix*, either because it fails to regulate the transcriptional machinery or because it can no longer bind to the FBE *in vivo*. However, in *ix* mutant males, DSX^M still functions to completely repress FBE expression (Garrett-Engele *et al.*, 2002), as it does in normal males. Thus, there are two independent mechanisms at work in regulating the *Yp* genes; DSX^F utilizes *ix* as a cofactor for activation of the *Yp* genes, while DSX^M utilizes an independent mechanism to repress these genes.

It is not clear whether the *Yp* gene model for sex-specific regulation by DSX will apply more generally to other DSX target genes, especially since some genes are female-specific, like the *Yp* genes, while others are male-specific. Directed studies of sexual dimorphism in specific tissues have identified some potential DSX targets (discussed in the following sections). In one example, *Wnt2* is expressed male-specifically in the genital disc (Kozopas *et al.*, 1998) and gonad (DeFalco *et al.*, 2008), and has DSX consensus binding sites in its putative upstream regulatory region that are conserved in other *Drosophila* species (DeFalco *et al.*, 2008). In *dsx* mutants, *Wnt2* is expressed in the gonad of both XX and XY animals (DeFalco *et al.*, 2008). Thus, if *Wnt2* is a direct DSX target, it is normally repressed by DSX^F in females, opposite of how DSX^F affects the *Yp* genes. Similarly, male-specific expression of *branchless* (*bnl*) in the genital disc is also repressed by DSX^F, and *bnl* has DSX consensus binding sites in its putative regulatory region (Ahmad and Baker, 2002). More comprehensive approaches have taken advantage of molecular and genomic analysis to identify sex-specifically expressed genes and study their changes in expression in different species (Andrews *et al.*, 2000; Arbeitman *et al.*, 2002, 2004; Dauwalder *et al.*, 2002; DiBenedetto *et al.*, 1987; Fujii and Amrein, 2002; Goldman and Arbeitman, 2007; Jin *et al.*, 2001; Parisi *et al.*, 2004; Ranz *et al.*, 2003; Xu *et al.*, 2002). A critical next step is to use similar genomic approaches to identify direct DSX target genes. Future work to understand DSX action at the level of the individual target gene, and the whole genome, promises to dramatically change our understanding of how DSX functions to control sexual dimorphism.

When and where does dsx act? The sex determination pathway usually acts at a particular time and place to control sexually dimorphic development of a specific tissue, and thus there are likely to be different classes of sex-specific genes. The *early* genes are those genes that are more-or-less directly regulated by the sex determination cascade. These include genes that are the direct targets of DSX (or FRU), along with the genes immediately downstream of these direct targets. Early genes will be expressed sex-specifically in a particular tissue at the time that *dsx* regulates sex-specific development of that tissue, and they may not be expressed at later stages of development. In contrast, the *late* genes are those genes that are expressed in a sex-specific manner because they are required for the terminal differentiation or function of a sex-specific tissue. Thus, the expression of these genes may be far downstream of the action of the sex determination pathway, and may be more dependent on the identity of a sex-specific tissue than on the action of *dsx*. Finally, some genes may be *continuously* regulated by the sex determination pathway. For some tissues, *dsx* may be required for both the initiation and maintenance of sexual dimorphism, and some genes can even be regulated by *dsx* in the adult fly. Knowing when *dsx* acts to control sex-specific gene expression in a particular tissue is essential for identifying direct DSX targets, and understanding how *dsx* controls sexual dimorphism.

Experiments in which the activity of the sex determination pathway is altered at different times in development, such as those utilizing a temperature-sensitive allele of *tra-2* (Belote and Baker, 1982), can provide an indication of when in development the sex determination pathway is important for a particular sexually dimorphic tissue. In most sex-specific tissues studied so far, the sex determination pathway is required at a discrete point (or points) in development, after which it is not required for the maintenance of that tissue (Baker and Ridge, 1980; Belote and Baker, 1982; Bownes *et al.*, 1990; Chapman and Wolfner, 1988; DiBenedetto *et al.*, 1987; Epper and Bryant, 1983; Feng *et al.*, 1991; Granadino *et al.*, 1992a; Jursnich and Burtis, 1993; Sanchez *et al.*, 2001; Taylor and Truman, 1992). In contrast, the sex determination pathway can regulate the *Yp* genes even in the adult (Belote *et al.*, 1985; Bownes *et al.*, 1987), providing an example of continuous regulation by the sex determination pathway. However, such continuous regulation is likely to be rare, since other genes that are sex-specifically expressed in adults are no longer under the influence of the sex determination pathway (Arbeitman *et al.*, 2004; Chapman and Wolfner, 1988). Thus, many of the sex-specific transcripts in adults may represent "late" genes that are not directly regulated by the sex determination pathway. In order to identify the "early" genes that DSX regulates to control sexual dimorphism, it will be necessary to search for sex-specific genes that are expressed at the time and place that *dsx* is controlling the sexually dimorphic development of a particular tissue.

Lastly, it is important to consider where *dsx* is acting to control sexual dimorphism. It has been assumed that the sex determination pathway is active in most or all cells. However, recent work has shown that the expression of DSX and FRU is highly tissue-specific, indicating that they are only acting to provide sexual identity to a subset of cells in the fly (Hempel and Oliver, 2007; Lee *et al.*, 2000, 2002; Sanders and Arbeitman, 2008; Usui-Aoki *et al.*, 2000). In addition, the sex determination pathway has been shown to act in a cell-autonomous manner to control sexual identity (Baker and Ridge, 1980). However, there are now several examples of nonautonomous sex determination in *Drosophila*, where a particular cell type does not decide for itself what its sexual identity should be, but instead relies on cues coming from another cell type (see the following sections). Since these findings are quite different from the traditional view of sex determination in *Drosophila*, we will discuss them further after we have reviewed how sexual dimorphism is created in some individual tissues.

2. THE CREATION OF SEXUAL DIMORPHISM

The primary purpose of sex determination is to create sexual dimorphism, the differences between the sexes. For all sexually dimorphic tissues, it is important to understand two principles: (1) how the tissue is specified and patterned, which gives its initial identity and developmental potential, and (2) how sex determination genes such as *dsx* interact with that initial pattern to alter the development of the tissue in a sex-specific manner. We will focus on somatic tissues that rely on *dsx* for their sexually dimorphic development. Other recent reviews focus on how sexual dimorphism is created in the germline (Hempel, *et al.*, this volume), and how *fru* regulates sex-specific behavior (Manoli *et al.*, 2006; Shirangi and McKeown, 2007). The development of each of the tissues we discuss in the following sections could be the topic of an entire review, and so we will limit our discussion largely to what is known about how *dsx* acts to create sexual dimorphism.

2.1. Sexual dimorphism in the gonad: Ovary vs. Testis

While many tissues develop differently in males versus females, sexual dimorphism is particularly important in the gonads, which must produce the sperm or eggs necessary for sexual reproduction. In addition, while many aspects of sexual dimorphism are specific for particular species, sexual dimorphism in the gonad is more universal, and thus more likely to be evolutionarily conserved. Here we discuss the formation of sexual dimorphism in the somatic gonad, which is essential for providing the structure of the testes and ovaries and for promoting sex-specific development of the germline.

2.1.1. Overview of gonad formation

The gonad forms as the germ cells coalesce with specialized somatic cells (somatic gonadal precursors, SGPs) to create the embryonic gonad (stage 14; reviewed by Starz-Gaiano and Lehmann, 2001). Both the soma and the germline show evidence of being sexually dimorphic at the time of gonad formation (Fig. 3.2). The testis begins to develop extensively during embryogenesis, while ovary morphogenesis is delayed until late larval and early pupal stages. In the embryo, testis development includes the recruitment of additional somatic cell types. The male-specific SGPs are initially observed in both sexes, but undergo programmed cell death in the female, and so only join the gonad in males. In addition, during late embryogenesis, the pigment cell precursors are recruited from the surrounding fat body to join the testis. Also, during late embryogenesis, a subset of SGPs in the male forms the "hub", a somatic structure that creates the niche and influences both the germline and somatic stem cells of the testis. Less is known about

Sexual Dimorphism in the *Drosophila* Soma

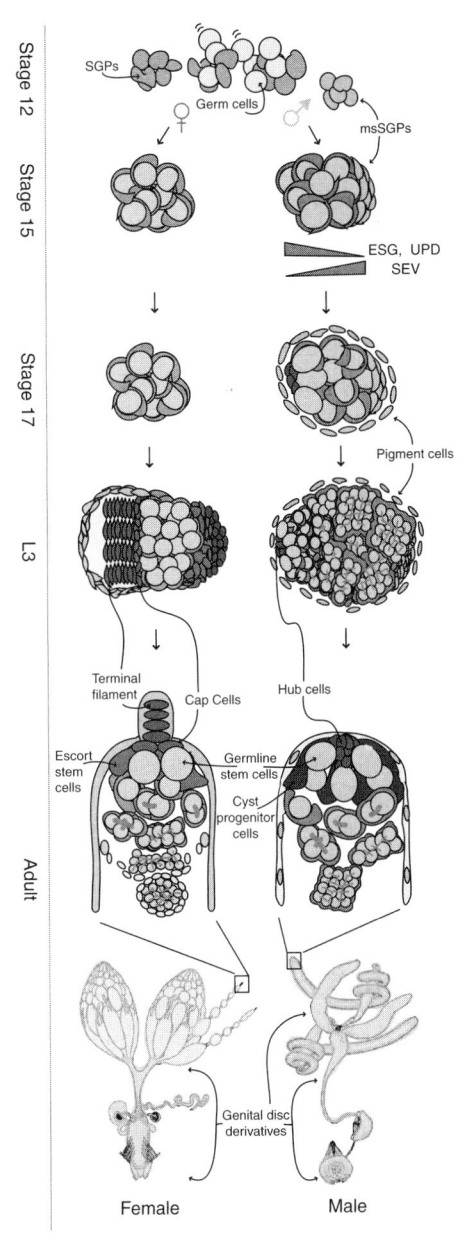

Figure 3.2 Development of sexual dimorphism in the gonad. Embryonic stages are as described (Campos-Ortega and Hartenstein, 1985). L3, 3rd instar larvae. The adult stage depicts the apical end of a single ovariole in the female and testis in the male, as indicated. The gonad begins as a bi-potential organ. Germ cells and somatic gonadal precursors (SGPs) interact and form an embryonic gonad by stage 15. Both the germline and somatic gonad are sexually dimorphic at this time. msSGPs are initially specified in

which embryonic cells give rise to specific cells of the adult ovary. Female-specific gene expression is observed in both the germline and the SGPs in the embryo, but structures such as the terminal filaments and cap cells, which create the stem cell niche in the ovary, only form during late larval stages. This is also the time when the ovary is divided into individual egg-producing units (ovarioles). Both the ovary and the testis connect to the rest of the reproductive tract, derived from the genital disc, during early pupal stages.

2.1.2. Somatic gonadal precursors

SGP development The SGPs first form within the mesoderm as three distinct clusters on either side of the embryo (reviewed in Van Doren, 2006). The key genes *eyes absent* (Boyle et al., 1997) and *Six4* (Clark et al., 2006; Kirby et al., 2001) likely act together to specify SGP identity. In addition, the SGP clusters have distinct anterior/posterior identities which are controlled by the homeotic genes. Anterior SGPs are specified by *abdominal-A* (*abd-A*), while posterior SGPs are specified by *abd-A* and *Abdominal-B* (*Abd-B*) (Boyle and DiNardo, 1995; DeFalco et al., 2004).

The earliest evidence for sexual dimorphism in the gonad is the upregulation of the JAK/STAT pathway in male germ cells, but not female germ cells, which is observed just prior to gonad coalescence (stage 13; Wawersik et al., 2005). The activation of this pathway in the germ cells requires contact with male SGPs, and even female germ cells can respond if they contact male SGPs. This is likely due to male-specific expression of ligand(s) for the JAK/STAT pathway in SGPs and one such ligand, *unpaired* (*upd*), is detected in anterior SGPs specifically in males at a slightly later stage (Wawersik et al., 2005). Other examples of male-specific gene expression in the SGPs at this stage include, the transcription factor *escargot* (*esg*), expressed in anterior SGPs (Gönczy et al., 1992; Le Bras and Van Doren, 2006), and the receptor tyrosine kinase *sevenless*, expressed in posterior SGPs (Kitadate et al., 2007). In addition, at least one molecular marker exhibits expression in female SGPs, but not male SGPs at this time (Casper and

---

both sexes and join only the posterior of the male gonad. In the male, pigment cells and the embryonic hub form during stage 17. The female gonad undergoes ovary morphogenesis during late L3 to make individual ovarioles and establish the female germline stem cell niche at the L3-pupal transition. In adults, germline stem cells (GSCs) contact the somatic niche formed by cap cells in females and hub cells in males. Somatic stem cells (cyst progenitor cells in male and escort stem cells in females) also contact the niche. germ cells (yellow at st 12, then light pink or blue); SGPs (green, st 12, then pink or blue; msSGPs (orange); hub cells (red); pigment cells (gray), terminal filaments (dark purple); cap cells (magenta); escort stem cells (purple); hub (red), cyst progenitor cells (navy blue). Adult ovaries (pink), testes (blue), and genital disk derivatives (gray). (See Color Insert.)

MVD, in preparation), indicating that both male and female SGPs have a unique sexual identity at the time of gonad formation.

Sexual dimorphism in the SGPs is important for many subsequent aspects of testis and ovary development. Sex-specific signaling from the SGPs controls the development of other cell types, including the male-specific somatic gonadal precursors (msSGPs), pigment cell precursors, and germ cells (see the following section and Hempel et al., this volume). In males, the anterior SGPs will form the testis stem cell niche or hub and it has been proposed that anterior SGPs also give rise to the stem cell niche in the ovary (Asaoka and Lin, 2004). One possibility for the remaining SGPs is that they give rise to those somatic cells in the adult that remain most closely associated with the germ cells: the cyst cells and cyst progenitor (somatic stem cells) in the male, and their counterparts in the female. Each of these later roles of the SGPs requires additional sex-specific gene expression.

Role of dsx *in early SGPs* In the embryonic gonad, *dsx* controls all known aspects of sexually dimorphic SGP cell fate. In mutants expressing a dominant masculinizing form of *dsx* (dsx^D), germ cells now exhibit upregulation of the JAK/STAT pathway in both sexes, demonstrating that this signal from the SGPs is under the control of *dsx* (Wawersik et al., 2005). Male-specific expression of *esg* in anterior SGPs and *sev* in posterior SGPs are dependent on *dsx* (Kitadate et al., 2007; Le Bras and Van Doren, 2006). In the absence of *dsx*, *esg*, and *sev* are expressed in both XX and XY gonads at levels that resemble wild type males (Kitadate et al., 2007; Le Bras and Van Doren, 2006). Thus, DSX^F is required (directly or indirectly) to repress these genes in female SGPs. Since these assays are not quantitative, a role for DSX^M in maximal expression of these genes in males is still possible. In the absence of *dsx* function, the embryonic gonad has a male phenotype in many respects, indicating that repression of male characteristics by DSX^F is a general principle for creating sexual dimorphism at this stage. However, since the gonads in *dsx* mutant adults do not resemble either testes or ovaries (Hildreth, 1965), *dsx* is clearly required at some point in gonad development in both sexes.

Surprisingly, the expression pattern for *dsx* in the gonad has only recently been investigated. The pattern of DSX^M is highly tissue-specific within the embryo, and it is only expressed in SGPs and msSGPs (Hempel and Oliver, 2007) (Note: this was examined with an antibody specific for DSX^M). Similarly, *dsx* RNA is restricted to the embryonic gonad, likely the SGPs and msSGPs (BDGP Gene Expression Database and DeFalco et al., 2008). Expression begins as the SGPs are first specified in the mesoderm, and is dependent on SGP identity. In *eya* mutants, SGP identity and DSX expression are greatly reduced, and only those few cells that retain SGP identity still express DSX (Hempel and Oliver, 2007). Since *dsx* is expressed in the SGPs, it may act in a cell-autonomous manner to regulate sexual identity in these cells. This appears to be the case for formation of the hub (see the following section), but has not been tested for other aspects of sex-specific SGP development. DSX^M is also expressed in several somatic cell types in the adult testis (Hempel and Oliver, 2007). Whether this represents a continuous role for *dsx* in the maintenance of sexual dimorphism in the adult gonad remains to be tested.

2.1.3. Male-specific somatic gonadal precursors

msSGP development The msSGPs are another early sign of sexual dimorphism in the gonad (DeFalco *et al.*, 2003). msSGPs are initially specified in both sexes, but only join the posterior of the male gonad. In females, they die via apoptosis before they reach the female gonad (DeFalco *et al.*, 2003). The msSGPs are specified as a single cluster of mesodermal cells on either side of the embryo, in a location posterior and ventral to where the SGPs form (DeFalco *et al.*, 2003). msSGP identity is dependent on *Abd-B*, but not *abd-A* (DeFalco *et al.*, 2004). While the msSGPs do not require many of the genes required for SGP specification, they do express *eya* and *Six4* (Clark *et al.*, 2007; DeFalco *et al.*, 2003), indicating that SGPs and msSGPs share a related identity.

Interestingly, the msSGPs express the transcription factor Sox100B (DeFalco *et al.*, 2003), which is the only fly homolog of mammalian Sox9 (Loh and Russell, 2000). Sox9 is expressed in the developing male gonad in a number of species examined, and is required for testis formation in mice and humans (Chaboissier *et al.*, 2004; Foster *et al.*, 1994; Kent *et al.*, 1996). Similarly, *Sox100B* mutants have a strong effect on testis development in *Drosophila*, while the ovary is unaffected (unpublished observations). Thus, there are common players acting to create sexual dimorphism in the gonads of diverse species.

In the adult, msSGPs give rise to the terminal epithelium at the base of the male testis (unpublished observations). The terminal epithelium is the region of the testis that makes contact with the genital disc to connect to the developing reproductive tract. It also plays an essential role in the last stages of spermatogenesis where it is responsible for anchoring sperm at the base of the testis and degrading the contents of the waste bag after sperm individualization (Tokuyasu, 1974).

Role of dsx *in the msSGPs* In *dsx* mutants, msSGPs survive and join the gonad in both sexes. Similarly, if apoptosis is blocked in the msSGPs, they are able to join the gonad in females and exhibit a similar pattern of gene expression as msSGPs in males (DeFalco *et al.*, 2003). This indicates that *dsx* primarily regulates sexual dimorphism in the msSGPs by regulating programmed cell death, rather than msSGP identity. Since apoptosis of the msSGPs in females requires the programmed cell death gene *head involution defective* (*hid*), this is a prime target for sex-specific regulation (DeFalco *et al.*, 2003).

Interestingly, *dsx* does not act in the msSGPs themselves to regulate programmed cell death. By manipulating the sex determination pathway in specific cells ("sexual mosaics"), it is possible to examine in which cells this pathway is acting to control sexual dimorphism. General expression of the feminizing factor TRA in otherwise male (XY) animals is sufficient to induce apoptosis of the msSGPs. However, msSGP development is unaffected when TRA is expressed in the msSGPs themselves; they still survive in males and die in females as they normally would (DeFalco *et al.*, 2008). Since *dsx* is only expressed in the SGPs and msSGPs at this time (Hempel and Oliver, 2007), this suggests that a signal from the SGPs regulates survival/death of the msSGPs. This is supported by the finding that msSGPs die in both sexes when the SGPs are absent (DeFalco *et al.*, 2008). Thus, *dsx* appears to regulate sexual dimorphism in the msSGPs through a nonautonomous mechanism; *dsx* acts in the SGPs to control male-specific expression of a survival signal for the msSGPs. The nature of this signal has not yet been identified. It is also interesting that, even though DSX is expressed in the msSGPs, it does not appear to act in these cells to control msSGP survival. It remains possible that there is an autonomous role for DSX in the msSGPs to regulate later aspects of their development.

2.1.4. Pigment cells

Pigment cell development The pigment cells are an epithelial cell layer that covers the outside of the testis and gives the testis its yellow color (Fuller, 1993). Even though the function of the pigment cells in gonad is not fully known, it is clear from mutants which lack pigment cells, that they are important for normal testis development (Kozopas *et al.*, 1998). They are also important for the development of a second cell layer that surrounds the testis, the muscle sheath, which is recruited to the testis from the genital disc (Kozopas *et al.*, 1998).

Pigment cells form during the last stage of embryogenesis (stage 17) and are recruited from the surrounding fat body to ensheath the developing testis (DeFalco *et al.*, 2008). Interestingly, these cells are a second male-specific cell type in the gonad that expresses the Sox9 homolog Sox100B (DeFalco *et al.*, 2008). These cells also express the transcription factor *empty spiracles* (*ems*). *ems* is required for initial pigment cell formation (DeFalco *et al.*, 2008) and *Sox100B* is required for later aspects of pigment cell development (unpublished observations).

Role of *dsx* ***in pigment cells*** Similar to the msSGPs, sexual identity of the pigment cells is regulated nonautonomously. Although pigment cells are normally formed by the male gonad (Stern and Hadorn, 1939), a male genital disc can induce the formation of pigment cells when transplanted into a female larva (Hadorn and Bertani, 1948). Thus, female cells are able to take on pigment cell identity. Using sexual mosaics in the embryonic gonad, it has been found that the sex determination pathway does not act in the pigment cells themselves to control their sexually dimorphic development (DeFalco et al., 2008); "female" cells are able to take on the male pigment cell identity as long as they are associated with a male gonad. Further, the pigment cell precursors do not even express DSX (Hempel and Oliver, 2007), and so *dsx* is unlikely to act directly in these cells.

Instead, *dsx* controls the male-specific development of the pigment cell precursors nonautonomously by regulating expression of the secreted ligand WNT2 in the gonad (DeFalco et al., 2008). *Wnt2* is initially expressed in the gonad in both sexes (Russell et al., 1992), although at higher levels in the male (DeFalco et al., 2003). However, at the time of pigment cell formation (stage 17), *Wnt2* expression in the gonad is entirely male-specific (DeFalco et al., 2008). *Wnt2* is necessary for the formation of pigment cells in males and expression of *Wnt2* alone is sufficient to induce pigment cell formation in females (DeFalco et al., 2008; Kozopas et al., 1998). *Wnt2* is also expressed in the male genital disc, likely accounting for the ability of this tissue to induce pigment cell formation (Kozopas et al., 1998). In *dsx* mutants, *Wnt2* expression is present in the gonads of all embryos and pigment cells form in both sexes (DeFalco et al., 2008). Thus, the action of DSX^F is critical for repressing *Wnt2* expression in females. As mentioned earlier, the presence of evolutionarily conserved DSX binding sites in the *Wnt2* upstream regulatory region suggests that regulation of *Wnt2* by DSX may be direct (DeFalco et al., 2008).

2.1.5. Stem cell niches In *Drosophila*, both the ovary and testis harbor stem cell populations that are necessary for continuous production of large numbers of gametes. These stem cells are maintained and controlled by their surrounding microenvironment, or niche, produced by specialized somatic cells. The stem cell niches exhibit some intriguing similarities, and also some clear differences, between the two sexes (Dansereau and Lasko, 2008; Fuller and Spradling, 2007; Gilboa and Lehmann, 2004, Lin, 2002). The gametes are derived from germline stem cells (GSCs), which also exhibit different properties in males and females, and clearly give rise to very distinct products, the sperm and egg. Finally, both the ovary and testis have distinct types of somatic stem cells that are also required for gametogenesis. Thus, the stem cell systems in the ovary and testis represent an interesting place to study how sexual dimorphism is created in both the soma and the germline.

Development of the male stem cell niche The male stem cell niche is created by the "hub", a tight cluster of somatic cells that interacts closely with the GSCs and somatic stem cells (cyst progenitor cells) of the testis (Aboïm, 1945; Hardy et al., 1979; Kiger et al., 2001; Tulina and Matunis,

2001). In addition, the cyst progenitor cells also regulate the GSCs, and therefore contribute to the niche (Leatherman and Dinardo, 2008). The hub forms during the last stage of embryogenesis (Gönczy et al., 1992; Le Bras and Van Doren, 2006) and, since spermatogenesis begins by early larval stages (Aboïm, 1945), it may already form a functioning stem cell niche at this time. Anterior SGPs in the embryonic gonad have a sex-specific identity as indicated by their expression of *esg* and *upd* in males (Le Bras and Van Doren, 2006; Wawersik et al., 2005). At the end of embryogenesis (stage 17), a subset of these male cells coalesce into a tight group to form the embryonic hub. These cells retain expression of *esg* and *upd*, and also initiate expression of a number of additional molecular markers that are expressed in the adult hub (Le Bras and Van Doren, 2006). Signaling from the germ cells to the SGPs through the *Boss/sevenless* receptor tyrosine kinase pathway is important for modulating the number of SGPs that take on hub cell identity (Kitadate et al., 2007). A number of cell–cell adhesion molecules (DE-cadherin, DN-cadherin, Fasciclin 3) are specifically expressed in the developing hub (Le Bras and Van Doren, 2006) and are likely to be important for hub formation and the interactions between hub cells and GSCs. In addition, integrin-mediated contact with the extracellular matrix is essential for maintaining the position of the hub at the anterior of the gonad (Tanentzapf et al., 2007).

The development of the GSCs and somatic stem cells that populate the testis niche has been less well documented. It is likely that those germ cells that contact the hub will go on to form GSCs, and the remaining germ cells will directly enter spermatogenesis (as "one shot" germ cells). Evidence for both GSC formation and "one shot" spermatogenesis has been observed when single labeled germ cells are transplanted into host testes (Asaoka and Lin, 2004). It is unknown from where the somatic stem cells (cyst progenitor cells) arise in the testis. However, since a subset of anterior SGPs form the hub, and SGPs do not give rise to other testis cell types (terminal epithelium, pigment cells, or muscle sheath), it is likely that somatic stem cells are derived from non-hub SGPs. One possibility is that anterior SGPs that do not form the hub become somatic stem cells, while posterior SGPs differentiate directly as cyst cells for the "one shot" germ cells, but this remains to be determined.

Ovary development and the formation of the female stem cell niche   Each ovary consists of approximately 16 individual ovarioles that are competent for egg production (reviewed by Spradling, 1993). As in the testis, each ovariole has a stem cell niche at its proximal end, created by specialized somatic cells (Kirilly and Xie, 2007). This niche regulates both a germline and somatic stem cell (escort stem cell) population (Fig. 3.2). However, the ovary has a second type of somatic stem cell, the follicle stem cells, located in a separate niche (Margolis and Spradling,

1995; Nystul and Spradling, 2007). Each ovariole is also surrounded by a muscle sheath and epithelial layer, analogous to the muscle layer and pigment cell layer on the testis. Thus, even though the later stages of oogenesis and spermatogenesis appear very different, there may be related cell types upon which the sex determination pathway acts to create an ovary versus testis.

While both the somatic cells and the germ cells have a sex-specific identity in the embryonic gonad, ovary morphogenesis does not happen until several days later during the third instar larval period (Fig. 3.2; King, 1970). At this time, the number of somatic cells and germ cells has increased and the germ cells, mixed with interstitial somatic cells, reside in the middle of the larval ovary, while additional somatic cells populate either end. The first sign of ovary morphogenesis is the formation of the terminal filaments, which assemble from anterior somatic cells into approximately 16 ordered rows, one for each ovariole (Godt and Laski, 1995; Sahut-Barnola et al., 1995). The *bric a brac* (*bab*) locus encodes related transcription factors (BAB1 and 2) that are expressed in the terminal filament cells and are required for terminal filament formation (Godt and Laski, 1995; Godt et al., 1993). *Cofilin/ADF* is also required for the movements of the terminal filament cells (Chen et al., 2001). After terminal filament formation, apical cells migrate between the terminal filaments and encapsulate subsets of germ cells to form the ovarioles (King, 1970), a process that requires the secreted ligand *Wnt4* (Cohen et al., 2002).

The GSC niche in the ovary is formed from cap cells, along with the terminal filaments (Xie and Spradling, 2000). Cap cells form adjacent to the terminal filaments beginning at the larval-pupal transition (Zhu and Xie, 2003), and cap cell number is regulated by the Notch pathway (Song et al., 2007; Ward et al., 2006). Soon after the cap cells form, DE-cadherin rich adherens junctions are observed between the cap cells and the adjacent germ cells, and these germ cells take on the characteristics of GSCs (Song et al., 2002; Zhu and Xie, 2003). Germ cells that reside in the anterior of the embryonic gonad preferentially form GSCs in the adult, while the others appear to directly differentiate (Asaoka and Lin, 2004).

Similarities and differences in the adult male and female stem cell niches The male and female niches are morphologically very different. The single, tight cluster of somatic cells making up the hub in the testis bears little resemblance to the multiple strings of terminal filaments and cap cells, one for each ovariole, that form in the ovary. Also, the timing of niche formation is quite different, with the testis hub being formed by the end of embryogenesis while the terminal filaments and cap cells form at the larval/pupal transition. Further, the germ cells also have distinct identities in males and females (Hempel, et al., this volume), and the GSCs express different molecular markers in the two sexes (e.g., Gönczy et al., 1992). However, the underlying mechanisms by which the male and female niches regulate the GSCs and somatic stem cells are surprisingly similar.

Recent work has highlighted three important features of how the gonad niches regulate their GSCs: adhesion to the GSCs, signaling to the GSCs, and the cell biology of the GSC divisions. These features are conserved in males and females, but with some important differences. The GSCs in both sexes adhere to their respective niches through cadherin-rich cell-cell contacts. The contacts between the female GSCs and cap cells have high levels of DE-cadherin (Song et al., 2002), as do the contacts between male GSCs and hub cells (Yamashita et al., 2003), and adherens junctions are present in both cases (Song et al., 2002; Yamashita et al., 2007). DE-cadherin is expressed in the niche as it initially forms and interacts with a subset of germ cells (Le Bras and Van Doren, 2006; Song et al., 2002). In females, GSCs mutant for DE-cadherin (*shotgun*) are lost from the niche when in competition with wild type GSCs, but not when all GSCs are mutant (Song et al., 2002). Thus, DE-cadherin contributes to niche-GSC interaction in females but is not required for GSC function (the role of DE-cadherin in the male GSCs has not been reported). DN-cadherin is also found at the niche-GSC interface in both sexes (Le Bras and Van Doren, 2006 and unpublished observations), suggesting that these cadherins may work together to promote niche-GSC adhesion.

The cells that adhere to the niche receive signals that promote stem cell identity. In both sexes, the TGF-β and JAK/STAT pathways are important within the niche, but in somewhat different ways. In females, TGF-β signaling plays the dominant role in regulating the GSCs (Xie and Spradling, 1998), while the JAK/STAT pathway is not required within the GSCs (Decotto and Spradling, 2005). In contrast, in males the JAK/STAT pathway is absolutely essential within in the GSC (Kiger et al., 2001; Tulina and Matunis, 2001), although the TGF-β is also required (Kawase et al., 2004; Schulz et al., 2004; Shivdasani and Ingham, 2003). Thus, while the same signaling pathways are employed in the niche in the two sexes, there are differences in how the GSCs respond. The JAK/STAT pathway is further required in the somatic cells in both sexes. It is required in the niche for its ability to signal to the GSCs through the TGF-β pathway (Lopez-Onieva et al., 2008; Wang et al., 2008). It is also required in the somatic stem cells (escort stem cells and cyst progenitor cells) for their maintenance (Decotto and Spradling, 2005; Leatherman and Dinardo, 2008). Therefore, the JAK/STAT pathway influences the GSCs directly only in males, but influences the GSCs indirectly in both sexes by regulating the niche and somatic stem cells.

Association with the niche is also essential for regulating the orientation of GSC division, so that one daughter cell remains in contact with the niche, and retains stem cell identity, while the other daughter leaves the niche and differentiates. In the male GSC, the mother centrosome always remains localized to the niche-GSC interface, while the daughter centrosome becomes positioned at the opposite side of the cell (Yamashita et al., 2003, 2007). In this way, the mitotic spindle is regulated so that one cell inherits the mother centrosome and remains at the niche as a GSC, while the other daughter enters differentiation. In females, a germline specific organelle, the spectrosome (Lin et al., 1994), is oriented at the niche-GSC interface, while this localization is not observed in males. The spectrosome in females is associated with centrosomes (Lin et al., 1994), and is required for orienting the mitotic spindle of the GSCs so that one daughter cell remains associated with the niche (Deng and Lin, 1997). Thus, the orientation of GSC divisions is tightly regulated in both males and females, though the detailed mechanisms may differ.

dsx *in niche formation* As is true for other aspects of sexual dimorphism in the somatic gonad, development of a sex-specific stem cell niche is regulated by *dsx*. Sexual dimorphism in the niche is initiated at the end of embryogenesis, as the hub forms from a subset of anterior SGPs in males but not in females. In *dsx* mutants, hubs form in both XX and XY embryos that are indistinguishable from hubs in normal males (Le Bras and Van Doren, 2006). Thus, DSX^F is essential for repressing hub formation in anterior SGPs in females. The sex determination pathway acts cell-autonomously to regulate hub cell formation, since expression of the feminizing factor TRA in subsets of anterior SGPs is sufficient to prevent them from contributing to the hub (DeFalco et al., 2008). Further, DSX is expressed in the anterior SGPs and in the hub itself (Hempel and Oliver, 2007), indicating that it can function directly in these cells. Several genes are expressed in the hub as it forms, such as *esg*, *upd*, and DE- and DN-cadherin (Gönczy et al., 1992; Le Bras and Van Doren, 2006), but it is not known if any of these are directly regulated by DSX.

How *dsx* regulates later aspects of niche development has not been reported. It has been proposed that both the hub and the terminal filaments/cap cells are formed from the same population of anterior SGPs. An enhancer trap that is expressed in anterior SGPs in the embryo is also expressed in the male and female niches at later stages (Asaoka and Lin, 2004), and these cell types also share other common molecular markers. By this model, the male and female niches would come from a common primordium, and *dsx* would act to determine whether these cells should form a hub at the end of embryogenesis, or wait and form terminal filaments and cap cells at the end of the larval period. Whether this model is correct, and how *dsx* acts to influence hub vs. terminal filament/cap cell formation, remains to be determined.

For most sexually dimorphic tissues, the sex determination pathway does not create wholly different organs in the two sexes. Instead, it takes a common primordium and creates organs with analogous cell types and structures that differ in more discrete ways between the sexes. This was less clear for the gonads, which, although derived from the same embryonic primordium, appeared to have little in common in males and females other than the germ cells. However, we now know that the testis and ovary are much more similar than they appear. As we have discussed, this is most apparent for the niches, GSCs and somatic stem cells (cyst progenitor cells and escort stem cells), but may also be true for other cell types, such as the muscle sheath and epithelial layer that surround both the testis and each ovariole. This leads to a model in which the testis and ovary are highly related structures with directly analogous cell types that develop from related precursors. Thus, here again, the sex determination pathway does not invent a completely different organ in males vs. females, but it instead takes a similar template and modifies it through less drastic changes. The difference in outcome is fairly dramatic to be sure: the single coiled testis versus multiple ovarioles, the differences in the niches and regulation of stem cells, the presence of follicle stem cells only in females, and the final outcome of creating sperm vs. eggs are all significant sexual dimorphisms. However, the more we learn about the underlying similarities between testis and ovary development, the more we understand how the sex determination pathway may only need to effect a small set of regulatory changes in a bi-potential tissue to create large difference between males and females.

2.2. The genital disc

The genital disc in *Drosophila* gives rise to the sex-specific genitalia (reproductive tract and external genitalia) and analia. These structures differ drastically in males versus females, and so the development of the genital disc provides an elegant model in which to study the creation of sexual dimorphism (Christiansen *et al.*, 2002; Estrada *et al.*, 2003; Sanchez and Guerrero, 2001). The disc is comprised of three primordia that derive from embryonic abdominal segments 8, 9 and 10 (or A8, A9, and A10, respectively) (Nothiger and Epper, 1977; Schupbach and Nothiger, 1978). The A10 domain (anal primordium, AP) gives rise to either male or female analia, depending on the sex of the animal. In contrast, it was originally thought that the A8 and A9 domains only developed in one of the sexes, with A8 (the female genital primordium, FGP) giving rise to the female genitalia, and A9 (the male genital primordium, MGP) to the male genitalia (Epper and Nothiger, 1982). In the opposite sex, the development of A8 or A9 was though to be completely repressed (the repressed female or male primordium, respectively). However, it is now known that, though their development is limited in one sex, the A8 and A9 domains do contribute to adult structures in both sexes; A8 gives rise to the 8th tergite in males and the A9 gives rise to the parovaria and part of the uterine wall in females (Keisman et al., 2001). In addition, cells are recruited from the surrounding mesoderm to become part of the genital disc in males, and these cells form the accessory glands (paragonia) and vas deferens of the testis (Ahmad and Baker, 2002).

Development of the genital disc As is the case for other imaginal discs, there are two main elements that provide patterning information for the genital disc; the homeotic genes determine the identity of the segment that gives rise to the disc and the "A/P organizer" controls growth and patterning within the disc. Unlike other discs which are derived from single segments of the fly's body plan, the genital disc derives from three segments: A8, A9, and A10. The homeotic genes *abd-A* and *Abd-B* are critical for providing segmental identity to different regions of the disc. *abd-A* is specifically expressed in the A8 primordium, whereas *Abd-B* shows expression in both the A8 and A9 primordium (Casares et al., 1997). *Abd-B* encodes two distinct transcripts, *Abd-B m* and *Abd-B r*, that encode different ABD-B isoforms (Kuziora and McGinnis, 1988; Sanchez-Herrero and Crosby, 1988). In the genital disc, *Abd-B m* is specifically expressed in the A8 primordium, whereas *Abd-B r* is expressed in the A9 primordium (Casares et al., 1997). Dramatically, in the absence of *Abd-B*, the male or female genitalia are transformed into legs or antennae (Estrada and Sanchez-Herrero, 2001). The development of the A10 primordium is regulated by *caudal* (*cad*), as it is required for formation of the analia and is expressed specifically in the A10 primordium (Casares et al., 1997; Macdonald et al., 1986; Moreno and Morata, 1999). When the precursors for the genital disc can first be detected in the embryo (stage 14/15), they already appear to be patterned by the homeotic genes (Chen et al., 2005; Foronda et al., 2006). In addition, it is clear that the different segmental primordia of the genital disc can influence each other's development, and signaling occurs between the primordia (Gorfinkiel et al., 2003).

The segments of the *Drosophila* body plan are divided into anterior and posterior compartments, and imaginal discs that derive from single segments have one anterior and one posterior compartment. The anterior/posterior compartment boundary establishes an important signaling center, the A/P organizer, which expresses secreted ligands such as Wingless (WG) and Decapentaplegic (DPP), members of the Wnt and TGF-β family, respectively (Estrada et al., 2003). The genital imaginal disc, however, appears to have three such anterior/posterior compartment boundaries, one for each of its segmental domains (Casares et al., 1997; Chen and Baker, 1997; Freeland and Kuhn, 1996). Further, each of these can establish an organizer that expresses WG and DPP in non-overlapping, complementary domains along the compartment boundary. Signaling by WG and DPP to other cells in the disc is a major point of regulation for genital disc development. *dsx* acts in concert with the homeotic genes to create differences in organizer signaling between the sexes in specific domains of the disc (e.g., A8 vs. A9).

The role of dsx *in the genital disc* In *dsx* mutants, both the A8 (FGP) and A9 (MGP) develop so that adults have aspects of both the female and male genitalia—the "double sex" phenotype that gives *doublesex* its name (Hildreth, 1965). In addition, the analia (A10) develop with an intersexual phenotype. A considerable amount is known about how *dsx* regulates sex-specific development of the A8 and A9 domains, but less is known about the sexually dimorphic development of A10. Interestingly, nonautonomous sex determination plays an important role in sex-specific development of the genital disc, similar to what was described in the previous sections for the gonad.

One mechanism by which *dsx* controls development of the genital disc involves sex-specific signaling from the organizers. Normally, in males the A9 (MGP) domain will grow extensively to give rise to the male genitalia, while in females the growth of the A9 domain is limited, and it gives rise only to the parovaria. DPP is expressed in the organizer of all three genital disc segments in males, but is absent from the A9 domain in females (Casares *et al.*, 1997; Chen and Baker, 1997; Freeland and Kuhn, 1996). The loss of DPP signaling from the A9 organizer is why this region fails to grow and develop in females. When "male" cells are created in an otherwise female tissue (e.g., *tra2* mutant clones), such clones can cause male-like growth of the A9 domain (Keisman and Baker, 2001; Sanchez *et al.*, 2001), and these are always associated with ectopic DPP expression (Sanchez *et al.*, 2001). Even if just the organizer region of A9 is transformed to a male identity, this is sufficient to induce growth and development of the A9 domain similar to what is seen in males (Keisman *et al.*, 2001). This occurs even though the other cells of the A9 domain still have a female identity; the sexual identity of the A9 organizer is sufficient to regulate whether or not DPP will be expressed, and nonautonomously regulate the growth and development of the surrounding cells. Thus, *dsx* regulates sexually dimorphic growth of the genital disc A9 segment by limiting DPP expression to the A9 organizer only in males.

Sex-specific control of organizer signaling may also control whether the A8 domain will grow extensively to form the female genitalia, or will remain repressed and give rise to the miniature 8th tergite in males. Unlike DPP, WG is expressed similarly in both sexes in organizer region of all three segmental domains. However, it has been proposed that the response to WG is blocked in the A8 domain in males but not females (Sanchez *et al.*, 2001). Thus, while *dsx* also acts to control sexually dimorphic growth of A8 by controlling signaling from the organizer, by this model it would do so cell-autonomously by influencing the ability of A8 cells to respond to WG in males.

In addition to regulating the growth and early development of the A8 and A9 domains, *dsx* is also required for later aspects of how these domains differentiate with the proper male vs. female character. Here again, *dsx* acts in a cell-autonomous manner to regulate the response of cells to the WG and DPP signals (Keisman and Baker, 2001; Sanchez *et al.*, 2001). *dachsund* (*dac*) is a transcription factor that is regulated by WG and DPP and is required for proper development of the male and female genitalia (Keisman and Baker, 2001). Interestingly, the regulation of *dac* is completely opposite in the two sexes; in females *dac* is activated by WG but repressed by DPP and in males it is activated by DPP and repressed by WG. Thus, *dsx* controls later aspects of sex-specific differentiation of the genital disc by regulating how cells interpret the WG and DPP signals.

Finally, *dsx* acts in a nonautonomous manner to regulate the recruitment of mesodermal cells into the male genital disc (Ahmad and Baker, 2002). *bnl*, which encodes an FGF ligand, is expressed in the A9 domain of the genital disc in males, but not females. *bnl* expression is regulated by *dsx*, and the presence of consensus DSX binding sites in the region upstream of the *bnl* start of transcription suggests this regulation could be direct (Ahmad and Baker, 2002). Male-specific expression of *bnl* acts nonautonomously to recruit surrounding mesodermal (*twist*-positive), adepithelial cells into the developing male genital disc. These cells go on to form the male accessory glands (paragonia) and vas deferens in the adult (Ahmad and Baker, 2002). It has also been observed that *twist*-expressing cells migrate from the genital disc to the gonad as the vas deferens contacts the developing gonad, and these cells form the muscle sheath around the testis (Kozopas *et al.*, 1998). However, since the mesodermal cells that are recruited into the genital disc are thought to lose *twist* expression (Ahmad and Baker, 2002), it is unclear whether a subset of these cells migrates to the testis to form the muscle sheath, or whether this sheath comes from other *twist*-positive adepithelial cells associated with the genital disc.

It is clear that *dsx* acts via several distinct mechanisms to control sexual dimorphism in the genital disc. However, in each case, *dsx* appears to regulate cell-cell signaling. Either *dsx* acts cell autonomously to control how a cell responds to a particular signal (e.g., whether *dac* is activated by WG or DPP), or it acts to regulate expression of the signal itself (e.g., DPP or *bnl*), to nonautonomously influence the sex-specific development of surrounding cells.

2.3. Sex combs and sexual dimorphism in the leg

Drosophila have three pairs of legs that develop from paired leg imaginal discs. The first (prothoracic) leg is sexually dimorphic in several respects, largely with respect to the pattern of sensory bristles. The most obvious is the presence of the male sex comb, a set of thick, dark bristles that are mechanosensory organs (touch receptors). Sex combs are used to grab the female's abdomen during intercourse (Spieth, 1952) and are important for male mating success (Ng and Kopp, 2008). Sex combs form on a particular region of the leg (the distal part of tarsal segment t1) in males, which gives rise to a transverse row of bristles in females. In *Drosophila melanogaster*, the male sex combs differ from the female bristles in three ways: sex combs have more bristles, the bristles are thicker and more pigmented, and the bristle row is rotated 90 degrees so that it is oriented longitudinally, rather than transversely across the leg. *dsx* regulates all of these characteristics; in *dsx* mutants each is intermediate between the normal male and female phenotype, including the orientation, which is in between the normal male (longitudinal) and female (transverse) positions (Baker and Ridge, 1980; Hildreth, 1965). The sex determination pathway affects sex comb bristle morphology in a cell-autonomous manner, but the orientation of the sex comb is a function of how many of the bristles have male vs. female identity, indicating that there is some nonautonomous interaction among the cells of the group with respect to their position (Baker and Ridge, 1980).

Sex combs only appear on the first leg, and only on the distal part of t1, demonstrating that both the segmental identity of the leg (first thoracic segment) and patterning of the leg along the proximal-distal axis, are essential for specifying the sex comb. The homeotic gene *Sex combs reduced* (*Scr*) is a key regulator of sex comb positioning and is required for sex comb formation (Struhl, 1982). Interestingly, *Scr* expression is sexually dimorphic, and it is expressed in the region of the leg in males where the sex comb forms, but not in the corresponding transverse bristle row in females (Barmina and Kopp, 2007). Thus, sexual identity can influence the expression of key patterning genes, such as the homeotic gene *Scr*. In males, ectopic expression of Scr is also sufficient to induce extra sex combs on additional tarsal segments (Barmina and Kopp, 2007). However, extra sex combs are not observed in females, indicating that sexual identity must provide additional information besides regulation of *Scr*. In addition, ectopic sex combs are only observed on the distal portion of tarsal segments, indicating that additional patterning information is also required. *bric a brac* (*bab1/2*) is also expressed in the developing leg disc, and behaves as a negative regulator of sex comb formation. Loss of *bab1/2* function leads to sex combs on additional tarsal segments in males, similar to gain of *Scr* function (Couderc et al., 2002; Godt et al., 1993).

In addition to the sex comb, other aspects of the sensory bristle pattern on the first leg are sex-specific, and dependent on *dsx* (Hildreth, 1965). Some of the bristles on the legs represent taste sensilla with chemoreceptive neurons. Both the number and type of taste sensilla are different in males versus females (Meunier et al., 2000; Nayak, 1983). Interestingly, Bray and Amrein (2003) identified a candidate pheromone receptor that is expressed in a subset of gustatory neurons on the prothoracic leg in males, but not females. This receptor is required for proper male mating behavior, and its male-specific expression is regulated by *dsx* (Bray and Amrein, 2003). Other regions of the fly (antennae and maxillary pulp) contain receptors for more volatile chemicals (odors) and show differences between the sexes (reviewed in Vosshall and Stocker, 2007). The chemical cues (pheromones) received by these taste and odorant receptors are also sexually dimorphic, and regulated by *dsx* (reviewed in Ferveur, 2005).

2.4. Sexual dimorphism in the abdomen

In addition to the terminal regions derived from the genital disc, other aspects of the posterior abdomen are sexually dimorphic in *Drosophila*. Perhaps the best known is the difference in pigmentation pattern in males versus females; in males, abdominal segments A5 and A6 are fully pigmented, while they only have a posterior stripe in females (the posterior stripe is not sex-specific and is regulated independently). Not all species of *Drosophila* have sexually dimorphic abdominal pigmentation, and the pigmented abdomen of *D. melanogaster* is what lends this species its name. In addition to the pigmentation pattern, the cuticular structures (tergites and sternites) of A6 and A7 are fused in males, but not females, and the A7 sternite has bristles in females that are absent from the fused A6/7 sternite in males. There are also sex-specific differences in the development of the nervous system in the abdomen (Taylor and Truman, 1992). Each of these dimorphic characteristics is regulated by *dsx* (Baker and Ridge, 1980; Hildreth, 1965; Taylor and Truman, 1992).

The sexually dimorphic development of these traits has been best studied for abdominal pigmentation. The abdominal cuticle forms from histoblast nests which, like the imaginal discs, are cells set aside in the embryo to form adult structures during metamorphosis. *dsx* acts in a cell-autonomous manner during pupal development to control sex-specific pigmentation (Baker and Ridge, 1980). *dsx* mutants exhibit pigmentation that is similar to wild type males, but does not extend as far into A5 (Baker and Ridge, 1980; Hildreth, 1965). Thus, while DSX^F is required to repress pigmentation in females, DSX^M appears to have some role in promoting the full male pattern.

dsx regulates the male pigmentation pattern by acting through *Abd-B* and *bab1/2* (Jeong et al., 2006; Kopp et al., 2000). *Abd-B* is a positive regulator of pigmentation, and *Abd-B* mutants lose male pigmentation while gain of *Abd-B* function results in ectopic pigmentation (Celniker and Lewis, 1993; Celniker et al., 1990; Hopmann et al., 1995). In contrast, *bab1/2* acts as a repressor of abdominal pigmentation, with *bab1/2* mutants exhibiting increased pigmentation while gain of *bab1/2* function represses male pigmentation (Couderc et al., 2002; Kopp et al., 2000). *Abd-B* regulates pigmentation by two mechanisms; it negatively regulates expression of *bab1/2* (Kopp *et al.*, 2000), and ABD-B is also a direct transcriptional activator of the pigmentation gene *yellow* (Jeong *et al.*, 2006). *bab1/2* and *Abd-B* also exhibit an antagonistic relationship in the development of other characteristics of the abdominal cuticle, indicating that these other sexually dimorphic traits may be regulated in a similar manner as pigmentation (Couderc *et al.*, 2002; Kopp *et al.*, 2000).

bab1/2 expression is repressed by *Abd-B*, and is lowest in the posterior regions (A5/6) where *Abd-B* is highest. *bab1/2* expression is also sexually dimorphic, being absent from A5/6 in males, where pigmentation is observed, but expressed at low levels in this domain in females, where pigmentation is repressed (Kopp *et al.*, 2000). Further, *bab1/2* expression is not sexually dimorphic in *Drosophila* species that lack sex-specific abdominal pigmentation, indicating that *bab1/2* is a principle regulator of the development and evolution of this sexually dimorphic trait (Kopp *et al.*, 2000). *Abd-B* expression appears the same in males and females, and is unchanged in *dsx* mutants, so this doesn't explain the sexually dimorphic expression of *bab1/2*. *bab1/2* expression is down-regulated in *dsx* mutant females, indicating that DSX^F normally promotes the higher levels of *bab1/2* observed in A5/6 in females. In addition, the fact that pigmentation in A5 is not complete in *dsx* mutants, suggests that DSX^M might also help *Abd-B* to repress *bab1/2* in males.

ABD-B also acts directly to regulate *yellow* expression (Jeong *et al.*, 2006). A defined enhancer in the *yellow* regulatory region is sufficient to recapitulate the normal male-specific expression in A5/6 in a manner that is genetically responsive to *bab1/2* and *Abd-B*. ABD-B binding sites in this enhancer are essential for this expression, indicating that ABD-B directly activates this gene (Jeong *et al.*, 2006). However, again, *Abd-B* is not expressed differently in males and females, and how sex-specific expression of this enhancer is regulated has not been elucidated.

It is interesting to note that, as discussed in the previous sections, *bab1/2* regulates many aspects of sexual dimorphism in *Drosophila*. *bab1/2* regulates abdominal pigmentation and cuticle pattern, sex comb formation in the leg, and ovary development. In each case, *bab1/2* acts to promote the female mode of development; it represses male abdominal pigmentation, negatively regulates male sex comb formation, and is required for formation of the terminal filaments of the ovary. In addition, *bab1/2* function is highly dose-dependent, suggesting that changes in *bab1/2* expression could lead to phenotypic variation in sexual traits and adaptive selection (discussed in Couderc *et al.*, 2002). Thus, there appears to be a common relationship between *bab1/2* and *dsx* in the creation of sexual dimorphism; whether *bab1/2* is a direct target for regulation by DSX remains to be determined (See note added in proof).

2.5. Sexual dimorphism in the fat body

Although the *Drosophila* fat body is involved in fat storage, it is also a dynamic signaling center that regulates such diverse processes and energy metabolism and growth, steroid hormone signaling, the immune system and aging. As discussed in the previous sections, the fat body produces yolk proteins (Yp) in females, but not males, which circulate through the hemolymph and eventually find their way into developing oocytes. The tandem genes *Yp1* and *Yp2* remain the only known direct transcriptional target of DSX, being activated by DSX^F and repressed by DSX^M (discussed in the previous sections).

Interestingly, proper sexual identity in the fat body is also required for normal courtship behavior, suggesting that signaling from the fat body to the nervous system can regulate sex-specific behavior (Lazareva *et al.*, 2007). *takeout* (*to*) is expressed in the head fat body in males, but not females, and encodes a secreted protein with homology to lipophilic hormone binding proteins (Dauwalder *et al.*, 2002; Lazareva *et al.*, 2007). *to* belongs to a family of related genes, and some other members of this family also exhibit sex-specific expression (Dauwalder *et al.*, 2002). *to* mutants show a subtle but clear reduction in male courtship, particularly in a sensitized (*fru*/+) background, indicating that *to* is required for proper sex-specific behavior (Dauwalder *et al.*, 2002).

Male-specific expression of *to* is positively regulated by both *dsx* and *fru* as *to* expression is decreased in males mutant for *dsx* or *fru*. *to* expression is also greatly increased in *dsx* mutant females, indicating that DSX^F negatively regulates *to* (*to* is unchanged in *fru* mutant females, consistent with *fru* producing no functional protein in females). Thus, the two genes acting downstream of TRA to regulate sexual dimorphism, *dsx* and *fru*, can cooperate to regulate common processes and target genes. *dsx* and *fru* are both required from proper sex-specific behavior (Shirangi *et al.*, 2006) and genomic analysis of sex-specific gene expression in the head and nervous system promises to reveal additional *dsx* and *fru* targets in the head fat body and other tissues (Goldman and Arbeitman, 2007).

2.6. Muscle of Lawrence

The muscle of Lawrence (MOL) a paired muscle that is found in the fifth abdominal regions of males but not in females (Lawrence and Johnston, 1984). The MOL is another example of nonautonomous sex determination in *Drosophila*; the sex-specific formation of the MOL relies on the sex of the nerves that innervate the muscle, and not the sex of the muscle itself (Currie and Bate, 1995; Lawrence and Johnston, 1986). However, this is regulated by *fru* and not *dsx* (Taylor, 1992; Taylor and Knittel, 1995; Usui-Aoki *et al.*,

2000). The MOL fails to develop properly in *fru* mutants (Taylor and Knittel, 1995), and ectopic expression of *fru* can induce MOL formation in females (Usui-Aoki *et al.*, 2000). Thus, neurons are able to influence the sex-specific development of the tissues they innervate, and *fru* can control nonautonomous sex determination, similar to *dsx*.

3. A New Look at an Old Problem

The recent work discussed in the previous sections has shed some surprising new light on the process of sex determination in *Drosophila*. These new ideas have changed some fundamental assumptions about how the process of sex determination works.

3.1. Not all cells decide their own sex

Somatic sex determination in *Drosophila* is commonly thought to be a cell autonomous process where each cell decides its own sexual fate based on its chromosomal constitution (Gilbert, 2003; Wolpert *et al.*, 2006). This idea is largely supported by work on gynandromorphs, organisms that are mosaic for patches of male and female tissue. Morgan and Bridges studied gynandromorphs and concluded that "male and female parts and their sex-linked characters are strictly self-determining, each developing according to its own aspiration...not interfered with the aspirations of its neighbors, nor is it overruled by the action of the gonads." (Morgan and Bridges, 1919). This indicates that flies do not have dominant hormones that determine the sex of many different cells in the body in a nonautonomous manner, as is observed in mammals. Further, initial studies indicated that the sex determination pathway primarily acts in a cell-autonomous manner to control sexual dimorphism (Baker and Ridge, 1980).

However, as discussed in the previous sections, this question has now been revisited for several different cell types with a very different answer. It is now clear that many cells decide their sex in a nonautonomous manner; local cell–cell interactions are important for determining their sex independent of their own sex chromosome constitution. Nonautonomous control has been observed for aspects of sex determination in several independent cell types in the gonad (the msSGPs, pigment cell precursors, and germ cells) and in the genital disc and muscle of Lawrence. Thus, a great deal of nonautonomous sex determination occurs in *Drosophila*, and the mechanism by which any particular cell type decides its sex must be assessed carefully for each cell type studied.

3.2. Not all cells even know their own sex

In addition to the finding that not all cells decide their own sex, we now know that not all cells even have the machinery necessary to assess their own sex. In order for a *Drosophila* cell to "know" its sex, it must have a mechanism for translating the number of X chromosomes into sexual identity. As discussed in the previous sections, this is the role of sex-specific transcription factors such as DSX and FRU. Recent observations indicate that the expression pattern of DSX and FRU are highly tissue-specific. For example, within the embryo, the only cells that express DSX are the SGPs and msSGP in the embryonic gonad (Hempel and Oliver, 2007), and within the nervous system, only a small fraction of neurons express FRU (Lee *et al.*, 2000; Usui-Aoki *et al.*, 2000). Since these factors are essential for allowing a cell to interpret information from the sex determination cascade, a cell cannot truly "know" its sex without expressing such a critical downstream factor.

In *Drosophila*, X chromosome number and *Sxl* also control the process of dosage compensation, where the expression from a single X chromosome in males is increased to match that of the two X chromosomes in females. In the absence of *Sxl*, this process is activated in XX animals, and is lethal. It is thought that most or all cells count their X chromosome number and initiate dosage compensation accordingly, in an autonomous manner (reviewed by Cline and Meyer, 1996). However, just because a cell "knows" its X chromosome number, this does not mean that it "knows" its sex. If a cell does not express a downstream factor such as DSX, it cannot interpret the information about X chromosome number in such a way that it can develop differently in males and females.

The idea that not all cells should know their sex makes perfect sense when one considers how sexual dimorphism is created. A common theme in development is that the same genes and signaling pathways are used again and again in multiple contexts. Thus, the genes that are important for development of sex-specific tissues are likely to also be used to control the development of other tissues in a non-sex specific manner. For example, the signaling pathways that are regulated by *dsx* to ensure proper sexual dimorphism in the gonad and genital disc (e.g., JAK/STAT, WNT, FGF, TGF-β, etc.) are the same core signaling pathways that are required in other tissues in both sexes. Clearly, it would be problematic for *dsx* to regulate the expression of such genes in tissues that are not sexually dimorphic. One way to avoid this is to simply not have *dsx* expressed in those tissues. Thus, cells are on a "need to know basis" with respect to their sexual identity. Further, it may even be dangerous for some cells to "know" their sex. Indeed, ectopic expression of *dsx* can cause the wrong structures to take on sex-specific traits, and can even kill the animal (Jursnich and Burtis, 1993).

3.3. Integrating sex and patterning

As discussed in the previous sections, sexual dimorphism is created by how patterning information, such as that provided by the homeotic genes or signals from an organizer, combines with sexual identity to allow a tissue to develop differently in males and females. There are several ways that sexual identity and patterning information can be integrated during development. First, sex and pattern can provide independent inputs into the regulation of a particular target. For example, *dsx* determines how *dac* expression will be regulated by WG and DPP expressed from the genital disc organizer (Keisman and Baker, 2001; Sanchez *et al.*, 2001), and *bab1/2* is regulated by combined inputs from *dsx* and *Abd-B* to control abdominal pigmentation (Kopp *et al.*, 2000). Alternatively, the sex determination pathway can directly alter the expression of patterning genes. For instance, *dsx* determines whether DPP is expressed in the A9 organizer of the genital disc (Keisman and Baker, 2001; Sanchez *et al.*, 2001), and the homeotic gene *Scr* is expressed differently in the leg disc of males and females (Barmina and Kopp, 2007). Lastly, we now know that patterning genes can also affect the expression of sex determination genes, since *dsx* and *fru* are expressed in a tissue-specific manner. For example, *dsx* expression within the embryonic gonad is dependent on *eya*, which specifies SGP identity (Hempel and Oliver, 2007). Thus, sex determination and patterning information can be integrated at multiple levels to control the creation of sexual dimorphism.

3.4. Of flies and men

These new ideas about how sex determination works in *Drosophila* strike at the heart of some of the main differences that were thought to exist between sex determination in flies and mammals. One difference between flies and mammals was thought to be the reliance on autonomous sex determination (in flies) versus nonautonomous sex determination (in mammals). With our new knowledge of nonautonomous sex determination in *Drosophila*, this difference is no longer so apparent. There is still no evidence for systemically acting sexual hormones in *Drosophila*, akin to testosterone in mammals. However, the local, nonautonomous, cell-cell interactions that regulate sexual dimorphism in the *Drosophila* gonad and genital disc are very similar to those that control testis vs. ovary formation in mammals (Capel *et al.*, this volume).

In addition, in mammals the processes of sex determination and dosage compensation are separate; the Y chromosome and *SRY* control sex determination, while the number of X chromosomes controls dosage compensation. Most or all cells undergo dosage compensation, but only a small subset of cells determines their own sex through the expression *SRY*. In particular, *SRY* is expressed in a subset of cells in the somatic gonad, which become Sertoli cells (Capel *et al.*, this volume). An analogous situation exists in *Drosophila* where, again, most or all cells undergo dosage compensation, but only those cells that express *dsx* or *fru* can determine their own sex. Interestingly, within the embryo, *dsx* expression is limited to the cells of the somatic gonad, similar to *SRY* expression in mammals. Finally, flies and mammals use common genes for the creation of sexual dimorphism. Homologs of *dsx*, the DMRTs, are essential for gonad development in mammals, and the *Sox9* homolog, *Sox100B*, is required for proper development of the testis in *Drosophila* (discussed in the previous sections). Thus, as we learn more about the mechanisms by which sexual dimorphism is created during development, we find a deeper level of evolutionary conservation. This is particularly apparent in the gonads, which have a common function, the production of the sex-specific gametes, across very different animal species.

Note added in proof

New work by the Carroll lab now demonstrates that the *bab1/2* locus is under direct regulation by both DSX and ABDB (Williams *et al.*, 2008).

REFERENCES

Aboïm, A. N. (1945). Développement embryonnaire et post-embryonnaire des gonades normales et agamétiques de *Drosophila melanogaster*. *Revue Suisse de Zoologie* **52,** 53–154.

Ahmad, S. M., and Baker, B. S. (2002). Sex-specific deployment of FGF signaling in *Drosophila* recruits mesodermal cells into the male genital imaginal disc. *Cell* **109,** 651–661.

Amrein, H., Gorman, M., and Nothiger, R. (1988). The sex-determining gene tra-2 of *Drosophila* encodes a putative RNA binding protein. *Cell* **55,** 1025–1035.

An, W., Cho, S., Ishii, H., and Wensink, P. C. (1996). Sex-specific and non-sex-specific oligomerization domains in both of the doublesex transcription factors from *Drosophila melanogaster*. *Mol. Cell Biol.* **16,** 3106–3111.

Andrews, J., Garcia-Estefania, D., Delon, I., Lu, J., Mevel-Ninio, M., Spierer, A., Payre, F., Pauli, D., and Oliver, B. (2000). Ovo Transcription factors function antagonistically in the *Drosophila* female germline. *Development* **127,** 881–892.

Arbeitman, M. N., Furlong, E. E., Imam, F., Johnson, E., Null, B. H., Baker, B. S., Krasnow, M. A., Scott, M. P., Davis, R. W., and White, K. P. (2002). Gene expression during the life cycle of *Drosophila melanogaster*. *Science* **297,** 2270–2275.

Arbeitman, M. N., Fleming, A. A., Siegal, M. L., Null, B. H., and Baker, B. S. (2004). A genomic analysis of *Drosophila* somatic sexual differentiation and its regulation. *Development* **131,** 2007–2021.

Asaoka, M., and Lin, H. (2004). Germline stem cells in the *Drosophila* ovary descend from pole cells in the anterior region of the embryonic gonad. *Development* **131,** 5079–5089.

Avila, F. W., and Erickson, J. W. (2007). *Drosophila* JAK/STAT pathway reveals distinct initiation and reinforcement steps in early transcription of Sxl. *Curr. Biol.* **17,** 643–648.

Baker, B. S., and Ridge, K. A. (1980). Sex and the single cell. I. On the action of major loci affecting sex determination in *Drosophila melanogaster*. *Genetics* **94,** 383–423.

Baker, B. S., and Wolfner, M. F. (1988). A molecular analysis of doublesex, a bifunctional gene that controls both male and female sexual differentiation in *Drosophila melanogaster*. *Genes Dev.* **2,** 477–489.

Barbash, D. A., and Cline, T. W. (1995). Genetic and molecular analysis of the autosomal component of the primary sex determination signal of *Drosophila melanogaster*. *Genetics* **141,** 1451–1471.

Barmina, O., and Kopp, A. (2007). Sex-specific expression of a HOX gene associated with rapid morphological evolution. *Dev. Biol.* **311,** 277–286.

Bell, L. R., Maine, E. M., Schedl, P., and Cline, T. W. (1988). Sex-lethal, a *Drosophila* sex determination switch gene, exhibits sex-specific RNA splicing and sequence similarity to RNA binding proteins. *Cell* **55,** 1037–1046.

Bell, L. R., Horabin, J. I., Schedl, P., and Cline, T. W. (1991). Positive autoregulation of Sex-lethal by alternative splicing maintains the female determined state in *Drosophila*. *Cell* **65,** 229–239.

Belote, J. M., and Baker, B. S. (1982). Sex determination in *Drosophila melanogaster*: Analysis of transformer-2, a sex-transforming locus. *Proc. Natl. Acad. Sci. USA* **79,** 1568–1572.

Belote, J. M., Handler, A. M., Wolfner, M. F., Livak, K. J., and Baker, B. S. (1985). Sex-specific regulation of yolk protein gene expression in *Drosophila*. *Cell* **40,** 339–348.

Boggs, R. T., Gregor, P., Idriss, S., Belote, J. M., and McKeown, M. (1987). Regulation of sexual differentiation in *D. melanogaster* via alternative splicing of RNA from the transformer gene. *Cell* **50,** 739–747.

Bownes, M. (1994). The regulation of the yolk protein genes, a family of sex differentiation genes in *Drosophila melanogaster*. *Bioessays* **16,** 745–752.

Bownes, M., Scott, A., and Blair, M. (1987). The use of an inhibitor of protein synthesis to investigate the roles of ecdysteroids and sex-determination genes on the expression of the genes encoding the *Drosophila* yolk proteins. *Development* **101,** 931–941.

Bownes, M., Steinmann-Zwicky, M., and Nothiger, R. (1990). Differential control of yolk protein gene expression in fat bodies and gonads by the sex-determining gene tra-2 of *Drosophila*. *EMBO J.* **9,** 3975–3980.

Boyle, M., and DiNardo, S. (1995). Specification, migration, and assembly of the somatic cells of the *Drosophila* gonad. *Development* **121,** 1815–1825.

Boyle, M., Bonini, N., and DiNardo, S. (1997). Expression and function of clift in the development of somatic gonadal precursors within the *Drosophila* mesoderm. *Development* **124,** 971–982.

Bray, S., and Amrein, H. (2003). A putative *Drosophila* pheromone receptor expressed in male-specific taste neurons is required for efficient courtship. *Neuron* **39,** 1019–1029.

Bridges, C. (1913). Non-disjunction of the sex chromosomes of *Drosophila*. *J. Exp. Zool.* **15,** 587.

Bridges, C. (1916). Non-disjunction as proof of the chromosome theory of heredity. *Genetics* **1,** 1–52107–163.

Bridges, C. (1921). Triploid intersexes in *Drosophila melanogaster*. *Science* **54,** 252–254.

Bridges, C. (1930). Haploid *Drosophila* and the theory of genic balance. *Science N. S.* **72,** 405.

Burtis, K. C., and Baker, B. S. (1989). *Drosophila doublesex* gene controls somatic sexual differentiation by producing alternatively spliced mrnas encoding related sex-specific polypeptides. *Cell* **56,** 997–1010.

Burtis, K. C., Coschigano, K. T., Baker, B. S., and Wensink, P. C. (1991). The doublesex proteins of *Drosophila melanogaster* bind directly to a sex-specific yolk protein gene enhancer. *EMBO J.* **10,** 2577–2582.

Butler, B., Pirrotta, V., Irminger-Finger, I., and Nothiger, R. (1986). The sex-determining gene tra of *Drosophila*: Molecular cloning and transformation studies. *Embo J.* **5,** 3607–3613.

Cabrera, C. V., and Alonso, M. C. (1991). Transcriptional activation by heterodimers of the achaete-scute and daughterless gene products of *Drosophila*. *Embo J.* **10,** 2965–2973.

Campos-Ortega, J. A., and Hartenstein, V. (1985). The Embryonic Development of *Drosophila melanogaster*. Springer-Verlag, Heidelberg.

Casares, F., Bender, W., Merriam, J., and Sanchez-Herrero, E. (1997). Interactions of *Drosophila* Ultrabithorax regulatory regions with native and foreign promoters. *Genetics* **145**, 123–137.

Caudy, M., Vassin, H., Brand, M., Tuma, R., Jan, L. Y., and Jan, Y. N. (1988). Daughterless, a *Drosophila* gene essential for both neurogenesis and sex determination, has sequence similarities to myc and the achaete-scute complex. *Cell* **55**, 1061–1067.

Celniker, S. E., and Lewis, E. B. (1993). Molecular basis of transabdominal–a sexually dimorphic mutant of the bithorax complex of *Drosophila*. *Proc. Natl. Acad. Sci. USA* **90**, 1566–1570.

Celniker, S. E., Sharma, S., Keelan, D. J., and Lewis, E. B. (1990). The molecular genetics of the bithorax complex of *Drosophila*: *Cis*-regulation in the abdominal-B domain. *EMBO J.* **9**, 4277–4286.

Chaboissier, M. C., Kobayashi, A., Vidal, V. I., Lutzkendorf, S., van de Kant, H. J., Wegner, M., de Rooij, D. G., Behringer, R. R., and Schedl, A. (2004). Functional analysis of Sox8 and Sox9 during sex determination in the mouse. *Development* **131**, 1891–1901.

Chapman, K. B., and Wolfner, M. F. (1988). Determination of male-specific gene expression in *Drosophila* accessory glands. *Dev. Biol.* **126**, 195–202.

Chase, B. A., and Baker, B. S. (1995). A genetic analysis of intersex, a gene regulating sexual differentiation in *Drosophila melanogaster* females. *Genetics* **139**, 1649–1661.

Chen, E. H., and Baker, B. S. (1997). Compartmental organization of the *Drosophila* genital imaginal discs. *Development* **124**, 205–218.

Chen, J., Godt, D., Gunsalus, K., Kiss, I., Goldberg, M., and Laski, F. A. (2001). Cofilin/ADF is required for cell motility during *Drosophila* ovary development and oogenesis. *Nat. Cell Biol.* **3**, 204–209.

Chen, E. H., Christiansen, A. E., and Baker, B. S. (2005). Allocation and specification of the genital disc precursor cells in *Drosophila*. *Dev. Biol.* **281**, 270–285.

Cho, S., and Wensink, P. C. (1996). Purification and physical properties of the male and female double sex proteins of *Drosophila*. *Proc. Natl. Acad. Sci. USA* **93**, 2043–2047.

Cho, S., and Wensink, P. C. (1997). DNA binding by the male and female doublesex proteins of *Drosophila melanogaster*. *J. Biol. Chem.* **272**, 3185–3189.

Christiansen, A. E., Keisman, E. L., Ahmad, S. M., and Baker, B. S. (2002). Sex comes in from the cold: The integration of sex and pattern. *Trends Genet.* **18**, 510–516.

Clark, I. B., Boyd, J., Hamilton, G., Finnegan, D. J., and Jarman, A. P. (2006). D-six4 plays a key role in patterning cell identities deriving from the *Drosophila* mesoderm. *Dev. Biol.* **294**, 220–231.

Clark, I. B., Jarman, A. P., and Finnegan, D. J. (2007). Live imaging of *Drosophila* gonad formation reveals roles for Six4 in regulating germline and somatic cell migration. *BMC Dev. Biol.* **7**, 52.

Cline, T. W. (1980). Maternal and zygotic sex-specific gene interactions in *Drosophila melanogaster*. *Genetics* **96**, 903–926.

Cline, T. W. (1984). Autoregulatory functioning of a *Drosophila* gene product that establishes and maintains the sexually determined state. *Genetics* **107**, 231–277.

Cline, T. W., and Meyer, B. J. (1996). Vive la difference: Males vs females in flies vs worms. *Annu. Rev. Genet.* **30**, 637–702.

Cohen, E. D., Mariol, M. C., Wallace, R. M., Weyers, J., Kamberov, Y. G., Pradel, J., and Wilder, E. L. (2002). Dwnt4 regulates cell movement and focal adhesion kinase during *Drosophila* ovarian morphogenesis. *Dev. Cell* **2**, 437–448.

Coschigano, K. T., and Wensink, P. C. (1993). Sex-specific transcriptional regulation by the male and female doublesex proteins of *Drosophila*. *Genes Dev.* **7**, 42–54.

Couderc, J. L., Godt, D., Zollman, S., Chen, J., Li, M., Tiong, S., Cramton, S. E., Sahut-Barnola, I., and Laski, F. A. (2002). The bric a brac locus consists of two paralogous genes encoding BTB/POZ domain proteins and acts as a homeotic and morphogenetic regulator of imaginal development in *Drosophila*. *Development* **129,** 2419–2433.

Cronmiller, C., Schedl, P., and Cline, T. W. (1988). Molecular characterization of daughterless, a *Drosophila* sex determination gene with multiple roles in development. *Genes Dev.* **2,** 1666–1676.

Currie, D. A., and Bate, M. (1995). Innervation is essential for the development and differentiation of a sex-specific adult muscle in *Drosophila melanogaster*. *Development* **121,** 2549–2557.

Dansereau, D. A., and Lasko, P. (2008). The development of germline stem cells in *Drosophila*. *Methods Mol. Biol.* **450,** 3–26.

Dauwalder, B., Tsujimoto, S., Moss, J., and Mattox, W. (2002). The *Drosophila* takeout gene is regulated by the somatic sex-determination pathway and affects male courtship behavior. *Genes Dev.* **16,** 2879–2892.

Decotto, E., and Spradling, A. C. (2005). The *Drosophila* ovarian and testis stem cell niches: Similar somatic stem cells and signals. *Dev. Cell* **9,** 501–510.

DeFalco, T. J., Verney, G., Jenkins, A. B., McCaffery, J. M., Russell, S., and Van Doren, M. (2003). Sex-specific apoptosis regulates sexual dimorphism in the *Drosophila* embryonic gonad. *Dev. Cell* **5,** 205–216.

DeFalco, T., Le Bras, S., and Van Doren, M. (2004). Abdominal-B is essential for proper sexually dimorphic development of the *Drosophila* gonad. *Mech. Dev.* **121,** 1323–1333.

DeFalco, T., Camara, N., Le Bras, S., and Van Doren, M. (2008). Nonautonomous sex determination controls sexually dimorphic development of the *Drosophila* gonad. *Dev. Cell* **14,** 275–286.

Deng, W., and Lin, H. (1997). Spectrosomes and fusomes anchor mitotic spindles during asymmetric germ cell divisions and facilitate the formation of a polarized microtubule array for oocyte specification in *Drosophila*. *Dev. Biol.* **189,** 79–94.

DiBenedetto, A. J., Lakich, D. M., Kruger, W. D., Belote, J. M., Baker, B. S., and Wolfner, M. F. (1987). Sequences expressed sex-specifically in *Drosophila melanogaster* adults. *Dev. Biol.* **119,** 242–251.

Dobzahnsky, T. S. J. (1931). Evidence for multiple sex factors in the x-chromosome of *Drosophila melanogaster*. *Proc. Natl. Acad. Sci. USA* **17,** 513.

Ellis, H. M., Spann, D. R., and Posakony, J. W. (1990). Extramacrochaetae, a negative regulator of sensory organ development in *Drosophila*, defines a new class of helix-loop-helix proteins. *Cell* **61,** 27–38.

Epper, F., and Bryant, P. J. (1983). Sex-specific control of growth and differentiation in the *Drosophila* genital disc, studied using a temperature-sensitive transformer-2 mutation. *Dev. Biol.* **100,** 294–307.

Epper, F., and Nothiger, R. (1982). Genetic and developmental evidence for a repressed genital primordium in *Drosophila melanogaster*. *Dev. Biol.* **94,** 163–175.

Erdman, S. E., and Burtis, K. C. (1993). The *Drosophila* doublesex proteins share a novel zinc finger related DNA binding domain. *EMBO J.* **12,** 527–535.

Erdman, S. E., Chen, H. J., and Burtis, K. C. (1996). Functional and genetic characterization of the oligomerization and DNA binding properties of the *Drosophila* doublesex proteins. *Genetics* **144,** 1639–1652.

Erickson, J. W., and Cline, T. W. (1993). A bzip protein, Sisterless-a, collaborates with bhlh transcription factors early in *Drosophila* development to determine sex. *Genes Dev.* **7,** 1688–1702.

Erickson, J. W., and Quintero, J. J. (2007). Indirect effects of ploidy suggest X chromosome dose, not the X:A ratio, signals sex in *Drosophila*. *PLoS Biol.* **5,** e332.

Estrada, B., and Sanchez-Herrero, E. (2001). The Hox gene Abdominal-B antagonizes appendage development in the genital disc of *Drosophila*. *Development* **128,** 331–339.

Estrada, B., Casares, F., and Sanchez-Herrero, E. (2003). Development of the genitalia in *Drosophila melanogaster*. *Differentiation* **71,** 299–310.

Evans, D. S., and Cline, T. W. (2007). *Drosophila melanogaster* male somatic cells feminized solely by tra² can collaborate with female germ cells to make functional eggs. *Genetics* **175,** 631–642.

Feng, Y., Schiff, N. M., and Cavener, D. R. (1991). Organ-specific patterns of gene expression in the reproductive tract of *Drosophila* are regulated by the sex-determination genes. *Dev. Biol.* **146,** 451–460.

Ferguson-Smith, M. (2007). The evolution of sex chromosomes and sex determination in vertebrates and the key role of DMRT1. *Sex Dev.* **1,** 2–11.

Ferveur, J. F. (2005). Cuticular hydrocarbons: Their evolution and roles in *Drosophila* pheromonal communication. *Behav. Genet.* **35,** 279–295.

Flickinger, T. W., and Salz, H. K. (1994). The *Drosophila* sex determination gene *snf* encodes a nuclear protein with sequence and functional similarity to the mammalian U1A snrnp protein. *Genes Dev.* **8,** 914–925.

Foronda, D., Estrada, B., de Navas, L., and Sanchez-Herrero, E. (2006). Requirement of Abdominal-A and Abdominal-B in the developing genitalia of *Drosophila* breaks the posterior downregulation rule. *Development* **133,** 117–127.

Foster, J. W., Dominguez-Steglich, M. A., Guioli, S., Kowk, G., Weller, P. A., Stevanovic, M., Weissenbach, J., Mansour, S., Young, I. D., Goodfellow, P. N., Brook, J. D., and Schafer, A. J. (1994). Campomelic dysplasia and autosomal sex reversal caused by mutations in an SRY-related gene. *Nature* **372,** 525–530.

Freeland, D. E., and Kuhn, D. T. (1996). Expression patterns of developmental genes reveal segment and parasegment organization of *D. melanogaster* genital discs. *Mech. Dev.* **56,** 61–72.

Fujihara, T., Kawabe, M., and Oishi, K. (1978). A sex-transformation gene in *Drosophila melanogaster*. *J. Hered.* **69,** 229–236.

Fujii, S., and Amrein, H. (2002). Genes expressed in the *Drosophila* head reveal a role for fat cells in sex-specific physiology. *EMBO J.* **21,** 5353–5363.

Fuller, M. (1993). Spermatogenesis. In "The Development of *Drosophila melanogaster*" (M. Bate and A. Martinez Arias, Eds.), pp. 71–147. Cold Spring Harbor Press, Cold Spring Harbor.

Fuller, M. T., and Spradling, A. C. (2007). Male and female *Drosophila* germline stem cells: Two versions of immortality. *Science* **316,** 402–404.

Garrell, J., and Modolell, J. (1990). The *Drosophila* extramacrochaetae locus, an antagonist of proneural genes that, like these genes, encodes a helix-loop-helix protein. *Cell* **61,** 39–48.

Garrett-Engele, C. M., Siegal, M. L., Manoli, D. S., Williams, B. C., Li, H., and Baker, B. S. (2002). Intersex, a gene required for female sexual development in *Drosophila*, is expressed in both sexes and functions together with doublesex to regulate terminal differentiation. *Development* **129,** 4661–4675.

Gilbert, S. F. (2003). Developmental Biology Sinauer Associates.

Gilboa, L., and Lehmann, R. (2004). How different is Venus from Mars? The genetics of germ-line stem cells in *Drosophila* females and males. *Development* **131,** 4895–4905.

Godt, D., and Laski, F. A. (1995). Mechanisms of cell rearrangement and cell recruitment in *Drosophila* ovary morphogenesis and the requirement of bric a brac. *Development* **121,** 173–187.

Godt, D., Couderc, J. L., Cramton, S. E., and Laski, F. A. (1993). Pattern formation in the limbs of *Drosophila: Bric à brac* is expressed in both a gradient and a wave-like pattern and is required for specification and proper segmentation of the tarsus. *Development* **119,** 799–812.

Goldman, T. D., and Arbeitman, M. N. (2007). Genomic and Functional Studies of *Drosophila* Sex Hierarchy Regulated Gene Expression in Adult Head and Nervous System Tissues. *PLoS Genet.* **3,** e216.

Gönczy, P., Viswanathan, S., and DiNardo, S. (1992). Probing spermatogenesis in *Drosophila* with P-element enhancer detectors. *Development* **114,** 89–98.

Goralski, T. J., Edström, J. E., and Baker, B. S. (1989). The sex determination locus transformer-2 of *Drosophila* encodes a polypeptide with similarity to RNA binding proteins. *Cell* **56,** 931–936.

Gorfinkiel, N., Sanchez, L., and Guerrero, I. (2003). Development of the *Drosophila* genital disc requires interactions between its segmental primordia. *Development* **130,** 295–305.

Graham, P., Penn, J. K., and Schedl, P. (2003). Masters change, slaves remain. *Bioessays* **25,** 1–4.

Granadino, B., San Juan, A., Santamaria, P., and Sanchez, L. (1992a). Evidence of a dual function in fl(2)d, a gene needed for Sex-lethal expression in *Drosophila melanogaster*. *Genetics* **130,** 597–612.

Granadino, B., San Juan, A., Santamaria, P., and Sánchez, L. (1992b). Evidence of a dual function in *fl(2)d*, a gene needed for *Sex-lethal* expression in *Drosophila melanogaster*. *Genetics* **130,** 597–612.

Hadorn, E., and Bertani, G. (1948). Induktion mannlicher Pigmentierung in somatischen Zellen von *Drosophila*-Ovarien. *Rev. Suisse Zool.* **55,** 232–248.

Hardy, R. W., Tokuyasu, K. T., Lindsley, D. L., and Garavito, M. (1979). The germinal proliferation center in the testis of *Drosophila melanogaster*. *J. Ultrastruct. Res.* **69,** 180–190.

Hempel, L. U., and Oliver, B. (2007). Sex-specific doublesexm expression in subsets of *Drosophila* somatic gonad cells. *BMC Dev. Biol.* **7,** 113.

Hildreth, P. E. (1965). Doublesex, a recessive gene that transforms both males and females of *Drosophila* into intersexes. *Genetics* **51,** 659–678.

Hilfiker-Kleiner, D., Dübendorfer, A., Hilfiker, A., and Nöthinger, R. (1994). Genetic control of sex determination in the germ line and soma of the housefly, *Musca domestica*. *Development* **120,** 2531–2538.

Hong, C. S., Park, B. Y., and Saint-Jeannet, J. P. (2007). The function of Dmrt genes in vertebrate development: It is not just about sex. *Dev. Biol.* **310,** 1–9.

Hopmann, R., Duncan, D., and Duncan, I. (1995). Transvection in the iab-5,6,7 region of the bithorax complex of *Drosophila*: Homology independent interactions in trans. *Genetics* **139,** 815–833.

Jeong, S., Rokas, A., and Carroll, S. B. (2006). Regulation of body pigmentation by the Abdominal-B Hox protein and its gain and loss in *Drosophila* evolution. *Cell* **125,** 1387–1399.

Jin, W., Riley, R. M., Wolfinger, R. D., White, K. P., Passador-Gurgel, G., and Gibson, G. (2001). The contributions of sex, genotype and age to transcriptional variance in *Drosophila melanogaster*. *Nat. Genet.* **29,** 389–395.

Jinks, T. M., Polydorides, A. D., Calhoun, G., and Schedl, P. (2000). The JAK/STAT signaling pathway is required for the initial choice of sexual identity in *Drosophila melanogaster*. *Mol. Cell* **5,** 581–587.

Jursnich, V. A., and Burtis, K. C. (1993). A positive role in differentiation for the male doublesex protein of *Drosophila*. *Dev. Biol.* **155,** 235–249.

Kania, M. A., Bonner, A. S., Duffy, J. B., and Gergen, J. P. (1990). The *Drosophila* segmentation gene *runt* encodes a novel nuclear regulatory protein that is also expressed in the developing nervous system. *Genes Dev.* **4,** 1701–1713.

Kawase, E., Wong, M. D., Ding, B. C., and Xie, T. (2004). Gbb/Bmp signaling is essential for maintaining germline stem cells and for repressing bam transcription in the *Drosophila* testis. *Development* **131**, 1365–1375.

Keisman, E. L., and Baker, B. S. (2001). The *Drosophila* sex determination hierarchy modulates wingless and decapentaplegic signaling to deploy dachshund sex-specifically in the genital imaginal disc. *Development* **128**, 1643–1656.

Keisman, E. L., Christiansen, A. E., and Baker, B. S. (2001). The sex determination gene doublesex regulates the A/P organizer to direct sex-specific patterns of growth in the *Drosophila* genital imaginal disc. *Dev. Cell* **1**, 215–225.

Kent, J., Wheatley, S. C., Andrews, J. E., Sinclair, A. H., and Koopman, P. (1996). A male-specific role for SOX9 in vertebrate sex determination. *Development* **122**, 2813–2822.

Keyes, L. N., Cline, T. W., and Schedl, P. (1992). The primary sex determination signal of *Drosophila* acts at the level of transcription. *Cell* **68**, 933–943.

Kiger, A. A., Jones, D. L., Schulz, C., Rogers, M. B., and Fuller, M. T. (2001). Stem cell self-renewal specified by JAK-STAT activation in response to a support cell cue. *Science* **294**, 2542–2545.

Kim, S., Kettlewell, J. R., Anderson, R. C., Bardwell, V. J., and Zarkower, D. (2003). Sexually dimorphic expression of multiple doublesex-related genes in the embryonic mouse gonad. *Gene Exp. Patterns* **3**, 77–82.

Kim, S., Bardwell, V. J., and Zarkower, D. (2007). Cell type-autonomous and non-autonomous requirements for Dmrt1 in postnatal testis differentiation. *Dev. Biol.* **307**, 314–327.

King, R. C. (1970). Ovarian Development in *Drosophila melanogaster*. Academic Press, New York.

Kirby, R. J., Hamilton, G. M., Finnegan, D. J., Johnson, K. J., and Jarman, A. P. (2001). *Drosophila* homolog of the myotonic dystrophy-associated gene, SIX5, is required for muscle and gonad development. *Curr. Biol.* **11**, 1044–1049.

Kirilly, D., and Xie, T. (2007). The *Drosophila* ovary: An active stem cell community. *Cell Res.* **17**, 15–25.

Kitadate, Y., Shigenobu, S., Arita, K., and Kobayashi, S. (2007). Boss/Sev signaling from germline to soma restricts germline-stem-cell-niche formation in the anterior region of *Drosophila* male gonads. *Dev. Cell* **13**, 151–159.

Kopp, A., Duncan, I., Godt, D., and Carroll, S. B. (2000). Genetic control and evolution of sexually dimorphic characters in *Drosophila*. *Nature* **408**, 553–559.

Kozopas, K. M., Samos, C. H., and Nusse, R. (1998). Dwnt-2, a *Drosophila* Wnt gene required for the development of the male reproductive tract, specifies a sexually dimorphic cell fate. *Genes Dev.* **12**, 1155–1165.

Kramer, S. G., Jinks, T. M., Schedl, P., and Gergen, J. P. (1999). Direct activation of Sex-lethal transcription by the *Drosophila* runt protein. *Development* **126**, 191–200.

Kuziora, M. A., and McGinnis, W. (1988). Different transcripts of the *Drosophila* Abd-B gene correlate with distinct genetic sub-functions. *EMBO J.* **7**, 3233–3244.

Lawrence, P. A., and Johnston, P. (1984). The genetic specification of pattern in a *Drosophila* muscle. *Cell* **36**, 775–782.

Lawrence, P. A., and Johnston, P. (1986). The muscle pattern of a segment of *Drosophila* may be determined by neurons and not by contributing myoblasts. *Cell* **45**, 505–513.

Lazareva, A. A., Roman, G., Mattox, W., Hardin, P. E., and Dauwalder, B. (2007). A role for the adult fat body in *Drosophila* male courtship behavior. *PLoS Genet.* **3**, e16.

Le Bras, S., and Van Doren, M. (2006). Development of the male germline stem cell niche in *Drosophila*. *Dev. Biol.* **294**, 92–103.

Leatherman, J. L., and Dinardo, S. (2008). Zfh-1 controls somatic stem cell self-renewal in the *Drosophila* testis and nonautonomously influences germline stem cell self-renewal. *Cell Stem Cell* **3**, 44–54.

Lee, G., Foss, M., Goodwin, S. F., Carlo, T., Taylor, B. J., and Hall, J. C. (2000). Spatial, temporal, and sexually dimorphic expression patterns of the fruitless gene in the Drosophila central nervous system. *J. Neurobiol.* **43**, 404–426.

Lee, G., Hall, J. C., and Park, J. H. (2002). Doublesex gene expression in the central nervous system of Drosophila melanogaster. *J. Neurogenet.* **16**, 229–248.

Lin, H. (2002). The stem-cell niche theory: Lessons from flies. *Nat. Rev. Genet.* **3**, 931–940.

Lin, H., Yue, L., and Spradling, A. C. (1994). The Drosophila fusome, a germline-specific organelle, contains membrane skeletal proteins and functions in cyst formation. *Development* **120**, 947–956.

Loh, S. H. Y., and Russell, S. (2000). A Drosophila group E Sox gene is dynamically expressed in the embryonic alimentary canal. *Mech. Dev.* **93**, 185–188.

Lopez-Onieva, L., Fernandez-Minan, A., and Gonzalez-Reyes, A. (2008). Jak/Stat signalling in niche support cells regulates dpp transcription to control germline stem cell maintenance in the Drosophila ovary. *Development* **135**, 533–540.

Lucchesi, J. C., Kelly, W. G., and Panning, B. (2005). Chromatin remodeling in dosage compensation. *Annu. Rev. Genet.* **39**, 615–651.

Macdonald, P. M., Ingham, P., and Struhl, G. (1986). Isolation, structure, and expression of even-skipped: A second pair-Rule gene of Drosophila containing a homeo box. *Cell* **47**, 721–734.

Manoli, D. S., Meissner, G. W., and Baker, B. S. (2006). Blueprints for behavior: Genetic specification of neural circuitry for innate behaviors. *Trends Neurosci.* **29**, 444–451.

Margolis, J., and Spradling, A. (1995). Identification and behavior of epithelial stem cells in the Drosophila ovary. *Development* **121**, 3797–3807.

Mattox, W., Palmer, M. J., and Baker, B. S. (1990). Alternative splicing of the sex determination gene transformer-2 is sex-specific in the germ line but not in the soma. *Genes Dev.* **4**, 789–805.

McKeown, M., Belote, J. M., and Boggs, R. T. (1988). Ectopic expression of the female transformer gene product leads to female differentiation of chromosomally male Drosophila. *Cell* **53**, 887–895.

Meller, V. H., and Kuroda, M. I. (2002). Sex and the single chromosome. *Adv. Genet.* **46**, 1–24.

Meunier, N., Ferveur, J. F., and Marion-Poll, F. (2000). Sex-specific non-pheromonal taste receptors in Drosophila. *Curr .Biol.* **10**, 1583–1586.

Moniot, B., Berta, P., Scherer, G., Sudbeck, P., and Poulat, F. (2000). Male specific expression suggests role of DMRT1 in human sex determination. *Mech. Dev.* **91**, 323–325.

Moreno, E., and Morata, G. (1999). Caudal is the Hox gene that specifies the most posterior Drosophile segment. *Nature* **400**, 873–877.

Morgan, T., and Bridges, C. (1919). The origin of gynandromoprhs. *Carnegie Institution of Washington* **278**, 1–122.

Nagoshi, R. N., and Baker, B. S. (1990). Regulation of sex-specific RNA splicing at the Drosophila doublesex gene: Cis-acting mutations in exon sequences alter sex-specific RNA splicing patterns. *Genes Dev.* **4**, 89–97.

Nagoshi, R. N., McKeown, M., Burtis, K. C., Belote, J. M., and Baker, B. S. (1988). The control of alternative splicing at genes regulating sexual differentiation in D. melanogaster. *Cell* **53**, 229–236.

Narendra, U., Zhu, L., Li, B., Wilken, J., and Weiss, M. A. (2002). Sex-specific gene regulation. The Doublesex DM motif is a bipartite DNA-binding domain. *J. Biol. Chem.* **277**, 43463–43473.

Nayak, S. S. R. (1983). Sensilla on the Tarsal Segments and Mouthparts of Adult Drosophila melanogaster meigen (Diptera: Drosophilidae). *Int. J. Insect Morphol. Embryol.* **12**, 273–291.

Ng, C. S., and Kopp, A. (2008). Sex combs are important for male mating success in *Drosophila melanogaster*. *Behav. Genet.* **38,** 195–201.

Nothiger, R. D. A., and Epper, F. (1977). Gynandromorphs reveal two separate primordia for male and female genitalia. *Roux's Arch. Dev. Biol.* **181,** 367–373.

Nystul, T., and Spradling, A. (2007). An epithelial niche in the *Drosophila* ovary undergoes long-range stem cell replacement. *Cell Stem Cell* **1,** 277–285.

Ottolenghi, C., Veitia, R., Quintana-Murci, L., Torchard, D., Scapoli, L., Souleyreau-Therville, N., Beckmann, J., Fellous, M., and McElreavey, K. (2000). The region on 9p associated with 46,XY sex reversal contains several transcripts expressed in the urogenital system and a novel doublesex-related domain. *Genomics* **64,** 170–178.

Parisi, M. N. R., Edwards, P., Minor, J., Naiman, D., Lu, J., Doctolero, M., Vainer, M., Chan, C., Malley, J., Eastman, S., and Oliver, B. (2004). A survey of ovary-, testis-, and soma-biased gene expression in *Drosophila melanogaster* adults. *Genome Biol.* **5,** R40.

Parkhurst, S. M., Bopp, D., and Ish-Horowicz, D. (1990). X:A ratio, the primary sex-determining signal in *Drosophila*, Is transduced by helix-loop-helix proteins. *Cell* **63,** 1179–1191.

Paroush, Z., Finley, R. L., Kidd, T., Wainwright, S. M., Ingham, P. W., Brent, R., and Ish-Horowicz, D. (1994). Groucho is required for *Drosophila* neurogenesis, segmentation, and sex determination and interacts with hairy-related bhlh proteins. *Cell* **79,** 805–815.

Patterson, J. T., Stone, W., and Bedichek, S. (1937). Further studies on X chromosome balance in *Drosophila*. *Genetics* **22,** 407–426.

Penalva, L. O., and Sanchez, L. (2003). RNA binding protein sex-lethal (Sxl) and control of *Drosophila* sex determination and dosage compensation. *Microbiol. Mol. Biol. Rev.* **67,** 343–359, Table of contents.

Penn, J. K., Graham, P., Deshpande, G., Calhoun, G., Chaouki, A. S., Salz, H. K., and Schedl, P. (2008). Functioning of the *Drosophila* Wilms'-tumor-1-associated protein homolog, Fl(2)d, in sex-lethal-dependent alternative splicing. *Genetics* **178,** 737–748.

Pomiankowski, A., Nothiger, R., and Wilkins, A. (2004). The evolution of the *Drosophila* sex-determination pathway. *Genetics* **166,** 1761–1773.

Ranz, J. M., Castillo-Davis, C. I., Meiklejohn, C. D., and Hartl, D. L. (2003). Sex-dependent gene expression and evolution of the *Drosophila* transcriptome. *Science* **300,** 1742–1745.

Raymond, C. S., Parker, E. D., Kettlewell, J. R., Brown, L. G., Page, D. C., Kusz, K., Jaruzelska, J., Reinberg, Y., Flejter, W. L., Bardwell, V. J., Hirsch, B., and Zarkower, D. (1999). A region of human chromosome 9p required for testis development contains two genes related to known sexual regulators. *Hum. Mol. Genet.* **8,** 989–996.

Raymond, C. S., Murphy, M. W., O'Sullivan, M. G., Bardwell, V. J., and Zarkower, D. (2000). Dmrt1, a gene related to worm and fly regulators, is required for mammalian testis differentiation. *Genes Dev.* **14,** 2587–2595.

Russell, J., Gennissen, A., and Nusse, R. (1992). Isolation and expression of two novel Wnt/wingless gene homologues in *Drosophila*. *Development* **115,** 475–485.

Sahut-Barnola, I., Godt, D., Laski, F. A., and Couderc, J. L. (1995). *Drosophila* ovary morphogenesis: Analysis of terminal filament formation and identification of a gene required for this process. *Dev. Biol.* **170,** 127–135.

Salz, H. K., Maine, E. M., Keyes, L. N., Samuels, M. E., Cline, T. W., and Schedl, P. (1989). The Drosophila female-specific sex-determination gene, *Sex lethal*, has stage-, tissue-, and sex-specific rnas suggesting multiple modes of regulation. *Genes Dev.* **3,** 708–719.

Sanchez, L., and Guerrero, I. (2001). The development of the *Drosophila* genital disc. *Bioessays* **23,** 698–707.

Sanchez, L., and Nothiger, R. (1983). Sex determination and dosage compensation in *Drosophila melanogaster*: Production of male clones in XX females. *Embo J.* **2,** 485–491.

Sanchez, L., Granadino, B., and Torres, M. (1994). Sex determination in *Drosophila melanogaster*: X-linked genes involved in the initial step of sex-lethal activation. *Dev. Genet.* **15,** 251–264.

Sanchez, L., Gorfinkiel, N., and Guerrero, I. (2001). Sex determination genes control the development of the *Drosophila* genital disc, modulating the response to Hedgehog, Wingless and Decapentaplegic signals. *Development* **128,** 1033–1043.

Sanchez-Herrero, E., and Crosby, M. A. (1988). The Abdominal-B gene of *Drosophila melanogaster*: Overlapping transcripts exhibit two different spatial distributions. *EMBO J.* **7,** 2163–2173.

Sanders, L. E., and Arbeitman, M. N. (2008). Doublesex establishes sexual dimorphism in the *Drosophila* central nervous system in an isoform-dependent manner by directing cell number. *Dev. Biol.* **320,** 378–390.

Sato, S., Tomomori-Sato, C., Banks, C. A., Parmely, T. J., Sorokina, I., Brower, C. S., Conaway, R. C., and Conaway, J. W. (2003). A mammalian homolog of *Drosophila melanogaster* transcriptional coactivator intersex is a subunit of the mammalian Mediator complex. *J. Biol. Chem.* **278,** 49671–49674.

Schulz, C., Kiger, A. A., Tazuke, S. I., Yamashita, Y. M., Pantalena-Filho, L. C., Jones, D. L., Wood, C. G., and Fuller, M. T. (2004). A misexpression screen reveals effects of bag-of-marbles and TGF beta class signaling on the *Drosophila* male germ-line stem cell lineage. *Genetics/Society* **167,** 707–723.

Schupbach, T. W. E., and Nothiger, R. (1978). The embryonic organization of the genital disc studied in genetic mosaics of *Drosophila melanogaster*. *Roux's Arch. Dev. Biol.* **185,** 249–270.

Schutt, C., and Nothiger, R. (2000). Structure, function and evolution of sex-determining systems in Dipteran insects. *Development* **127,** 667–677.

Sefton, L., Timmer, J. R., Zhang, Y., Beranger, F., and Cline, T. W. (2000). An extracellular activator of the *Drosophila* JAK/STAT pathway is a sex-determination signal element. *Nature* **405,** 970–973.

Shirangi, T. R., and McKeown, M. (2007). Sex in flies: What 'body-mind' dichotomy? *Dev. Biol.* **306,** 10–19.

Shirangi, T. R., Taylor, B. J., and McKeown, M. (2006). A double-switch system regulates male courtship behavior in male and female *Drosophila melanogaster*. *Nat. Genet.* **38,** 1435–1439.

Shivdasani, A. A., and Ingham, P. W. (2003). Regulation of stem cell maintenance and transit amplifying cell proliferation by tgf-beta signaling in *Drosophila* spermatogenesis. *Curr. Biol.* **13,** 2065–2072.

Song, X., Zhu, C. H., Doan, C., and Xie, T. (2002). Germline stem cells anchored by adherens junctions in the *Drosophila* ovary niches. *Science* **296,** 1855–1857.

Song, X., Call, G. B., Kirilly, D., and Xie, T. (2007). Notch signaling controls germline stem cell niche formation in the *Drosophila* ovary. *Development* **134,** 1071–1080.

Spieth, H. (1952). Mating behavior within the genus *Drosophila* (Diptera). *Bull. Am. Mus. Nat. Hist.* **99,** 395–474.

Spradling, A. C. (1993). Developmental genetics of oogenesis. In "The Development of *Drosophila melanogaster*" (M. Bate and A. Martinez Arias, Eds.), pp. 1–70. Cold Spring Harbor Press, Cold Spring Harbor.

Starz-Gaiano, M., and Lehmann, R. (2001). Moving towards the next generation. *Mech. Dev.* **105,** 5–18.

Steinmann-Zwicky, M. (1994). Sex determination of the *Drosophila* germ line: *tra* and *dsx* control somatic inductive signals. *Development* **120,** 707–716.

Stern, C., and Hadorn, E. (1939). The Relation between the Color of Testes and Vasa Efferentia in *Drosophila*. *Genetics* **24,** 162–179.

Stevens, N. (1908). A study of germ cells of certain Diptera, with reference to the heterochromosomes and the phenomena of synapsis. *J. Exp. Zool.* **5,** 359.

Struhl, G. (1982). Genes controlling segmental specification in the *Drosophila* thorax. *Proc. Natl. Acad. Sci. USA* **79,** 7380–7384.
Sturtevant, A. H. (1945). A gene in *Drosophila melanogaster* that transforms females into males. *Genetics* **30,** 297–299.
Tanentzapf, G., Devenport, D., Godt, D., and Brown, N. H. (2007). Integrin-dependent anchoring of a stem-cell niche. *Nat. Cell Biol.* **9,** 1413–1418.
Taylor, B. J. (1992). Differentiation of a male-specific muscle in *Drosophila melanogaster* does not require the sex-determining genes doublesex or intersex. *Genetics* **132,** 179–191.
Taylor, B. J., and Truman, J. W. (1992). Commitment of abdominal neuroblasts in *Drosophila* to a male or female fate is dependent on genes of the sex-determining hierarchy. *Development* **114,** 625–642.
Taylor, B. J., and Knittel, L. M. (1995). Sex-specific differentiation of a male-specific abdominal muscle, the Muscle of Lawrence, is abnormal in hydroxyurea-treated and in fruitless male flies. *Development* **121,** 3079–3088.
Tokuyasu, K. T. (1974). Dynamics of spermiogenesis in *Drosophila melanogaster*. III. Relation between axoneme and mitochondrial derivatives. *Exp. Cell Res.* **84,** 239–250.
Tulina, N., and Matunis, E. (2001). Control of stem cell self-renewal in *Drosophila* spermatogenesis by JAK-STAT signaling. *Science* **294,** 2546–2549.
Usui-Aoki, K., Ito, H., Ui-Tei, K., Takahashi, K., Lukacsovich, T., Awano, W., Nakata, H., Piao, Z. F., Nilsson, E. E., Tomida, J., and Yamamoto, D. (2000). Formation of the male-specific muscle in female *Drosophila* by ectopic fruitless expression. *Nat. Cell Biol.* **2,** 500–506.
Van Doren, M. (2006). The development of the somatic gonad and fat bodies. *In* "Muscle Development in *Drosophila*" (H. Sink, Ed.). Landes Bioscience, Austin.
Van Doren, M., Ellis, H. M., and Posakony, J. W. (1991). The *Drosophila* extramacrochaetae protein antagonizes sequence-specific DNA binding by daughterless/achaete-scute protein complexes. *Development* **113,** 245–255.
Villares, R., and Cabrera, C. V. (1987). The *achaete-scute* gene complex of *D. melanogaster*: Conserved domains in a subset of genes required for neurogenesis and their homology to *myc*. *Cell* **50,** 415–424.
Vosshall, L. B., and Stocker, R. F. (2007). Molecular architecture of smell and taste in *Drosophila*. *Annu. Rev. Neurosci.* **30,** 505–533.
Wang, L., Li, Z., and Cai, Y. (2008). The JAK/STAT pathway positively regulates DPP signaling in the *Drosophila* germline stem cell niche. *J. Cell Biol.* **180,** 721–728.
Ward, E. J., Shcherbata, H. R., Reynolds, S. H., Fischer, K. A., Hatfield, S. D., and Ruohola-Baker, H. (2006). Stem cells signal to the niche through the Notch pathway in the *Drosophila* ovary. *Curr. Biol.* **16,** 2352–2358.
Waterbury, J. A., Jackson, L. L., and Schedl, P. (1999). Analysis of the doublesex female protein in *Drosophila melanogaster*: Role on sexual differentiation and behavior and dependence on intersex. *Genetics* **152,** 1653–1667.
Wawersik, M., Milutinovich, A., Casper, A. L., Matunis, E., Williams, B., and Van Doren, M. (2005). Somatic control of germline sexual development is mediated by the JAK/STAT pathway. *Nature* **436,** 563–567.
Williams, T. M., Selegue, J. E., Werner, T., Gompel, N., Kopp, A., and Carroll, S. B. (2008). The regulation and evolution of a genetic switch controlling sexually dimorphic traits in *Drosophila*. *Cell* **134,** 610–623.
Wolpert, L. S. J., Jessell, T., Lawrence, P., Roberston, E., and Meyerowitz, E. (2006). Principles of Development, 3rd ed. Oxford University Press, Oxford.
Xie, T., and Spradling, A. C. (1998). Decapentaplegic is essential for the maintenance and division of germline stem cells in the *Drosophila* ovary. *Cell* **94,** 251–260.
Xie, T., and Spradling, A. C. (2000). A niche maintaining germ line stem cells in the *Drosophila* ovary. *Science* **290,** 328–330.

Xu, A., Park, S. K., D'Mello, S., Kim, E., Wang, Q., and Pikielny, C. W. (2002). Novel genes expressed in subsets of chemosensory sensilla on the front legs of male *Drosophila melanogaster*. *Cell Tissue Res.* **307,** 381–392.

Yamashita, Y. M., Jones, D. L., and Fuller, M. T. (2003). Orientation of asymmetric stem cell division by the APC tumor suppressor and centrosome. *Science* **301,** 1547–1550.

Yamashita, Y. M., Mahowald, A. P., Perlin, J. R., and Fuller, M. T. (2007). Asymmetric inheritance of mother versus daughter centrosome in stem cell division. *Science* **315,** 518–521.

Yang, D., Lu, H., Hong, Y., Jinks, T. M., Estes, P. A., and Erickson, J. W. (2001). Interpretation of X chromosome dose at Sex-lethal requires non-E-box sites for the basic helix-loop-helix proteins SISB and daughterless. *Mol. Cell Biol.* **21,** 1581–1592.

Younger-Shepherd, S., Vaessin, H., Bier, E., Jan, L. Y., and Jan, Y. N. (1992). *Deadpan*, an essential pan-neural gene encoding an HLH protein, acts as a denominator in *Drosophila* sex determination. *Cell* **70,** 911–922.

Zarkower, D. (2002). Invertebrates may not be so different after all. *Novartis Found Symp* **244,** 115–26; Discussion in 126–235, 203–206, 253–257.

Zhu, C. H., and Xie, T. (2003). Clonal expansion of ovarian germline stem cells during niche formation in *Drosophila*. *Development* **130,** 2579–2588.

CHAPTER FOUR

Drosophila Germline Sex Determination: Integration of Germline Autonomous Cues and Somatic Signals

Leonie U. Hempel,[*] Rasika Kalamegham,[*] John E. Smith III,[*] *and* Brian Oliver

Contents

1. Introduction	110
2. Sexually Dimorphic Niches Regulate Germ Cells	110
2.1. The male niche and proliferative zone	110
2.2. Signaling pathways	115
2.3. The female niche and proliferative zone	118
2.4. Signaling pathways	121
3. Sex Determination in the Germ Cells	122
3.1. Somatic sex determination	122
3.2. Germline sex determination	125
4. Germ Cells in a Sex-Mismatched Soma	128
4.1. Male germ cells within a female soma	128
4.2. Female germ cells within a male soma	138
5. Concluding Remarks	142
Acknowledgments	143
References	143

Abstract

The *Drosophila* testis and ovary are major genetically tractable systems for studying stem cells and their regulation. This has resulted in a deep understanding of germline stem cell regulation by the microenvironment, or niche. The male and female germline niches differ. Since sex is determined through different mechanisms in the soma than in the germline, genetic or physical manipulations can be used to experimentally mismatch somatic and germline sexual identities. The phenotypic consequences of these mismatches have

[*] Equal contributing authors
Laboratory of Cellular and Developmental Biology, National Institute of Diabetes and Digestive and Kidney Diseases, National Institutes of Health, 50 South Drive, Bethesda, Maryland 20892

Current Topics in Developmental Biology, Volume 83
ISSN 0070-2153, DOI: 10.1016/S0070-2153(08)00404-3

striking similarities to those resulting from manipulations of signals within the niche. A critical role of the germline sex determination pathway may therefore be to ensure the proper receipt and processing of signals from the niche.

1. INTRODUCTION

In *Drosophila*, germline stem cells (GSCs) are situated in a niche at the anterior end of the adult gonad while mature gametes are localized to the posterior, such that an anatomical axis of germ cell proliferation and differentiation is established in both sexes. The somatic cells adjacent to the stem cells constitute and provide the physical microenvironment, the stromal niche, that supports the stem cells. Stem cell niches exist in several adult tissues such as the gut epithelium, skin, neural tissues, hematopoietic system, and the germline (Ohlstein et al., 2004). Short-range signals from the niche prevent differentiation while maintaining stem cell fate. The sex-specific germline serves as a valuable system for understanding the errors that result from perturbation of this important microenvironment.

In this chapter, we outline the differences in cellular architecture between male and female gonads and discuss major signaling events that influence development of the male versus female germline. We also highlight the importance of sex-specific germline niches in influencing germline sex as can be inferred from the dramatic consequences observed when germ cells are within a sex-mismatched somatic environment.

2. SEXUALLY DIMORPHIC NICHES REGULATE GERM CELLS

Drosophila gonads are sexually dimorphic (Compare Figs. 4.1 and 4.3). Despite their stark morphological differences there are analogous mechanisms employed during gametogenesis by both sexes while crucial differences remain.

2.1. The male niche and proliferative zone

2.1.1. Cellular architecture

The testis is a long coiled tube containing germ cells in varying stages of spermatogenesis. At the apex of the testis are the niche and the mitotic proliferative zone (reviewed in Fuller, 1993). At the core of the niche is a cluster of 10–12 small densely packed somatic cells called the hub (Hardy et al., 1979). There are 6–9 germline stem cells (GSCs) arranged around the hub in a characteristic rosette pattern. Each GSC is enclosed by a pair of

Germline Sex Determination in *Drosophila* 111

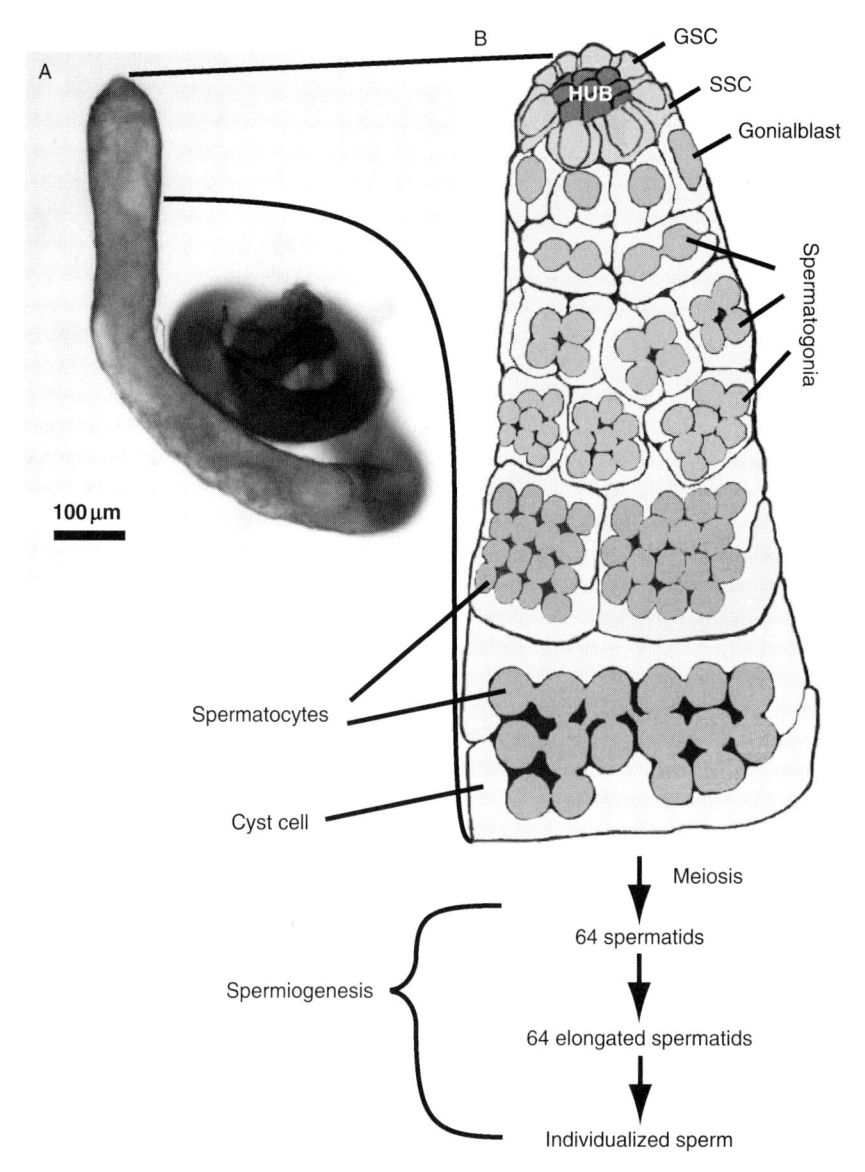

Figure 4.1 Male germ cell development. (A) Photomicrograph of a single adult testis. (B) Diagram of the apical tip of the testis. For details see text section 2.1—the male niche. Anterior top.

somatic stem cells (SSCs) also known as cyst progenitor cells. The GSCs and the SSCs maintain close contact with the apical hub through thin cytoplasmic extensions (Fig. 4.1). GSCs divide asymmetrically to give rise to two daughter cells, one of which remains adjacent to the hub and retains its GSC identity while the other is displaced away from the hub. The daughter that

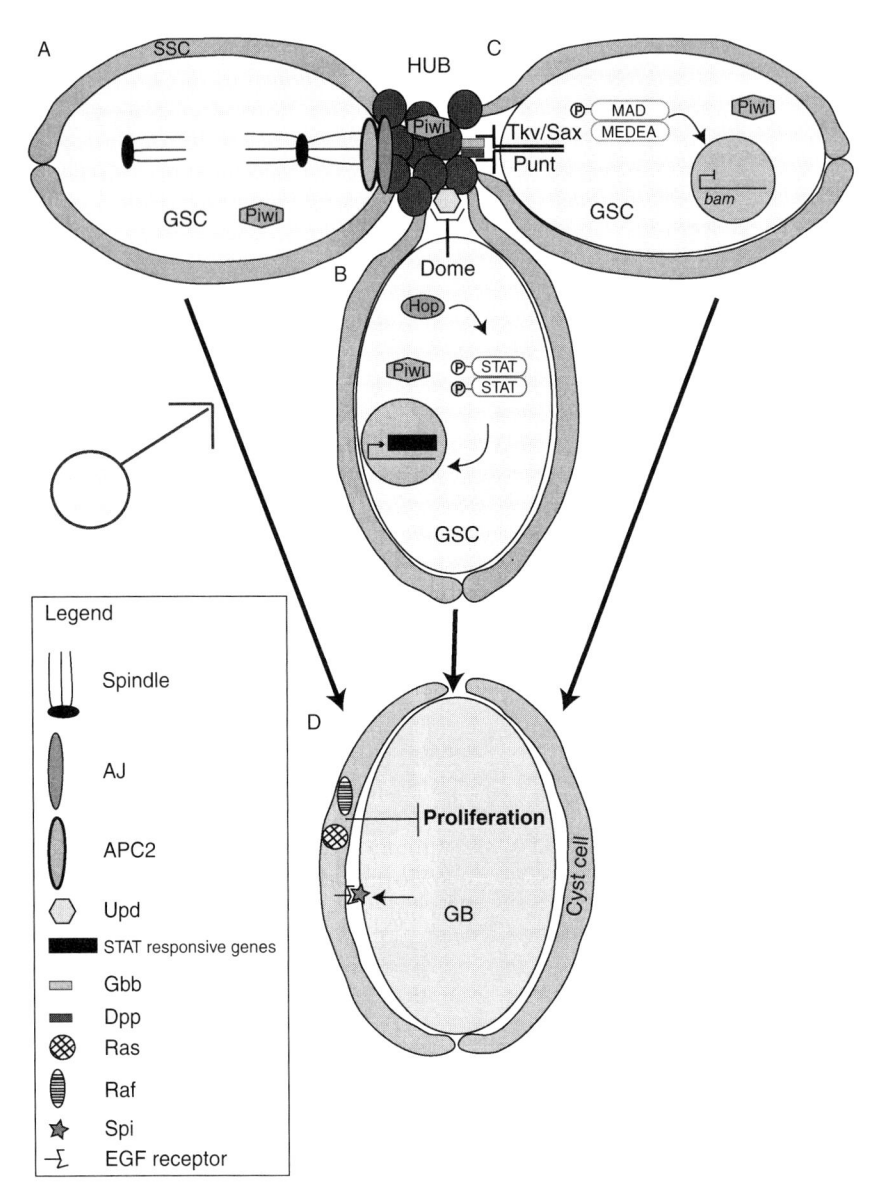

Figure 4.2 Male Niche. The germinal proliferation center of the male consists of a central core of somatic cells (the HUB) around which germline stem cells (GSC) are arranged radially. Each GSC is surrounded by two somatic stem cells (SSCs). Signaling cascades are shown separately for clarity as are spaces between the various cell types. (A) GSCs attach to the hub via adherens junctions (AJ). Orthogonally oriented mitotic spindles anchor the GSC cortex to the hub via APC2 and help ensure asymmetric division of the GSCs. (B) Secretion of the ligand Upd from hub cells activates JAK-STAT signaling in GSCs, which is required for their self-renewal. STAT phosphorylation leads to transcriptional activation of STAT responsive genes (shown as a black box since they

Germline Sex Determination in *Drosophila* 113

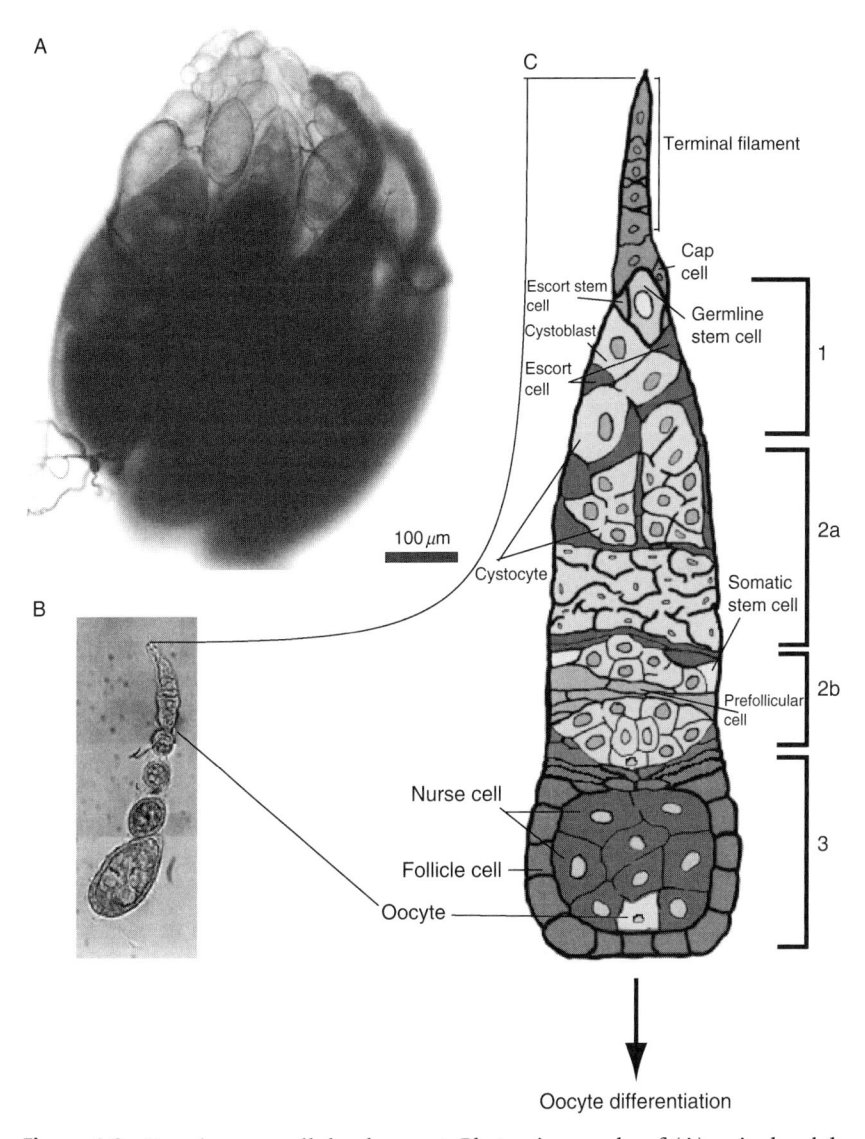

Figure 4.3 Female germ cell development. Photomicrographs of (A) a single adult ovary and (B) a single ovariole. Scale bar shown. (C) Diagram of germarium. (For details see text section 2.2—the female niche). (Image adapted from King, 1970). Anterior top.

---

are unknown). (C) Gbb and Dpp expressed in hub cells control GSC self-renewal by activation of BMP signaling, which leads to repression of *bam* transcription through the Mad/Med complexes. (D) EGF signaling is activated in somatic cyst cells by the ligand Spi secreted by the gonialblast (GB). Activated EGF signaling (through Ras and Raf) is required for the differentiation of GBs. (See Color Insert.)

exits from the niche, the gonialblast, is enveloped by two cyst cells (nondividing progeny of SSCs). Gonialblasts divide to generate spermatogonia, which undergo four rounds of synchronous mitotic divisions to generate a cohort of 16 cells. Spermatogonia undergo premeiotic DNA replication and after premeiotic S-phase, the 16 cells are referred to as primary spermatocytes. Primary spermatocytes undergo meiosis (after meiosis I, the cells are called secondary spermatocytes) to generate 64 spermatids and ultimately sperm (for details see Fuller, 1993). The asymmetry of GSC division is the key to stem cell maintenance—the cell maintaining contact with the hub remains a GSC while the gonialblast, which has no contact with the hub begins a program of cell division.

Much work has been undertaken to characterize the mechanism underlying this asymmetric division. The orientation of the mitotic spindle relative to an anchoring complex between the hub and the GSC regulates, at least in part, the asymmetric division of male GSCs (Fig. 4.2A). The mitotic spindle lies perpendicular to the hub and is set up by the positioning of the centrosomes during interphase (Yamashita et al., 2007). Upon centrosome duplication, the mother centrosome remains anchored to the hub-GSC interface and is inherited by the new GSC, while the daughter centrosome moves away from the hub and is inherited by the gonialblast. The integral centrosomal protein Centrosomin (Cnn) is required for centrosome positioning and spindle orientation (Yamashita et al., 2003). Since Cnn is required to anchor astral microtubules to the centrosome, it has been postulated that astral microtubules might link the centrosome and the cell cortex to properly orient the spindle. Indeed, in *cnn* partial loss-of-function mutants there is an increase in GSC number because instead of dividing asymmetrically, both daughters remain adjacent to the hub and thus maintain a GSC identity (Yamashita et al., 2003). This suggests that hub contact is a critical determinant of GSC identity.

GSCs are attached to the hub cells via adherens junctions (Yamashita et al., 2005). Adenomatous Polyposis Coli 2 (APC2) is present at the interface between the hub and the GSCs where it colocalizes with Shotgun (Shg, the *Drosophila* E-Cadherin homolog), an adherens junction component and Armadillo (Arm, the *Drosophila* β-Catenin homolog) (Yamashita et al., 2003). GSCs of *apc2* loss-of-function mutants have misoriented centrosomes and spindles, similar to those observed in *cnn* mutants and the number of GSCs around the hub is also increased (Yamashita et al., 2003). Thus, the astral microtubules that emanate from the centrosome appear to be captured by a protein complex containing APC2 and anchored to the interface of the GSC cortex with the hub in order to promote asymmetric divisions.

In addition to being anchored at the hub, GSCs and their progeny are also directly associated with SSCs and cyst cells. The SSCs that surround each GSC divide asymmetrically and concomitantly with the associated

GSC to produce a pair of new SSCs as well as a pair of squamous nondividing daughters (cyst cells) that enclose the gonialblast and its progeny. During the course of spermatogenesis, the somatic cyst cells stretch around the developing cohort of germ cells to accommodate about a 20-fold increase in germ cell volume. *Zero population growth* (*zpg*) encodes a germline-specific gap-junction protein homologous to Innexin-4 (Inx4) and is localized at the interface between spermatogonia and somatic cyst cells as well as between primary spermatocytes and somatic cyst cells (Tazuke *et al.*, 2002). Only a few germ cells (GSCs or gonialblasts) are present in *zpg* mutant testes, suggesting that gap junctions are required for the survival of the germ cells. *zpg* null mutant spermatogonia seem unable to progress in development suggesting that communication between spermatogonia and cyst cells via gap junctions is an important regulator of germ cell development (Gilboa *et al.*, 2003; Tazuke *et al.*, 2002).

The gonialblast undergoes four rounds of mitotic divisions with incomplete cytokinesis, generating a cluster of 16 interconnected spermatogonia (Rasmussen, 1973). During mitosis stable intercellular bridges called ring canals form at the site of incomplete cytokinesis. Ring canals connect the mitotic cells within a cyst, allowing them to share a common cytoplasm. As spermatogonia progress through mitotic divisions, the spectrosome (a spherical, spectrin-rich cytoplasmic organelle present in GSCs and gonialblasts) elongates into a rod-shaped structure called the fusome, which ultimately forms a highly branched vesicular structure interconnecting the cohort of mitotic germ cells via the ring canals (Hime *et al.*, 1996). Spermatogonia next enter meiotic prophase I, which lasts about 3.5 days and are then designated as primary spermatocytes. Extensive transcription leading to about a 20-fold increase in size of the nuclei is characteristic of this stage. After the long meiotic prophase, most transcription is shut down and the cells complete meiosis to generate haploid gametes. Autoradiographical studies investigating gene expression indicate that there is little transcription postmeiotically in males, although the mechanism of transcription cessation is still unclear (Olivieri and Olivieri, 1965). The process of postmeiotic sperm morphogenesis from an immotile cell to a highly specialized motile one is called spermiogenesis. The transformations include chromatin condensation, cell and axoneme elongation, mitochondrial rearrangement, and sperm individualization. These postmeiotic differentiation processes have been extensively reviewed elsewhere (see Fuller, 1993; Renkawitz-Pohl *et al.*, 2005).

2.2. Signaling pathways

Signaling between the germline and the niche plays a crucial role in the self-renewal, proliferation, and differentiation of the germline in both sexes. In the following sections, we will focus on key pathways conserved in both males and females, but employed differently in the two sexes. Janus kinase (Jak)/Signal Transducer and Activator of Transcription (STAT) signaling defines GSC fate, while Transforming Growth Factor β (TGF-β) signaling shields GSCs from precocious expression of genes that lead to proliferation. Finally, Epidermal Growth Factor Receptor (EGFR) signaling regulates encapsulation of germ cells by the somatic cyst cells, a process that ensures normal differentiation.

Stem cell self-renewal in males is controlled by the Jak/STAT signaling pathway which includes an extracellular cytokine ligand, Unpaired (Upd), a cytokine receptor, Domeless (Dome), a receptor activated kinase, Janus Kinase (Jak) encoded by the *hopscotch* locus (*hop*), and the transcription factor STAT92E (Kiger *et al.*, 2001; Rawlings *et al.*, 2004; Tulina and Matunis, 2001). Upd is specifically expressed in the hub cells. Upon glycosylation and subsequent secretion, it becomes tightly associated with the extracellular matrix, which limits its diffusion to a hub-proximal region (Harrison *et al.*, 1998). Dome is on the surface of the neighboring GSCs and the Jak/STAT pathway is activated in GSCs following Upd binding to it. Hop dimerizes, and phosphorylates cytoplasmic Stat92E, which translocates to the nucleus to activate specific targets (Fig. 4.2B) (reviewed in Arbouzova and Zeidler, 2006). The activated Jak/STAT pathway confers stem cell identity (Brown *et al.*, 2001; Kiger *et al.*, 2001; Tulina and Matunis, 2001). Failure to activate the pathway results in loss of GSCs (Kiger *et al.*, 2001; Tulina and Matunis, 2001). Conversely, ectopic expression of Upd in GSCs and SSCs results in a dramatic increase in the number of GSCs (and possibly gonialblasts) and failure to progress through spermatogonial divisions (Tulina and Matunis, 2001). Active STAT is detectable in GSCs in contact with the hub, but not in the gonialblast daughter cells resulting from asymmetric divisions of GSCs. This suggests that gonialblasts do not receive enough Upd to activate the Jak/STAT pathway, (although the situation may be more complex as loss of GSCs allows gonialblasts to assume GSC identity to repopulate the niche (Brawley and Matunis, 2004)). Thus, STAT signaling is highly restricted to cells abutting the hub and is essential for GSC self-renewal.

TGF-β signaling from the hub is required in GSCs and gonialblasts to repress differentiation (Fig. 4.2C). The pathway includes the extracellular ligands Glass bottom boat (Gbb) and Decapentaplegic (Dpp) and two-component receptor complexes which phosphorylate and activate a cytosolic R-Smad (receptor-regulated class of Smads), which then translocate to the nucleus to regulate gene expression (Reviewed in Raftery and Sutherland, 1999). The hub and somatic cyst cells express Gbb and Dpp, which act co-operatively to maintain GSC fate (Kawase *et al.*, 2004). Loss of *gbb* activity results in a failure to renew the GSC population, resulting in a single cohort of spermatogonial cysts. While reduced *gbb* dose and loss of *dpp* function results in a similar phenotype, loss of *dpp* function alone does not result in this phenotype. This suggests that Gbb is critical while Dpp is an ancillary ligand (Kawase et al., 2004). An important repressive target of TGF-β/BMP signaling is *bag of marbles (bam)*. Bam is expressed in spermatogonia and when it accumulates to high levels, spermatagonia cease mitotic amplification divisions and initiate the spermatocyte differentiation program (Gonczy et al., 1997; McKearin and Ohlstein, 1995; Schulz et al., 2004). Spermatogonia of loss-of-function *bam* mutant males do not cease amplification of mitosis and instead produce cysts of 32, 64, or more spermatogonia. Conversely, precocious expression of *bam* in GSCs and gonialblasts prevents amplification divisions leading to accumulation of single germ cells (GSCs and gonialblasts), which eventually die (Schulz et al., 2004). In *gbb* (but not *dpp*) mutant testes, GSCs inappropriately express Bam indicating that *gbb* is required for repressing *bam* transcription in GSCs. Interestingly, overexpression of *dpp* but not *gbb* leads to transcriptional repression of *bam* in all germ cells (Kawase et al., 2004). Gbb and (to a lesser extent) Dpp may act as short-range signals capable of blocking the spermatogonial program. In this model, spermatogonia undergoing mitotic divisions move away from the hub and therefore TGF-β signals, consequently Bam levels increase. Bam levels accumulate above a threshold leading to spermatogonia ending mitotic proliferation and entering meiosis (Shivdasani and Ingham, 2003). As a general rule, GSCs may need to be shielded from expression of any gene that is involved in differentiation, thus, repression of *bam* in GSCs might play an important permissive role in maintaining GSC identity.

EGFR signaling from the germ cells is required for proper encapsulation of gonialblasts by somatic cyst cells (Fig. 4.2D) (Schulz et al., 2002). The isolated microenvironment around the gonialblasts and spermatogonia provided by the surrounding pairs of cyst cells may help regulate the differentiation of the enclosed germ cells since EGF signaling is necessary for the spermatogonium to spermatocyte transition. EGFR is a tyrosine kinase receptor, which activates the mitogen activated protein kinase (MAPK) cascade by regulating the activities of the monomeric GTPase Raf and the serine-threonine protein kinase Ras (reviewed in Freeman, 2002). Germ cells signal via the ligand Spitz (Spi) to the surrounding cyst cells (Sarkar et al., 2007). Spi is expressed as an inactive transmembrane protein and is cleaved into a functional soluble fragment (Wasserman and Freeman, 1998). Proteolytic cleavage and activation of Spi requires the proteins Rhomboid-1 (Rho-1) and Star (Bier et al., 1990; Kolodkin et al., 1994). An additional Rhomboid protein (Rho-2) encoded by *stem cell tumor (stet)* is expressed in adult testes and is required in germ cells for their association with somatic cyst cells as well as for normal differentiation of germ cells (Guichard et al., 2000; Schulz et al., 2002; Wasserman et al., 2000). Star (*Star*) is transcribed at high levels in germ cells at the apical tip where GSCs, gonialblasts, and spermatogonia reside (Schulz et al., 2002). Genetic mosaic analyses have shown that the functions of EGFR and its downstream effector Raf are required in somatic cyst progenitor cells and somatic cyst cells but not in germ cells (Kiger *et al.*, 2000; Tran *et al.*, 2000). The current working model posits that Stet and Star activate Spi in germ cells, which subsequently binds to EGFR on somatic cyst cells (Schulz *et al.*, 2002). This is supported by the presence of activated MAPK in cyst cells (Kiger *et al.*, 2000). In *stet* loss-of-function testis MAPK expression is restricted to the hub cells and cyst progenitor cells, as a result mitotically active spermatogonia are interconnected by short branched fusomes suggesting failure to complete the full four rounds of mitosis (Schulz *et al.*, 2002). There are fewer somatic cyst cells in *stet* mutant testis when compared to wild-type and they are round instead of bean shaped, indicating a failure to envelop the gonialblasts and subsequent spermatogonial cysts (Schulz *et al.*, 2002). A similar phenotype is observed in *Egfr*, *spi*, and *raf* loss-of-function testis where GSCs, gonialblasts, and early spermatogonia accumulate (Kiger *et al.*, 2000; Sarkar *et al.*, 2007; Tran *et al.*, 2000). Interestingly, in addition to accumulation of early spermatogonia, there are also spermatogonia that undergo excessive mitotic divisions. Thus, EGFR activity in somatic cyst cells is not only required to restrict self-renewal or proliferation of germ cells, but also to permit differentiation of stem cells to gonialblasts and the transition from the spermatogonia to the spermatocyte stage.

2.3. The female niche and proliferative zone

2.3.1. Cellular architecture

An ovary is subdivided into about 18 individual ovarioles, each of which functions as an independent assembly line to produce eggs (Fig. 4.3) (for details see Spradling, 1993). The germarium is located at the anterior tip of each ovariole and consists of the niche and the proliferative germ cells (Fig. 4.4C) (Brown and King, 1962; Wieschaus and Szabad, 1979). The ovarian niche is at the anterior tip of each germarium and is composed of a stack of 8–10 disc-like somatic cells, called the terminal filament, and 5–7 squamous epithelial cells, the cap cells that literally cap the underlying 2–3 GSCs. The GSCs divide asymmetrically such that the anteriorly positioned daughter cell remains in contact with the cap cells and maintains GSC identity, while the posteriorly displaced daughter cell leaves the niche, and differentiates into a cystoblast (Lin and Spradling, 1995). Interspersed between the GSCs are 4–6 escort stem cells. Escort stem cells are akin to the SSCs of males; they are interspersed between the GSCs and surround them while maintaining contact with the niche (Decotto and Spradling, 2005). The differentiated daughters of the escort stem cells, called escort cells, are displaced away from the niche after cell division and enclose the cystoblast destined for differentiation. Like their male counterparts, female cystoblasts undergo four rounds of synchronous divisions with incomplete cytokinesis to give rise to 16 cystocytes. Unlike in males where cyst cells remain in contact with the spermatocytes, in females after the 16-cell cyst forms the associated escort cells undergo apoptosis (Decotto and Spradling, 2005). The germ cell cyst then becomes tightly associated with another somatic cell type—the somatic prefollicle cells derived from two follicle stem cells maintained in two stroma-free niches at the region 2a/2b boundary (Nystul and Spradling, 2007). If there are insufficient prefollicle cells to replace the escort cells the apoptotic germline cyst progression checkpoint is activated and the germline cyst itself undergoes apoptosis (Smith *et al.*, 2002). This suggests that female cystocyte viability requires envelopment by either escort or follicle cells, or a sequential envelopment by the two somatic cell types. Thus, while the gross cellular anatomy of the apex and the germarium are different, many principles of germarial organization are quite analogous to the testis apex.

Stem cell divisions are asymmetric by definition. This asymmetry depends on mitotic spindle orientation perpendicular to the interface of the GSCs with the niche (Deng and Lin, 1997; Yamashita *et al.*, 2003). The spectrosome in female GSCs is localized to the apical interface with the cap cells and serves as an anchor for one of the spindle poles (Fig. 4.4A). This is in contrast to the situation in males, where the spectrosome is unanchored and does not appear to play a role in establishing spindle orientation. The female spectrosome is also closely associated with the centrosome, however, the centrosome is not required for spindle orientation (or later oogenesis) (Lin *et al.*, 1994; Stevens *et al.*, 2007). Thus, the asymmetric divisions of GSCs within the female niche are very reminiscent of the process in males, although there are clear differences in the mechanisms used to accomplish the task.

Both male and female GSCs are anchored to the overlying niche via adherens junctions containing a complex composed of Shotgun (Shg) and Armadillo (Arm) (Song and Xie, 2002). Adherens junctions between GSCs and the niche are not identical in the two sexes. At least one gene (*gef26*) regulates niche E-cadherin function in males but not in females (Wang *et al.*, 2006). Loss of *shg* or *arm* function in GSCs results in a loss of stem cells from the ovarian niche indicating a loss of GSC anchoring (Song and Xie, 2002). Thus, adherens junctions are required to anchor GSCs to their niches in both sexes and this anchoring is necessary for GSC maintenance and self-renewal (Song and Xie, 2002).

Further, direct coupling of cap cells and GSCs via gap junctions is also important. Zpg is localized in the germarium at the interface between germ cells and somatic cells including GSCs and cap cells and GSCs and escort cells. Loss-of-function *zpg* mutant ovaries contain few GSCs and/or cystoblasts (Gilboa *et al.*, 2003). In contrast to females, Zpg is not detected at the interface between GSCs and SSCs or between GSCs and the hub in males, indicating that the composition of gap junctions between GSCs and the

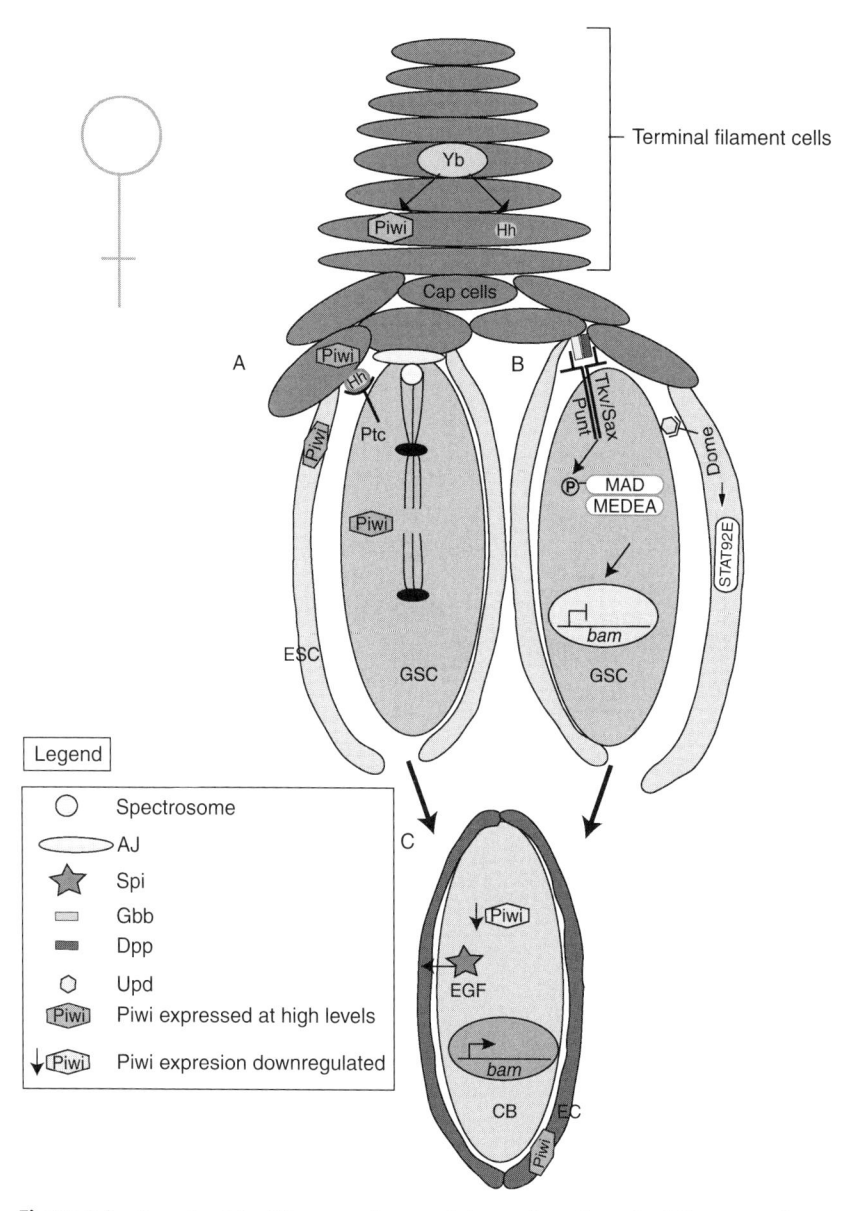

Figure 4.4 Female niche. The apex of an ovariole consists of terminal filament cells and cap cells which constitute the niche. GSCs are surrounded by escort stem cells (ESCs). Signaling cascades are shown separately for clarity as are spaces between the various cell types. (A) GSCs attach to the niche cells via adherens junctions (AJ). Orthogonally oriented mitotic spindles anchor the GSC cortex to the cap cells by the spectrosome. Yb, Piwi, Hh, Gbb and Dpp are expressed in the niche and act to control GSC self-renewal extrinsically. Piwi is also expressed in the GSCs and controls GSC division intrinsically. (B) Dpp/Gbb signaling leads to repression of *bam* in GSCs through Mad/Med complexes. The JAK-STAT pathway is activated in ESCs and controls their maintenance. (C) Bam is expressed in cystoblasts (CBs) and is required for their differentiation. (See Color Insert.)

overlying niche is not the same in males and females (Tazuke *et al.*, 2002). Further it has been shown that different gap-junctional proteins allow different small molecules to pass through (Stebbings *et al.*, 2000). This suggests that traffic between somatic and germ cells in the sexes could differ. Thus, while both adherence to the niche and gap-junctional intercellular communication are required for GSC self-renewal and for progression to mitotic stages in the two sexes, there are differences in the composition of both.

2.4. Signaling pathways

The same pathways utilized in the testis apex are deployed in the germarium, but not always to the same effect.

Jak/STAT signaling is required in both the testis apex and the germarium, but STAT activation is required in the germline of males as opposed to the soma of females (Fig. 4.4B). STAT activity is directly required within male GSCs where it confers stem cell identity. In contrast, the absence of Jak (due to loss-of-function in *hop*) in female GSCs is inconsequential (Gilboa and Lehmann, 2004). Nevertheless, loss of STAT function (through temperature sensitive alleles) results in drastically reduced GSC and escort stem cell numbers while increased Jak/STAT signaling, through overexpression of *upd* in germarial somatic cells, results in elevated numbers of cystoblasts and early cysts (Decotto and Spradling, 2005). STAT mutant germline clones do not affect normal development of germ cells but mutant ESCs and escort cells affect the development of closely associated germ cells (Decotto and Spradling, 2005). These germaria lack normal patterns of developing germline cysts and contain fewer GSCs with mislocalized spectrosomes (Decotto and Spradling, 2005). This suggests that female GSCs, cystoblasts, and cystocytes require a signal from the escort stem cells, not Jak/STAT *per se*. Activated STAT is thus required in escort stem cells, which in turn influence the female GSCs and affect their self-renewal.

TGF-β signaling is required in both male and female germ cells and has a much more conserved function. In both sexes, loss of TGF-β signaling leads to failed GSC self-renewal due at least in part to precocious derepression of *bam* expression. In both sexes, *bam* is also required for the cessation of mitotic divisions at the 16-cell stage and for progression to the next stage of differentiation. In females as in males, Dpp and Gbb ligands function synergistically to maintain GSCs by directly repressing transcription of the *bam* gene in GSCs (Fig. 4.4B and C) (Song *et al.*, 2004; Xie and Spradling, 1998, 2000). *dpp* is transcribed at low levels in cap cells and escort cells and at higher levels in prefollicle cells, which are located more posteriorly in the germarium (Xie and Spradling, 2000). *gbb* expression is detected more generally in the soma including escort cells and early follicle cells and it may also be expressed in the terminal filament cells and cap cells (Song *et al.*,

2004). The transcription pattern of these genes suggests that GSCs and cystoblasts are exposed to equivalent amounts of Dpp protein however, unlike in males where *bam* is repressed in both GSCs and gonialblasts; in females *bam* is repressed only in GSCs (Xie and Spradling, 2000). Therefore, intracellular modulation of sensitivity to Dpp signal, relieving Dpp-dependent transcriptional repression of *bam*, appears to be more important in the female germline (Casanueva and Ferguson, 2004).

In both sexes, the EGFR pathway is required for encapsulation of germ cells by somatic cells, which might be a prerequisite for proper interactions and reciprocal signaling between the two cell types (Fig. 4.4B). The somatic stem cells that are maintained in the male and female niches, interspersed between the GSCs are morphologically similar and produce nondividing daughter cells (cyst cells in males and escort cells in females), which enclose gonialblasts or cystoblasts and their progeny respectively. Male and female mutants lacking *stet* function show a similar phenotype: somatic cells fail to enclose germ cells and germ cells accumulate at early stages of differentiation. In females the accumulating cells resemble stem cells and cystoblasts, but in males the cells progress to the spermatogonial stage and mitotic interconnected cells accumulate indicating that EGFR signaling is required at a later step of differentiation in males than in females.

A comparison of the signaling pathways employed by the two sexes in their respective niches is outlined in Table 4.1.

3. Sex Determination in the Germ Cells

The sexual dimorphism in the gonads detailed above is a manifestation of the divergent sexual identity of cells within the gonads. The process of sex determination within the soma differs from that within the germline. Somatic sex determination is a cell-autonomous process, however, germline sex determination is not strictly cell-autonomous, incorporating not only information about the number of X chromosomes (germline karyotype) but also information about the sexual identity of the associated somatic cells.

3.1. Somatic sex determination

The sexual identity of the somatic gonad depends on the splicing autoregulation of *Sex-lethal* (*Sxl*) to produce Sxl protein in females, but not in males (Bell *et al.*, 1991). In the soma, the decision to be female or male is established in the early embryo (during mitotic cycles 12–14) when *Sxl* transcription is initiated in XX embryos but not in XY (or XO) embryos (Salz *et al.*, 1989). Sxl is required for female-specific splicing of the *transformer (tra)* pre-mRNA (McKeown *et al.*, 1987). Tra and the constitutively

Table 4.1 Comparison of signaling pathways between male and female germ cell niche

	Male	Female
Spectrosome positioning	Spectrosome is unanchored and not localized with respect to the hub	Spectrosome is always anchored at one pole of the spindle. Zpg-containing gap junction is present at the interface of the GSC and cap cell and is associated with the spectrosome, likely anchoring it
Establishment of spindle orientation	Spindle is oriented by centrosome position and anchored to the interface between GSCs and the hub by a complex consisting of APC2, Arm and Shg	Spindle is oriented by the spectrosome, which is anchored at the interface between the GSC and cap cell
Gap junctions	GSCs do not express Zpg	Zpg-containing gap junction is present at the interface of the GSC and cap cell
	Germ cells lacking *zpg* function undergo differentiation, with spermatogonial cells accumulating	In females lacking Zpg, differentiation is arrested at the cystoblast stage
	Zpg is needed for communication between somatic cyst cells and spermatogonia	Zpg is required at an earlier stage than in males
JAK/STAT signaling	Primary pathway controlling GSC self-renewal in males. STAT activity is cell-autonomously required in male GSCs where it confers stem cell identity, allowing stem cell self-renewal	STAT activity is not required in female GSCs but is required in escort stem cells which in turn influence self-renewal of GSCs. Thus, the requirement is indirect

(*continued*)

Table 4.1 (continued)

	Male	Female
TGF-β/BMP signaling	Required to repress *bam* transcription in the GSCs. Loss of *gbb/dpp* function leads to loss of GSCs as the cells differentiate but fail to self-renew. *bam* is repressed in GSCs, gonialblasts and 2-cell spermatogonia and is required for cessation of mitotic divisions and differentiation into primary spermatocytes. *bam⁻* male germ cells progress further in differentiation compared to *bam⁻* female germ cells with spermatogonial cells accumulating	*bam* is repressed in GSCs but BamC is expressed in cystoblasts and mitotically-active cystocytes. Bam is required for differentiation from GSCs into cystoblasts. In *bam⁻* females differentiation is arrested at the cystoblast stage
EGFR pathway	Male germ cells mutant for *stet* progress to the spermatogonial stage until mitotic interconnected cells accumulate. EGFR signaling is required at a later step of differentiation	Cells resembling stem cells and cystoblasts accumulate in female *stet* mutants indicating EGFR signaling is needed during early differentiation

produced Transformer-2 (Tra2) protein are splicing enhancers that dictate female specific splicing of the *doublesex (dsx)* pre-mRNA, leading to production of the female-specific Dsx^F protein isoform (Burtis and Baker, 1989; Inoue *et al.*, 1990, 1992; Nagoshi *et al.*, 1988; Sosnowski *et al.*, 1989). In males, Sxl and Tra are absent and *dsx^M* is produced by default splicing, leading to production of male-specific Dsx^M (Bell *et al.*, 1988; Salz *et al.*, 1989). For details of somatic sex determination see the earlier chapter.

Dsx^F and Dsx^M are transcription factors that regulate development of multiple sexually dimorphic somatic structures including elaboration of abdominal pigmentation, development of external genitalia, sex combs, and abdominal neuroblasts (Baker and Ridge, 1980; Baker *et al.*, 1989). Dsx^M is strongly expressed in all the somatic cells closely associated with premeiotic germ cells in males—the hub cells, the SSCs and cyst cells (Hempel and Oliver, 2007). This observation suggests that the male soma may continue to actively instruct the adjacent germline either by secreting positive "male" signals or by repressing "female" signals. By analogy, a similar mechanism may be in place in the female gonad where Dsx^F may be expressed in the somatic cells closely associated with premeiotic germ cells.

3.2. Germline sex determination

Key female germline sex determination genes are expressed at higher levels in female GSCs, cystoblasts, and cystocytes relative to the corresponding male germ cells at the equivalent stages (Fig. 4.5). For example, the *ovo* gene has two germline-specific transcripts, *ovo-A* and *ovo-B* (derived from alternative promoters), which encode C$_2$H$_2$ transcription factors with the same DNA-binding specificity but with opposite effects on targets (Andrews *et al.*, 1998, 2000). In females, *ovo* transcripts are expressed in all germ cells, however, in males the transcripts are restricted to the apical tip, which likely corresponds with GSC, gonialblasts, and spermatogonia (Andrews and Oliver, 2002). Although both transcripts are expressed at higher levels in females when compared to males, *ovo-B* expression is higher when compared to *ovo-A* in females. The expression levels of both *ovo-A* and *ovo-B* are increased by the number of X chromosomes, thus the germ cell karyotype is a cell-autonomous cue resulting in higher levels of expression in females (Andrews and Oliver, 2002). A female soma specifically enhances expression of *ovo-B*, which becomes apparent in genotypes where (by specifically manipulating the somatic sex determination pathway) the XY soma alone is transformed from male to female without affecting the germline (Andrews and Oliver, 2002; Bielinska *et al.*, 2005). Ovo isoforms bind to their own promoters in an autoregulatory loop and act on the *ovarian tumor (otu)* promoter resulting in higher *otu* levels in female germ cells when compared to male germ cells (Andrews and Oliver, 2002; Bielinska *et al.*, 2005; Lu *et al.*, 1998; Sass *et al.*, 1995; Steinhauer and Kalfayan, 1992; Van

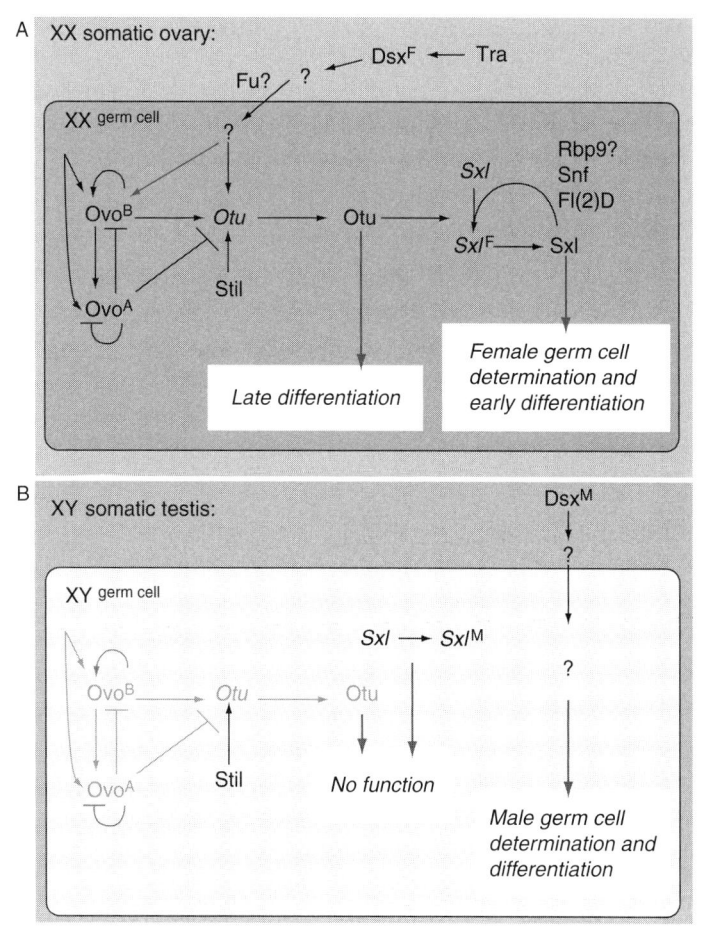

Figure 4.5 Germline sex determination pathway. Complete germ cell differentiation requires the evaluation of germ cell karyotype and integration of signal(s) from the somatic gonad to determine germline sex. (A) In females, *ovarian tumor (otu)* is a central player in integrating these two inputs. The 2 germline-specific *ovo* transcripts are expressed at higher levels in XX germ cells (than in XY germ cells) and thus the *ovo* locus appears to assess the germ cell karyotype. The proteins encoded by *ovo-A* and *ovo-B* are autoregulatory with opposite effects on transcription of their own promoters and the *otu* promoter. Somatic signal(s) from the female soma, downstream of *tra*, have a mild enhancement of *ovo-B* transcription and a more pronounced enhancement on *otu* transcription. The *standstill (stil)* gene is also required for robust *otu* expression. Downstream of *otu*, *Sxl* is sex-specifically spliced to produce protein in female germ cells. Along with *sans fille (snf)* and *female lethal d (fl(2)d)*, Sxl autoregulates its own splicing. *Ribosomal binding protein 9 (Rbp9)* is upstream of *Sxl* splicing regulation although it is not known precisely where it functions in the pathway. The *fused (fu)* gene likely functions in the soma to regulate female germline sex determination. (B) In males, little is known about germline sex determination. Both *ovo* and *otu* are expressed at lower levels (compared to female germ cells) but have no function. By default, *Sxl* is spliced into a male-specific isoform that does not encode a functional protein. It is not known how an XY karyotype is assessed or what may be downstream of the somatic signal which influences male germline sex determination. (Italics font denotes transcripts, Roman font denotes protein, gray color indicates lower levels than corresponding black color)

Buskirk and Schüpbach, 2002). This transcriptional regulation matches an "incoherent type 1 feed-forward circuit model" (Mangan and Alon, 2003) for the transcriptional regulation of *otu* consistent with basal levels of *ovo-A* and *ovo-B* being determined by the number of X chromosomes and an autoregulatory loop, while sensitivity to a positive input from a female soma on *ovo-B* expression results in an enhancement of *otu* transcription levels. Input from the soma also boosts *otu* expression independent of *ovo* because *otu* expression is higher in *ovo* mutant XY germ cells within a female soma than in wild-type testis (Hinson and Nagoshi, 1999). In addition, high levels of *otu* expression also require the *stand still (stil)* gene which is expressed robustly in all female germ cells (Sahut-Barnola and Pauli, 1999).

Sxl is spliced into a female-specific protein-encoding transcript downstream of *otu* regulation. This is based upon two lines of evidence: partial rescue of *ovo* and *otu* mutants by *Sxl* alleles that constitutively produce the female-specific transcript and the inappropriate expression of male-specific Sxl transcripts in *ovo* and *otu* mutant ovaries (Nagoshi et al., 1995; Oliver and Pauli, 1998; Oliver et al., 1993; Pauli et al., 1993). Sxl protein is found throughout the cytosol in the GSCs, cystoblasts and earliest cystocytes; thereafter Sxl protein levels abruptly decrease and nuclear foci appear (Bopp et al., 1993). The cytoplasmic localization of Sxl in GSCs and cystoblasts suggests that it may predominantly be functioning as a translational repressor in these cells, although that would not preclude its acting on targets with its better-defined role as a splicing factor (Bopp et al., 1993). Two genes required for *Sxl* splicing and autoregulation in the soma also function in the female germline: *sans fille (snf)* and *female lethal(2)d (fl(2)d)* (Hager and Cline, 1997; Ortega, 2005).

There are two additional genes whose mutant phenotypes place them in the same category as the other germline sex determination loci but whose precise roles remain uncertain. *RNA-binding protein 9 (Rbp9)*, which is allelic to *female sterile of Bridges (fes* or *fs(2)B)*, may be involved in *Sxl* autoregulation or otherwise upstream of *Sxl* regulation since mutant ovaries exhibit male-specific splicing of *Sxl* (Kim-Ha et al., 1999)(deCuevas, personal communication). Analysis of the *fused (fu)* gene has been confounded by its requirement as a component of the Hedgehog signal transduction cascade, which regulates the envelopment of germline cysts by the follicle cell layer (Besse et al., 2002). The Hedgehog pathway is not involved in germline sex determination, however, *fu* mutant ovaries express male-specific *Sxl* transcripts (Narbonne-Reveau et al., 2006; Oliver et al., 1993). Since mitotic clones in neither the germline nor the soma (in the follicle stem cell lineage) completely recapitulate the *fu* mutant ovary phenotype which has been shown to be ovary-autonomous by transplantation experiments, *fu* may function in the nonmitotic cells of the ovarian niche (the terminal filament and cap cells) or the escort stem cell lineage (Narbonne-Reveau et al., 2006; Smith et al., 1965).

Both *ovo* and *stil* are necessary for transcriptional activation of *otu* in the male germline; however, mutations in any of these genes have no effect on the male germline. *Sxl* is transcribed in the male germline but is not spliced into a protein-producing form. Even ectopic expression of Sxl protein in the male germline has no effect on spermatogenesis (Hager and Cline, 1997). Thus, the male germline sex determination pathway is distinct from the female germline sex determination pathway. Male germline sex cannot simply be a default fate since XY cells in a female soma have distinct phenotypic consequences (discussed below) strongly indicating that somatic signals must be incorporated into the germline sex decision.

4. Germ Cells in a Sex-Mismatched Soma

Germline sex and subsequent germ cell development is affected in germ cells within a sex-mismatched somatic environment. This observation highlights the function of the niche in influencing the sex of the germline and also the differences between the male and female niches. The developmental consequences of male germ cells within a female soma differ from those of female germ cells within a male soma. Male germ cells within a female soma overproliferate, generally without differentiation (Fig. 4.6). Female GSCs within a male soma are not maintained—initially germ cells begin proliferating and generate cells resembling spermatogonia, but are unable to complete differentiation to produce functional sperm. Ultimately the cells undergo cell death (Fig. 4.7).

4.1. Male germ cells within a female soma

The phenotypic consequences of male germ cells within a female soma can be observed in "mosaic intersexes," which can be generated by various techniques (described in Fig. 4.8). Some mosaic delineations are very clear such as XY or XO germ cells transplanted into XX hosts with or without an endogenous germline whereas chromosomal mosaics such as gynandromorphs and triploid intersexes are not as clearly delineated (Janzer and Steinmann-Zwicky, 2001; Laugé, 1966, 1969a; Laugé and King, 1979; Marsh and Wieschaus, 1978; Steinmann-Zwicky *et al.*, 1989; Van Deusen, 1977). Since the somatic sex determination regulatory cascade is well defined and the genes *tra*, *tra2*, and *dsx* are dispensable in the germline, XY or XO adults with a soma transformed from male to female by genetic manipulation of these genes produce clearly delineated sexual mosaics where the soma is sexually transformed from male to female but the germline is unaffected, at least autonomously (Andrews and Oliver, 2002; Bielinska *et al.*, 2005; Hinson *et al.*, 1999; Nagoshi *et al.*, 1995; Waterbury

Germline Sex Determination in *Drosophila* 129

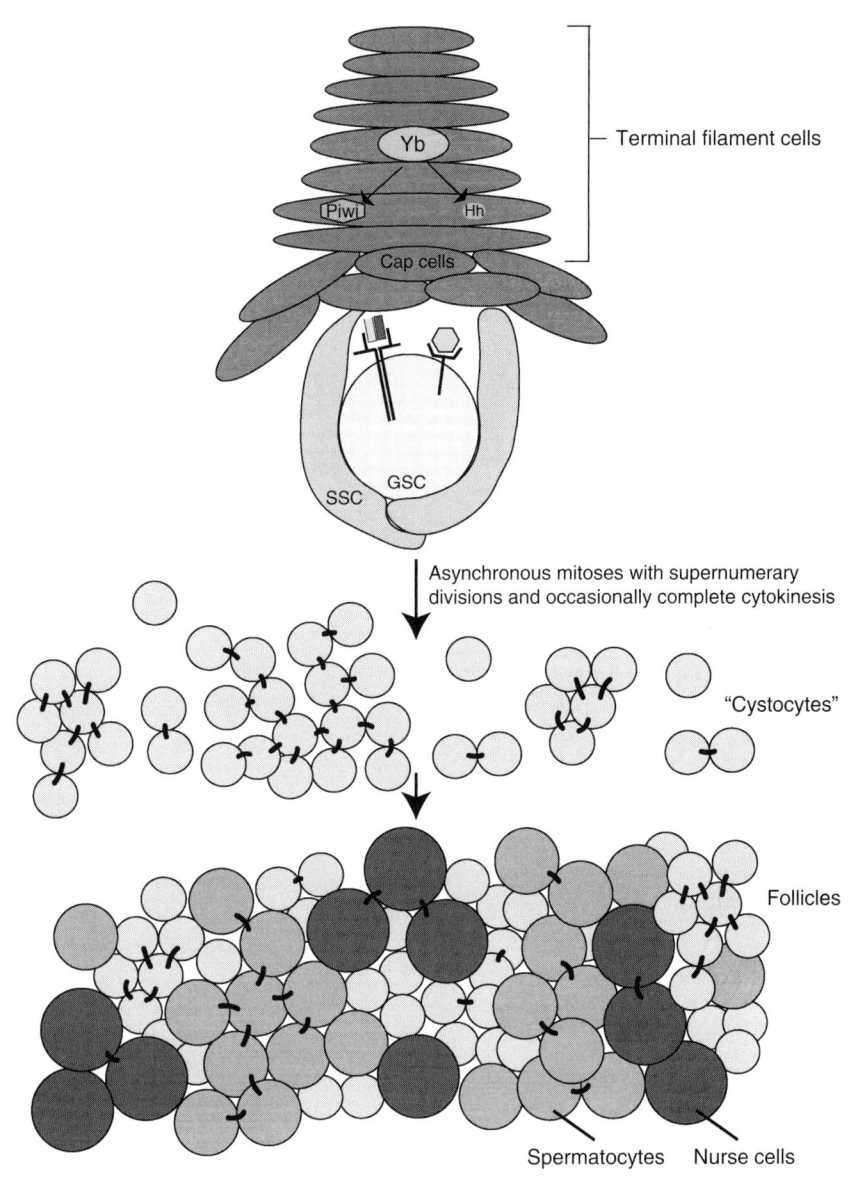

Figure 4.6 Aberrant development of male germ cells within a female soma. Graphic representing postulated fate of a male GSC in a mismatched female somatic environment. The end result of this mismatch is over-proliferation of early germ cells that develop up to the spermatocyte and nurse cell stage. The follicles contain nurse cells, spermatocytes and small cells. Many of these are single cells others are connected by ring canals in small clusters. It is not yet known if any of these cells are truly intersexual.

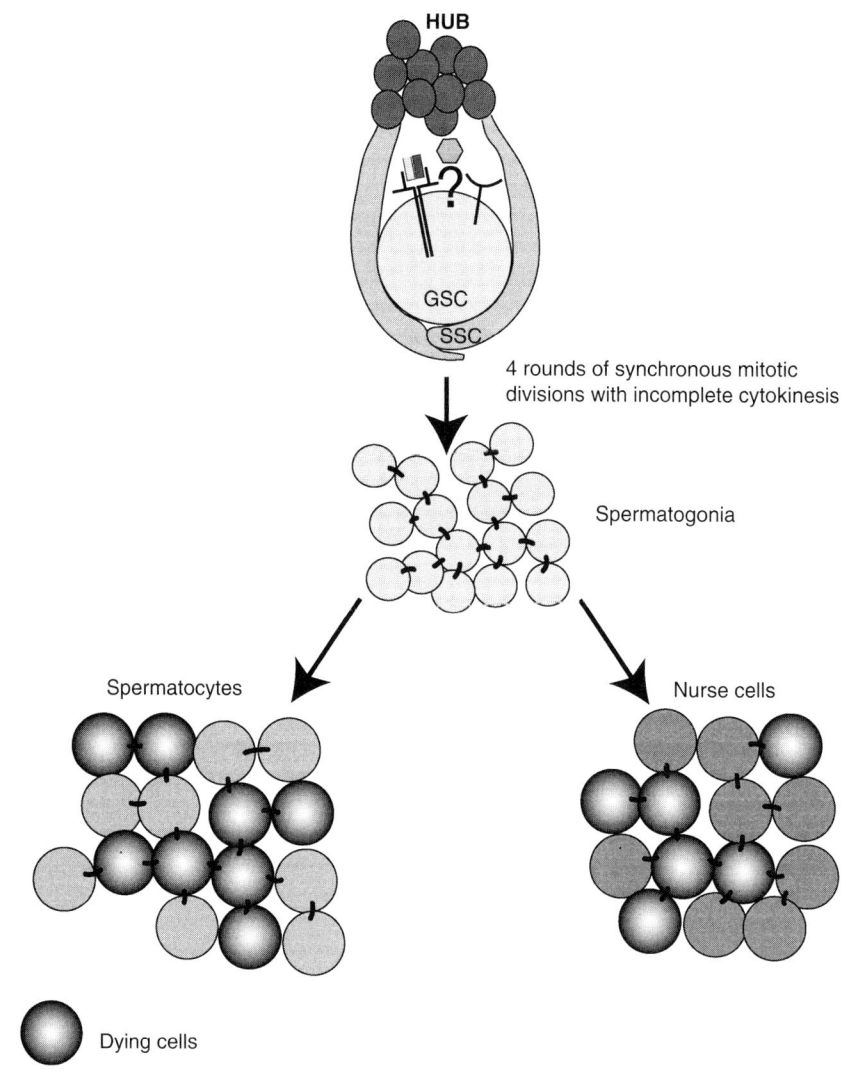

Figure 4.7 Aberrant development of female germ cells within a male soma. Graphic representing postulated fate of a female GSC in a mismatched male somatic environment. Some signaling pathways are shown to be disrupted (Jak-STAT and Hh) although the roles of other pathways can not be discounted. Development of a female germline within a male soma results in loss of germ cells. The germ cells develop up to the spermatocyte or nurse cell stage and finally degenerate.

et al., 2000). Such an XO soma transformed from male to female can completely support production of eggs from XX germ cells in gynandromorphs (Evans and Cline, 2007). In flies bearing mutations in germline sex determination genes, within an unaffected female soma there is an XX germline transformed from female to male (Gollin and King, 1981;

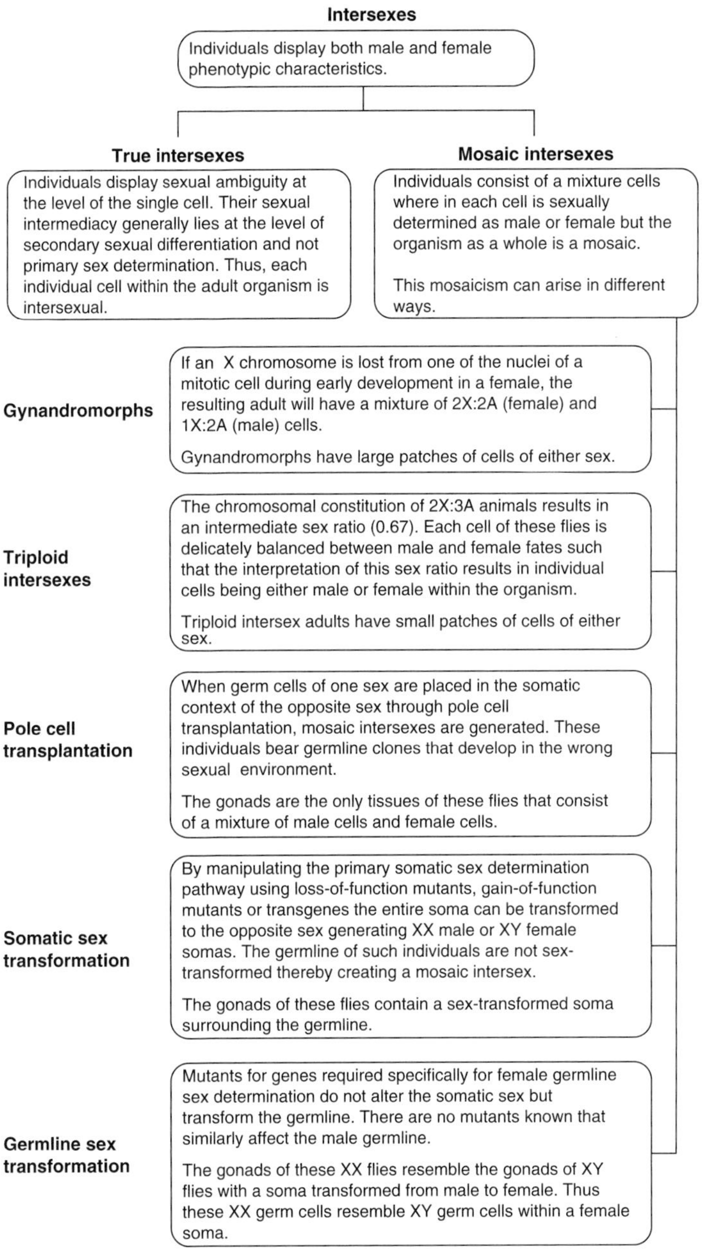

Germline Sex Determination in *Drosophila*

Intersexes
Individuals display both male and female phenotypic characteristics.

True intersexes
Individuals display sexual ambiguity at the level of the single cell. Their sexual intermediacy generally lies at the level of secondary sexual differentiation and not primary sex determination. Thus, each individual cell within the adult organism is intersexual.

Mosaic intersexes
Individuals consist of a mixture cells where in each cell is sexually determined as male or female but the organism as a whole is a mosaic.

This mosaicism can arise in different ways.

Gynandromorphs
If an X chromosome is lost from one of the nuclei of a mitotic cell during early development in a female, the resulting adult will have a mixture of 2X:2A (female) and 1X:2A (male) cells.

Gynandromorphs have large patches of cells of either sex.

Triploid intersexes
The chromosomal constitution of 2X:3A animals results in an intermediate sex ratio (0.67). Each cell of these flies is delicately balanced between male and female fates such that the interpretation of this sex ratio results in individual cells being either male or female within the organism.

Triploid intersex adults have small patches of cells of either sex.

Pole cell transplantation
When germ cells of one sex are placed in the somatic context of the opposite sex through pole cell transplantation, mosaic intersexes are generated. These individuals bear germline clones that develop in the wrong sexual environment.

The gonads are the only tissues of these flies that consist of a mixture of male cells and female cells.

Somatic sex transformation
By manipulating the primary somatic sex determination pathway using loss-of-function mutants, gain-of-function mutants or transgenes the entire soma can be transformed to the opposite sex generating XX male or XY female somas. The germline of such individuals are not sex-transformed thereby creating a mosaic intersex.

The gonads of these flies contain a sex-transformed soma surrounding the germline.

Germline sex transformation
Mutants for genes required specifically for female germline sex determination do not alter the somatic sex but transform the germline. There are no mutants known that similarly affect the male germline.

The gonads of these XX flies resemble the gonads of XY flies with a soma transformed from male to female. Thus these XX germ cells resemble XY germ cells within a female soma.

Figure 4.8 Intersexes can broadly be divided into either "true intersexes" or "mosaic intersexes". A number of different genetic or experimental manipulations can be used to generate mosaic intersexes useful for studying germline sex determination.

Granadino *et al.*, 1992; Kim-Ha *et al.*, 1999; Oliver *et al.*, 1990; Pauli *et al.*, 1993; Schüpbach, 1985). Strikingly, all of the above mosaics have very similar phenotypes; the resulting XX "male" germ cells are indistinguishable from XY or XO "male" germ cells within a female soma. The germ cells within these somatic females never fully differentiate into eggs, instead, they remain mitotically active and poorly differentiated. While specific differences between mutants for discrete loci are important, clear trends emerge when considering all of these "mosaic intersexes" as a single class. In an effort to integrate what is known about GSC regulation and early germ cell differentiation with the phenotypic observations of male germ cells within a female soma (and vice versa in the section below), we have focused on literature with clear representation of specific germ cell developmental stages.

Within the female germarial niche, both GSCs and cystoblasts of male germ cells within a female niche resemble wild-type female GSCs and cystoblasts in morphology and probably number. A germarium houses 2–3 GSCs while the apical tip of the testis, due to its larger size, houses 6–9 GSCs (compare schema in Figs. 4.1B and 4.3C). The large GSCs are recognizable in the most apical position beneath the terminal filament in camera lucida drawings from EM sections of *otu* and *fu* mutants and in triploid intersexes with gonads morphologically resembling ovaries (King *et al.*, 1978; Koch *et al.*, 1967; Laugé and King, 1979). Further, the GSC has a spectrosome localized to the apex of the cell abutting the niche in *Rbp9* and *fu* mutants as well as in germ cells of XY flies with a soma transformed from male to female (Janzer and Steinmann-Zwicky, 2001; Kim-Ha *et al.*, 1999; Narbonne-Reveau *et al.*, 2006). Anchoring of the spectrosome in GSCs is female-specific and is associated with a gap junction suggesting a cap cell may direct the formation of a gap junction with a GSC regardless of sex. Thus, there is a potential for miscommunication between male germ cells and female somatic cells. A female niche directs the early cell biology of male GSCs. In a wild-type female GSC, the spectrosome orients the mitotic spindles, even in the absence of centrosomes, whereas a wild-type male GSC requires centrosomes for spindle orientation (Stevens *et al.*, 2007; Yamashita *et al.*, 2003); but, it has not yet been tested if centrosomes or spectrosomes are required for GSC mitotic spindle orientation in male germ cells within a female soma. Any potential defects resulting from spindle orientation do not appear to affect the asymmetric division of the male GSCs in a female niche. In EM sections of germaria from females lacking *fu* and *Rbp9*, cystoblasts are discernable as single large germ cells subapically apposed to the GSCs, as in wild-type females, the caveat being that wild-type female cystoblasts and male gonialblasts are morphlogically indistiguishable (Johnson and King, 1972; Koch *et al.*, 1967). Cystoblasts of *Rbp9* and *fu* mutants also show BamC expression which is consistent with either *bam* derepression in these "male germ cells" because they respond like female germ cells or because they respond like male germ cells but derepress *bam* prematurely due to weaker TGF-β signal from an ovarian versus testis niche (Janzer and Steinmann-Zwicky, 2001; Kim-Ha *et al.*, 1999; Narbonne-Reveau *et al.*, 2006). Based on germ cell size and immunostaining for spectrosomes, the number of GSCs and cystoblasts may be increased relative to wild-type females in *fu* and *otu* mutants (Koch *et al.*, 1967; Narbonne-Reveau *et al.*, 2006; Rodesch *et al.*, 1997). Excessive GSC numbers might be a male germline characteristic, as there are more GSCs in the testis apex than in the germarium although such a difference in GSC number may be the result of the difference in size of the stromal niche (surface area of hub cells versus cap cells) to which the GSCs are attached. Protein traps with enriched expression in GSCs and cystoblasts are now available and could be used to resolve this issue (Buszczak *et al.*, 2007). Thus, no difference has been demonstrated between wild-type female GSCs or cystoblasts and those from "mosaic intersexes" of male germ cells within a female soma.

There are clear differences in the mitotic cystocyte divisions between wild-type female germ cells and male germ cells within a female soma. Wild-type female or male germ cells undergo four rounds of synchronous mitotic divisions with incomplete cytokinesis to yield a 16-cell cyst, however, when male germ cells are within a female soma the mitotic divisions become asynchronous, cytokinesis is occasionally complete and mitosis does not cease after four cell cycles. Examination of colchicine-pretreated germaria show equal numbers of 2, 4, or 8 cell clusters of cells with mitotic figures, however, in *Rbp9* mutants there were excessive numbers of single cells or 2 cell clusters of mitotic figures and clusters of odd numbers of mitotic cells (up to 11) indicating that the mitoses are not synchronous and that supernumerary divisions occur (Johnson and King, 1972). While single germ cells (GSCs and cystoblasts) are normally located within 15 μm of the terminal filament, in *Rbp9* mutants single germ cells can be found within the germarium as far as 100 μm away from the terminal filament consistent with some mitotic divisions culminating in complete cytokinesis (Johnson and King, 1972). In wild-type female cystocytes, the fusome grows by the formation of a new fusome plug within each newly forming ring canal and migrates (along with the ring canal) to join the central fusome (originially inherited in the cystoblast from the asymmetric division of the GSC). The central fusome pulls the ring canals into a rosette cluster, the mitotic spindle orients orthogonally to the central fusome for the next mitotic division resulting in an invariant "maximal branching" pattern of cystocyte connections (Storto and King, 1989). In XX flies with a germline transformed from female to male (*Rbp9* mutant), cystocyte branching patterns suggest there may be a defect in mitotic spindle orientation in dividing cystocytes (Johnson and King, 1972; Koch *et al.*, 1967). The fusome, which is derived from the endoplasmic reticulum and connects a common ER between the cystocytes, likely serves to facilitate communication between cystocytes allowing synchronization of the cell divisions (Snapp et al., 2004). Loss of synchronized mitotic divisions, odd numbers of cystocytes, lack of rosette configuration, and irregular cystocyte connection are all observed in *hts hu li tai shao* (*ht3*) mutant ovaries in which the fusome does not form (Grieder et al., 2000; Lin and Spradling, 1995). The fusomes in XY flies with a soma transformed from male to female or in XX flies with a germline transformed from female to male (*otu, Rbp9*, or *fu* mutants) are often spherical or dumbell-shaped similar to those of 2-cell cysts, or they are larger and either unbranched or irregularly branched (Hinson et al., 1999; Kim-Ha et al., 1999; Lin et al., 1994; Narbonne-Reveau et al., 2006; Rodesch et al., 1997). The lack of synchronized mitotic divisions and irregular cystocyte connections may both be consequences of defects in fusome function when male germ cells are within a female soma.

Sexual dimorphisms in the fusome and ring canal may be useful in the analysis of male germ cells within a female soma. With the cessation of mitotic divisions in wild-type female cystocytes, the fusome begins to break down and centrosomes travel along the fusome into the oocyte helping to establish the oocyte-nurse cell polarity within the cyst (Grieder et al., 2000). At the same time, female germline-specific ring canal growth begins which includes the concurrent appearance on the inner rim of the ring canal of Hts-RC and F-actin (which is relocalized from the fusome). Hts-RC is encoded by a female germline-specific transcript (Petrella et al., 2007). In wild-type male spermatogonia and spermatocytes, the cell cycle synchrony is maintained by the fusome, which does not break down until after the meiotic divisions (Wilson, 2005). There are dire consequences if the fusome is not maintained in spermatogonia: in *hts* mutant testes (which lack a fusome) spermatocytes frequently have too few or too many centrosomes and are sterile due to meiotic spindle defects and resulting aneuploidy (Wilson, 2005). While F-actin is present on the female fusome, it requires strong fixation protocols for reliable detection unlike the male fusome where F-actin is a major component and is easily detected (Huynh, 2006). Ring canals in wild-type male germ cells lack F-actin and Hts-RC but incorporate three septins (Pnut, Sep1, Sep2) not present in wild-type female germ cells (Hime et al., 1996). We have already described how male germ cells within a female soma often have spherical or dumbell-shaped fusomes or larger fusomes that are either unbranched or have an irregular branching pattern; these fusomes persist in germ cells well past the region 2a/2b transition and are even found within follicles. In *otu* mutant germ cells (in contrast to wild-type female cystocytes), substantial levels of actin filaments were seen in all fusomes examined and sometimes ring canals lacking F-actin encircled the fusomes—a combination seen in wild-type male ring canals and fusomes (Rodesch et al., 1997). Hts-RC is found in a few *otu* mutant germ cells and EM studies show the presence of outer rim but lack of inner rim deposits on the ring canals of *otu* and *Rbp9* mutants (Johnson and King, 1972; King et al., 1978; Rodesch et al., 1997). Ring canals of XX germ cells transformed from female to male resemble spermatogonial ring canals, however, the presence of septins on the ring canals of male germ cells within a female soma has not yet been explored. Since the female germline-specific breakdown of the fusome is associated with active centrosome redistribution and inappropriate loss of fusome function results in centrosome redistribution in male germ cells it might be informative to investigate centrosome dynamics in male germ cells within a female soma. The examination of fusomes and ring canals suggests that male germ cells within a female soma may be more like wild-type male germ cells than wild-type female germ cells.

There is an increase of somatic escort cells in germaria of females with an XX germline transformed from female to male. In wild-type germaria, mitotically active germline cysts are enveloped by a pair of somatic escort cells that travel with the germline cysts through regions 1 and 2a of the germarium and undergo apoptosis at the border of region 2b (Decotto and Spradling, 2005). It is likely that wild-type escort cell apoptosis occurs in response to either loss of a prosurvival signal or initiation of a proapoptotic signal from the postmitotic germline cyst since in *bam* mutants, in which germ cells are arrested prior to the cystoblast stage and remain mitotically active, escort cells are present throughout the ovariole (Decotto and Spradling, 2005). However, the long-range signaling which maintains the follicle stem cells within their nonstromal niches at the same location, or interaction with the follicle stem cell lineage could regulate escort cell apoptosis. The only "mosaic intersex" for which expression of escort cell markers has been reported is in *Sxl* mutant germaria where there are an increased number of escort cells, however, marker expression does not extend throughout the ovariole but stops some distance from the terminal filament (Decotto and Spradling, 2005). The increase in escort cell number may be related to the aberrant cell divisions observed in male germ cells within a female soma where sometimes cystocyte divisions go to completion. Perhaps escort cells proliferate to envelop germ cells that are no longer connected by cytoplasmic bridges, or alternately they persist in association with male germ cells within a female soma longer than with wild-type female germ cells; the male equivalent of escort cells, the somatic cyst cells, remain in association with wild-type male germ cells throughout spermatogenesis. Since the cyst cells in males (and presumably escort cells in females) express sex-specific Dsx isoforms, and are in close association with the germline cysts at a time when aberrant phenotypes are manifest, further investigations into the association of escort cells with male germ cells within a female soma are warranted.

Similar to wild-type females, follicles form in region 3 of the germarium in flies that have male germ cells within a female soma but the continued mitotic divisions of the germ cells result in the eponymous phenotype "ovarian tumor." Follicles can be found with hundreds to thousands of small germ cells, many of which are single cells or are connected by ring canals in small clusters of 2 or 3 cells (although in older flies there may be clusters with as many as 24 cells) (Gollin and King, 1981; King et al., 1957, 1978). Within these follicles, mitotically active cells can be found either as single cells or as clusters of germ cells interconnected by a fusome (King et al., 1957; Narbonne-Reveau et al., 2006). There are patches of *bam* expressing cells in tumorous follicles of *fu* mutants: fusome-interconnected clusters with high expression of *bam*, germ cells with low expression of *bam* and single cells that lack *bam* expression and contain a dot fusome (Narbonne-Reveau et al., 2006). The dot fusome could be indicative of either a GSC or cystoblast-like cell. Expression of *bam* within a wild-type germarium can be used to distinguish GSCs from cystoblasts; however, since *bam* should not be expressed in follicles at all, it is not clear if in these tumorous follicles it is a reliable indicator of GSC *versus* cystoblast identity. Use of other markers or short-term clonal lineage analysis may be useful in determining if some cells within these tumors divide asymmetrically like GSCs.

Within the tumorous follicles, some male germ cells within a female soma appear to begin differentiation along a male-specific or female-specific pathway. Germ cells found within ovaries containing XY transplanted pole cells or follicles of *otu*, *stil*, *Sxl*, or *ovo* mutants (as well as *ovo-otu* or *ovo-snf* interaction mutants) or triploid intersexes can morphologically resemble primary spermatocytes in nuclear positioning with distinct nucleoli and polar mitochondrial clouds visible by phase contrast microscopy in gonad squashes (Cline, 1984; Laugé and King, 1979; Oliver et al., 1990; Pauli et al., 1993; Pennetta and Pauli, 1997; Steinmann-Zwicky et al., 1989). The dense polar cloud of mitochondria is particularly diagnostic of primary spermatocytes prior to formation of the mitochondrial derivative (which goes on to form an important part of the sperm midpiece and tail during spermiogenesis), which suggests these cells have begun a male-specific differentiation pathway. This is consistent with the observation that male-specific Sxl^M transcripts are present in mutant ovaries (Lee et al., 2000; Oliver et al., 1993). Germ cells can also be found with "pseudo nurse cell" morphology with polytene rather than polyploid chromosomes—a chromatin state transiently seen in nurse cells in stage 4 follicles (King, 1970; Oliver et al., 1990, 1993; Pennetta and Pauli, 1997). In pseudonurse cells, *bam* expression is not detected, nor are fusomes present (Narbonne-Reveau et al., 2006). This suggests other germ cells begin a female-specific differentiation pathway. Germ cells with either male primary spermatocyte or female pseudonurse cell morphologies can even be found within the same ovariole (Pennetta and Pauli, 1997). This mosaicism is also evident from several sex-specific enhancer traps, which generally show patches of staining thoughout the ovariole with either male- or female-specific markers (reviewed in Oliver, 2002). By these criteria, germ cells with either male or female characteristics are present, but without concurrently looking for both male- and female-specific features at a cellular resolution, it is not possible to know if any of these germ cells represent true intersexes (initiating both sex-specific differentiation pathways). A possible reason for the patches of sex-specific morphologies and patchy expression of either male- or female-specific markers may be that male germ cells within a female soma are near a threshold of some binary switch between male versus female differentiation.

If the phenotypes (e.g., asynchronous germline divisions and continued mitotic divisions away from the apical niche) seen when male germ cells are within a female soma are due to the sex-mismatch between germline and soma, then one would expect that simultaneously transforming the sex of both soma and germline from female to male should restore male germ cell development. Indeed, when *otu* or *snf* mutant XX germ cells are within a soma transformed from female to male, the germ cells no longer over-proliferate (Andrews and Oliver, 2002). Thus the *otu* and *snf* mutant phenotype is dependent upon signaling from a female soma. Interestingly, the germ cells in this situation do not produce sperm like XO germ cells within a soma transformed from female to male but rather resemble wild-type XX germ cells within a male soma where the few germline cysts do not progress past primary spermatocyte development and eventually degenerate (see Section 4.2 for more discussion of this phenotypic class) (Andrews and Oliver, 2002; Seidel, 1963). Despite the male germ cell characteristics of XX germ cells of female germline mutants, it is clear that these XX "male" germ cells are not capable of successfully completing male germline differentiation.

While we do not know what molecules signal from the somatic gonad to the germ cells to influence germline sex determination, it is worth considering whether any of the known soma-to-germline signaling pathways that we have discussed could result in the phenotypes seen when male germ cells are within a female soma.

The TGF-β pathway signals from the niche to the GSCs to repress *bam* and allow GSC self-renewal. Since male GSCs are anchored to the cap cells they should receive TGF-β signals similar to wild-type female GSCs. If the level of TGF-β signaling were higher (perhaps due to male germ cells being inherently more sensitive to the ligands than female germ cells), one might expect *bam* repression in the cystoblast and earliest cystocyte divisions leading to excessive rounds of mitotic divisions without differentiation. In other words, an excessive proliferation of small, undifferentiated male germ cells within the gonad, a phenotype strikingly similar to that observed when male germ cells are within a female soma. However, *bam* transcription is not repressed in male cystoblasts. Intriguingly, in tumorous follicles of *fu* mutants a reporter for receipt of TGF-β signaling (*dad-lacZ*) is expressed in germ cells which do not express *bam* (or do so at low levels); in wild-type female germ cells this reporter is only expressed in GSCs and cystoblasts

(Narbonne-Reveau et al., 2006). Perhaps, away from the niche, male germ cells within a female soma are sensitive to low levels of TGF-β signaling whereas wild-type female germ cells are not. Alternatively, perhaps gene misexpression in male germ cells within a female soma results in inappropriate activation of TGF-β targets and overproliferation.

The major role of the Jak-STAT signaling pathway in the wild-type male gonad is to confer GSC self-renewal within the niche, but the major role of Jak-STAT signaling within the wild-type female gonad is to signal among somatic cells to refine subpopulations of follicle cells as the germline cyst becomes enveloped by follicular layer (Hombria and Brown, 2002). Perhaps male germ cells within a female soma are more sensitive to Jak-STAT ligands present in the germarium and follicles than wild-type female germ cells. In this scenario a male germ cell response would be adoption of GSC self-renewal characteristics, even away from the niche. This could be consistent with the phenotypes seen when male germ cells are within a female soma.

4.2. Female germ cells within a male soma

When female germ cells develop within a male soma (techniques outlined in Fig. 4.8), the resulting phenotypes are all very similar suggesting that the underlying defects and therefore the signaling pathways affected are essentially the same. Germ cells are observed only in newly eclosed XX adults with a soma transformed from female to male (tra^- or $tra2^-$ loss-of-function mutants or dsx^D gain-of-function mutants). However, gonads of XX first and second instar larvae with a soma transformed from female to male (tra^-) are filled with dividing spermatogonial cysts which are surrounded by somatic cells, characteristic of male germ cell development (Seidel, 1963). These larval spermatogonial cells are mitotically active and divide to form germline cysts of up to 16 cells, germline cysts with greater than 16 germ cells are never seen (Seidel, 1963). Most gonads (60–70%) of XX adults with a soma transformed from female to male (tra^- and $tra2^-$) are devoid of germ cells; however, gonads of adults that contain germ cells have been examined and offer vital clues about the behavior of XX germ cells within a male soma (Hinson and Nagoshi, 1999, 2002; Nöthiger et al., 1989; Seidel, 1963).

In XX adults with a soma transformed from female to male (tra^-), the GSCs are not attached to the hub; the unattached GSCs start to differentiate leaving behind an empty apical tip (Seidel, 1963). This is similar to the phenotype observed in gonads of wild-type female flies with mutations in components of the TGF-β or Jak/STAT pathways which are required for GSC self-renewal (directly or indirectly) suggesting that these pathways may be disturbed. If female GSCs receive insufficient Dpp and Gbb signals from the male niche, Bam would be expressed too early i.e. in GSCs, resulting in a *bam* overexpression phenotype (loss of the GSC population due to precocious differentiation). Insufficient Jak/STAT signaling would also lead to drastically reduced GSC numbers because the male SSCs may not be able to activate the same signaling pathway as ESCs, which would normally signal to a wild-type female GSC and help maintain its self-renewing capacity. Although Jak/STAT and TGF-β pathways are active in the niches of both sexes, female GSCs might receive insufficient signals due to differences in the expression levels between the sexes or due to an abolished or impaired cell–cell communication between female GSCs and the male hub and between female GSCs and male cyst cells. Furthermore, female GSCs might not be able to anchor properly within the male niche if there were an incompatibility of cell adhesion molecules and consequently would not be able to receive high enough levels of Jak/STAT or TGF-β signals and as a result fail to self-renew.

Cell–cell communication between GSCs and the niche may be abolished or impaired when germ cells of one sex are in contact with somatic cells of the opposite sex because the proteins composing gap-junction channels between germ cells and somatic cells in one sex may be different from those in the other. Zpg, for instance, is not detected at the interface between male GSCs and hub cells whereas in females Zpg is detected in foci at the interface between GSCs and the cap cells (Tazuke et al., 2002). This indicates that Zpg is perhaps required only by female GSCs to form functional gap junctions with the overlying cap cells. The putative partner to Zpg is thought to be Innexin 2 (*inx2*); its transcripts have been detected in female somatic cells immediately adjacent to germ cells expressing Zpg (Stebbings et al., 2002). Of the eight innexins in the *Drosophila* genome, *ogre* (*inx1*), *inx2*, and *inx3* have been detected in female follicle cells and in addition *inx2* is also expressed at low levels in the germline (Stebbings et al., 2002); however, in the male niche instead of Zpg a different innexin may be expressed. Testis ESTs for several innexins have been identified including *ogre*, *inx2*, and *inx5* suggesting that different combinations of innexins may be required for the formation of gap junctions between GSCs and the niche in one sex versus the other (Tazuke et al., 2002). Since gap junctions are formed by a pair of hemichannels, one from each cell opposed in the narrow intercellular gap between neighboring cell membranes, incompatibilities between subunits would affect the assembly of the cell–cell channel (Curtin et al., 2002). Studies have shown that fly innexins, in general, cannot functionally substitute for one another (Curtin et al., 2002; Stebbings et al., 2000). It has also been proposed that when different heteromeric complexes form gap junctions, it may allow different signaling molecules to pass through the gap-junction channel suggesting that gap junctions between male hub cells and female GSCs may not allow passage of the same molecules as between male hub cells and male GSCs (Panchin, 2005). Thus, female GSCs within a male soma may not be able to receive the correct signaling molecules or the right levels of signaling molecules from the male hub required for their self-renewal.

Another reason for the loss of female GSC self-renewing capacity might be the fact that the male niche does not express *female sterile (1)Yb (yb)* and only very low levels of *hedgehog (hh)*, genes required for female GSC self-renewal (Compare Figs. 4.2 and 4.4). *yb* regulates both *piwi* and *hh* (King *et al.*, 2001). The *yb* locus encodes a novel protein expressed specifically in the cap and terminal filament cells and loss-of-function mutants exhibit significantly reduced expression of Piwi in cap cells and somewhat reduced expression of Piwi in terminal filament cells (King *et al.*, 2001). Like loss-of-function mutations in *piwi*, loss-of-function mutations in *yb* lead to a depletion of GSCs in females likely due to differentiation of GSCs without self-renewing divisions (King and Lin, 1999). Yb is required for *hh* expression in the female niche (Fig. 4.4) (Forbes *et al.*, 1996a,b). *hh* is expressed strongly in terminal filament cells and cap cells and at much lower levels in escort cells and plays a role in maintenance of GSC numbers, although its role may be partially redundant with Piwi (Forbes *et al.*, 1996a; King *et al.*, 2001). Consequently, absence of *yb* and *hh* in the male hub might lead to loss of female GSC self-renewal. Thus, an empty apical niche may be caused by one or more of the following: failure of female GSCs to attach to a male niche, failure to form functional gap junctions between the male niche and female GSCs and/or insufficient levels of signaling ligands from the sex-transformed male niche to the XX germ cells.

Most germ cells at later stages of differentiation in testes with XX germ cells within a male soma resemble spermatogonia (Laugé, 1969b; Nöthiger *et al.*, 1989; Seidel, 1963; Steinmann-Zwicky *et al.*, 1989). Although identical in size, cystocytes can be distinguished from spermatogonia by the structure and positioning of the nucleolus—in cystocytes, the nucleolous abutts the nuclear membrane and is consistently dark, while it is centrally positioned in spermatogonia and is round with a dark periphery and a bright center (Seidel, 1963). This indicates that the male soma has a strong influence on the XX germ cells directing them to develop such that they are highly biased towards the male pathway. Spectrosomes that elongate into fusomes connecting cells within germline cysts are seen in the spermatogonia of XX adults with a soma transformed from female to male (tra^-). Male and female flies show a difference in their fusomes—F-actin is a major component of the male fusome but is not easily detectable in the fusome of *Drosophila* ovaries (Hime *et al.*, 1996). F-actin is readily detected in the branched fusomes of XX adults with a soma transformed from female to male (tra^-) further suggesting that these germ cells are highly male-biased in their development (Hinson and Nagoshi, 1999).

A few adult gonads also contain cells resembling primary spermatocytes. These cells have polar nuclei, mitochondrial clouds, and crystals in their cytoplasm (Laugé, 1969b; Seidel, 1963). Crystals are formed when spermatocytes develop without a Y chromosome (Meyer *et al.*, 1961). In male host embryos transplanted with pole cells, spermatocytes with crystals are observed indicating that they are of the genotype XX or XO. The formation of spermatocytes in XX animals with a soma transformed from female to male or in males that received XX germ cells by pole cell transplantation also suggests that XX germ cells within a male soma differentiate with great bias toward the male sex differentiation pathway. Interestingly, most spermatocytes in XX animals with a soma transformed from female to male do not increase significantly in size in contrast to wild-type male germ cell development wherein spermatocytes undergo a 20-fold increase in cell volume (Laugé, 1969b). Eventually these arrested spermatocytes degenerate, further indicating an inability to complete spermatogenesis (Laugé and King, 1979; Seidel, 1963).

Although most germ cells in gonads of XX animals with a soma transformed from female to male (loss-of-function tra^- or $tra2^-$, gain-of-function dsx^D) appear to resemble male germline stages, in some gonads germ cell clusters that resemble female germline stages are observed. These cysts were identified as female because they possess ring canals that incorporate the female-specific Hts-RC (Hinson and Nagoshi, 1999). Pseudonurse cells of varying sizes are also observed, although none reaching the size of true nurse cells (King et al., 1968; Nöthiger et al., 1989; Oliver et al., 1993; Seidel, 1963). These pseudonurse cells are found in groups but never form a cyst of up to 16 cells like a spermatogonial cyst. The fact that mitotic divisions were never observed in these cells supports the assumption that these cells are nurse cells, as polytene cells cannot divide (Brown and King, 1962; Seidel, 1963). Further, female-specific Sxl protein is also detected in some germ cell clusters suggesting that these clusters have a female identity (Oliver et al., 1993). The caveat is the lack of double labeling experiments with male and female germ cell specific markers, which would indicate whether germ cells exclusively express markers of one or the other sex indicating mosaic intersex identity or alternately, whether they simultaneously express markers of both sexes indicating a true intersex identity.

There are two possible explanations for the presence of female germ cell stages in XX animals with a soma transformed from female to male. XX germ cells may have an inherent tendency towards female development. In XX animals with a soma transformed from female to male this tendency might be suppressed by signals from the male soma, which actively secretes "male" signals. Only cells that do not receive a certain level of this masculinizing signal would be able to "escape" the suppression of female development. Consistent with this hypothesis, spermatogenic cell types are found in the absence of female cell types, but the reverse is never true (Nöthiger et al., 1989). Alternately, XX germ cells in absence of a positive feminizing signal from the soma may follow the male developmental pathway by default. In either scenario, the strength or extent of somatic sexual transformation of XX animals with a soma transformed from female to male (tra^-, $tra2-$, dsx^D) would determine the extent of germline sex transformation.

Consistent with this hypothesis is the fact that XX germ cells can differentiate into egg chambers and mature eggs if the sexual transformation is only partial as in true intersexes (XX *dsx*) (Hildreth, 1965); whereas a stronger somatic sexual transformation as in gynandromorphs with a soma transformed from female to male due to loss of *tra* function leads to germ cell differentiation highly biased towards the male pathway (Nöthiger *et al.*, 1989; Seidel, 1963). Furthermore, in testis of males transplanted with female pole cells, germ cells that exclusively follow the male developmental pathway are found and in 10–20% of these cases even immotile sperm are produced (Steinmann-Zwicky *et al.*, 1989). This bolsters the hypothesis that the strength or extent of the somatic sex transformation determines the extent of germline sex transformation. The reason for some germ cells in gonads of XX animals with a soma transformed from female to male not receiving a high enough threshold of "male" signal may be because they are not properly surrounded by somatic cyst cells. An examination for cyst cell markers would be helpful to examine if pseudonurse cells in XX animals with a soma transformed from female to male indeed lack contact with somatic cyst cells.

Taken together, the data show that the majority of XX GSC that develop within a male soma are unable to self-renew and therefore die or differentiate. The few GSCs that do differentiate develop such that they are highly biased towards the male germline differentiation pathway. In a few gonads of XX animals with a soma transformed from female to male, in addition to spermatogenic stages, germ cells resembling pseudonurse cells are observed indicating that some cells initiate female germ differentiation. However, it is still unclear if these female germ cells are an indicator of the gonads having a mixture of true male and female germ cells constituting a mosaic intersex germline, or whether germ cells in these gonads simultaneously follow both male and female germline developmental pathways indicating a true intersex identity.

5. Concluding Remarks

Three major events mark adult germ cell development: first, GSC self-renewal, second, mitotic proliferation of germline cysts until they reach the 16 cell stage, and final, differentiation of the 16 cell cohorts such that they undergo synchronous meiosis in males and differentiate into 15 nurse cells and one oocyte in females. Germ cells may integrate signals from the soma with intrinsic cues at each one of these steps to progress to the next stage of gametogenesis. It is not yet clear whether germline sex determination requires continuous inputs from the surrounding soma or whether it is determined at a specific time-point during development and is irreversible.

Different stages of germ cell development are affected when male germ cells are within a female soma and vice versa. For instance, GSC self renewal is only affected if female germ cells are within a male soma, not if male germ cells are within a female soma. This may indicate that male germ cells are more tolerant or capable of buffering somatic signals to a greater extent than female germ cells. Ultimately, neither male nor female germ cells can progress through gametogenesis in a sex-mismatched somatic environment to successfully generate functional gametes.

Unraveling the integration of somatic signals with germline autonomous cues would not only disclose the cause of the aberrant germ cell phenotypes observed when germ cells develop in a sexually mismatched soma, but would also lead to a detailed understanding of the differences between male and female niches in general. The pathway of cell autonomous cues required for germline sex determination in females is well on its way to being completely understood. However, the same is not true for males. Male germ cells must necessarily generate cell autonomous cues because they are clearly not exclusively dependent on the soma to determine their sex, as is obvious by the consequences of male germ cells within a female soma. The obvious next step in understanding germline sex determination in *Drosophila* is therefore to screen for XY germ cell autonomous cues needed for sex determination.

Elucidating the behavior of germ cells within their niche is fundamental to understanding germ cell development in *Drosophila* and, by extension, the requirement for and behavior of stem cells within their native niches in general.

ACKNOWLEDGMENTS

We thank members of the Oliver lab, Mary Lilly, Elissa Lei, and Jurrien Dean for helpful comments. This research was supported by the Intramural Research Program of the NIH, NIDDK.

REFERENCES

Andrews, J., and Oliver, B. (2002). Sex determination signals control ovo-B transcription in *Drosophila melanogaster* germ cells. *Genetics* **160,** 537–545.

Andrews, J., Levenson, I., and Oliver, B. (1998). New AUG initiation codons in a long 5' UTR create four dominant negative alleles of the *Drosophila* C2H2 zinc-finger gene ovo. *Dev. Genes Evol.* **207,** 482–487.

Andrews, J., Garcia-Estefania, D., Delon, I., Lu, J., Mevel-Ninio, M., Spierer, A., Payre, F., Pauli, D., and Oliver, B. (2000). OVO transcription factors function antagonistically in the *Drosophila* female germline. *Development* **127,** 881–892.

Arbouzova, N. I., and Zeidler, M. P. (2006). JAK/STAT signalling in *Drosophila*: Insights into conserved regulatory and cellular functions. *Development* **133,** 2605–16.

Baker, B. S., and Ridge, K. A. (1980). Sex and the single cell. I. On the action of major loci affecting sex determination in *Drosophila melanogaster*. *Genetics* **94**, 383–423.

Baker, B. S., Burtis, K., Goralski, T., Mattox, W., and Nagoshi, R. (1989). Molecular genetic aspects of sex determination in *Drosophila melanogaster*. *Genome* **31**, 638–645.

Bell, L. R., Maine, E. M., Schedl, P., and Cline, T. W. (1988). Sex-lethal, a *Drosophila* sex determination switch gene, exhibits sex-specific RNA splicing and sequence similarity to RNA binding proteins. *Cell* **55**, 1037–1046.

Bell, L. R., Horabin, J. I., Schedl, P., and Cline, T. W. (1991). Positive autoregulation of sex-lethal by alternative splicing maintains the female determined state in *Drosophila*. *Cell* **65**, 229–239.

Besse, F., Busson, D., and Pret, A. M. (2002). Fused-dependent Hedgehog signal transduction is required for somatic cell differentiation during *Drosophila* egg chamber formation. *Development* **129**, 4111–4124.

Bielinska, B., Lu, J., Sturgill, D., and Oliver, B. (2005). Core promoter sequences contribute to ovo-B regulation in the *Drosophila melanogaster* germline. *Genetics* **169**, 161–172.

Bier, E., Jan, L. Y., and Jan, Y. N. (1990). Rhomboid, a gene required for dorsoventral axis establishment and peripheral nervous system development in *Drosophila melanogaster*. *Genes Dev.* **4**, 190–203.

Bopp, D., Horabin, J. I., Lersch, R. A., Cline, T. W., and Schedl, P. (1993). Expression of the *Sex-lethal* gene is controlled at multiple levels during *Drosophila* oogenesis. *Development* **118**, 797–812.

Brawley, C., and Matunis, E. (2004). Regeneration of male germline stem cells by spermatogonial dedifferentiation *in vivo*. *Science* **304**, 1331–1334.

Brown, S., Hu, N., and Hombria, J. C. (2001). Identification of the first invertebrate interleukin JAK/STAT receptor, the *Drosophila* gene domeless. *Curr. Biol.* **11**, 1700–1705.

Brown, E. H., and King, R. C. (1962). Oogonial and spermatogonial differentiation within a mosaic gonad of *Drosophila melanogaster*. *Growth* **26**, 53–69.

Burtis, K. C., and Baker, B. S. (1989). *Drosophila* doublesex gene controls somatic sexual differentiation by producing alternatively spliced mRNAs encoding related sex-specific polypeptides. *Cell* **56**, 997–1010.

Buszczak, M., *et al.* (2007). The carnegie protein trap library: A versatile tool for *Drosophila* developmental studies. *Genetics* **175**, 1505–1531.

Casanueva, M. O., and Ferguson, E. L. (2004). Germline stem cell number in the *Drosophila* ovary is regulated by redundant mechanisms that control Dpp signaling. *Development* **131**, 1881–1890.

Cline, T. W. (1984). Autoregulatory functioning of a *Drosophila* gene product that establishes and maintains the sexually determined state. *Genetics* **107**, 231–277.

Curtin, K. D., Zhang, Z., and Wyman, R. J. (2002). Gap junction proteins are not interchangeable in development of neural function in the *Drosophila* visual system. *J. Cell Sci.* **115**, 3379–3388.

Decotto, E., and Spradling, A. C. (2005). The *Drosophila* ovarian and testis stem cell niches: Similar somatic stem cells and signals. *Dev. Cell* **9**, 501–510.

Deng, W., and Lin, H. (1997). Spectrosomes and fusomes anchor mitotic spindles during asymmetric germ cell divisions and facilitate the formation of a polarized microtubule array for oocyte specification in *Drosophila*. *Dev. Biol.* **189**, 79–94.

Evans, D. S., and Cline, T. W. (2007). *Drosophila melanogaster* male somatic cells feminized solely by traf can collaborate with female germ cells to make functional eggs. *Genetics* **175**, 631–642.

Forbes, A. J., Lin, H., Ingham, P. W., and Spradling, A. C. (1996a). Hedgehog is required for the proliferation and specification of ovarian somatic cells prior to egg chamber formation in *Drosophila*. *Development* **122**, 1125–1135.

Forbes, A. J., Spradling, A. C., Ingham, P. W., and Lin, H. (1996b). The role of segment polarity genes during early oogenesis in *Drosophila*. *Development* **122,** 3283–3294.
Freeman, M. (2002). A fly's eye view of EGF receptor signalling. *EMBO J.* **21,** 6635–42.
Fuller, M. T. (1993). Spermatogenesis. *In* "The Development of *Drosophila melanogaster*" (M. Bate and A. M. Arias, eds.), pp. 71–147. Cold Spring Harbor Press, New York.
Gilboa, L., and Lehmann, R. (2004). How different is Venus from Mars? The genetics of germ-line stem cells in *Drosophila* females and males. *Development* **131,** 4895–4905.
Gilboa, L., Forbes, A., Tazuke, S. I., Fuller, M. T., and Lehmann, R. (2003). Germ line stem cell differentiation in *Drosophila* requires gap junctions and proceeds via an intermediate state. *Development* **130,** 6625–6634.
Gollin, S. M., and King, R. C. (1981). Studies on *fs(1)1621*, a mutation producing ovarian tumors in *Drosophila melanogaster*. *Dev. Genet.* **2,** 203–218.
Gonczy, P., Matunis, E., and DiNardo, S. (1997). Bag-of-marbles and benign gonial cell neoplasm act in the germline to restrict proliferation during *Drosophila* spermatogenesis. *Development* **124,** 4361–4371.
Granadino, B., San Juán, A. B., Santamaria, P., and Sánchez, L. (1992). Evidence of a dual function in *fl(2)d*, a gene needed for *Sex-lethal* expression in *Drosophila melanogaster*. *Genetics* **130,** 597–612.
Grieder, N. C., de Cuevas, M., and Spradling, A. C. (2000). The fusome organizes the microtubule network during oocyte differentiation in *Drosophila*. *Development* **127,** 4253–4264.
Guichard, A., Roark, M., Ronshaugen, M., and Bier, E. (2000). Brother of rhomboid, a rhomboid-related gene expressed during early *Drosophila* oogenesis, promotes EGF-R/MAPK signaling. *Dev. Biol.* **226,** 255–266.
Hager, J. H., and Cline, T. W. (1997). Induction of female *Sex-lethal* RNA splicing in male germ cells: implications for *Drosophila* germline sex determination. *Development* **124,** 5033–5048.
Hardy, R. W., Tokuyasu, K. T., Lindsley, D. L., and Garavito, M. (1979). The germinal proliferation center in the testis of *Drosophila melanogaster*. *J. Ultrastruct. Res.* **69,** 180–190.
Harrison, D. A., McCoon, P. E., Binari, R., Gilman, M., and Perrimon, N. (1998). *Drosophila* unpaired encodes a secreted protein that activates the JAK signaling pathway. *Genes Dev.* **12,** 3252–3263.
Hempel, L. U., and Oliver, B. (2007). Sex-specific Doublesex[M] expression in subsets of *Drosophila* somatic gonad cells. *BMC Dev. Biol.* **7,** 113.
Hildreth, P. E. (1965). Doublesex, recessive gene that transforms both males and females of *Drosophila* into intersexes. *Genetics* **51,** 659–678.
Hime, G. R., Brill, J. A., and Fuller, M. T. (1996). Assembly of ring canals in the male germ line from structural components of the contractile ring. *J. Cell Sci.* **109,** 2779–2788.
Hinson, S., and Nagoshi, R. N. (1999). Regulatory and functional interactions between the somatic sex regulatory gene transformer and the germline genes ovo and ovarian tumor. *Development* **126,** 861–871.
Hinson, S., and Nagoshi, R. N. (2002). The involvement of ovarian tumour in the intracellular localization of Sex-lethal protein. *Insect Mol. Biol.* **11,** 241–248.
Hinson, S., Pettus, J., and Nagoshi, R. N. (1999). Regulatory and functional interactions between ovarian tumor and ovo during *Drosophila* oogenesis. *Mech. Dev.* **88,** 3–14.
Hombria, J. C., and Brown, S. (2002). The fertile field of *Drosophila* Jak/STAT signalling. *Curr. Biol.* **12,** R569–R575.
Huynh, J. R. (2006). Fusome as a cell–cell communication channel of *Drosophila* ovarian cyst. *In* "Cell–Cell Channels" (F. Baluska, *et al.*, Eds.), pp. 217–235. Springer, New York.
Inoue, K., Hoshijima, K., Sakamoto, H., and Shimura, Y. (1990). Binding of the *Drosophila* sex-lethal gene product to the alternative splice site of transformer primary transcript. *Nature* **344,** 461–463.

Inoue, K., Hoshijima, K., Higuchi, I., Sakamoto, H., and Shimura, Y. (1992). Binding of the *Drosophila* transformer and transformer-2 proteins to the regulatory elements of doublesex primary transcript for sex-specific RNA processing. *Proc. Natl. Acad. Sci. USA* **89,** 8092–8096.

Janzer, B., and Steinmann-Zwicky, M. (2001). Cell-autonomous and somatic signals control sex-specific gene expression in XY germ cells of *Drosophila*. *Mech. Dev.* **100,** 3–13.

Johnson, J. H., and King, R. C. (1972). Studies on fes, a mutation affecting cystocyte cytokinesis, in *Drosophila melanogaster*. *Biol. Bull.* **143,** 525–547.

Kawase, E., Wong, M. D., Ding, B. C., and Xie, T. (2004). Gbb/Bmp signaling is essential for maintaining germline stem cells and for repressing bam transcription in the *Drosophila* testis. *Development* **131,** 1365–1375.

Kiger, A. A., White-Cooper, H., and Fuller, M. T. (2000). Somatic support cells restrict germline stem cell self-renewal and promote differentiation. *Nature* **407,** 750–754.

Kiger, A. A., Jones, D. L., Schulz, C., Rogers, M. B., and Fuller, M. T. (2001). Stem cell self-renewal specified by JAK-STAT activation in response to a support cell cue. *Science* **294,** 2542–2545.

Kim-Ha, J., Kim, J., and Kim, Y. J. (1999). Requirement of RBP9, a *Drosophila* Hu homolog, for regulation of cystocyte differentiation and oocyte determination during oogenesis. *Mol. Cell. Biol.* **19,** 2505–2514.

King, R. C. (1970). "Ovarian Development in *Drosophila melanogaster*." Academic Press, New York.

King, F. J., and Lin, H. (1999). Somatic signaling mediated by fs(1)Yb is essential for germline stem cell maintenance during *Drosophila* oogenesis. *Development* **126,** 1833–1844.

King, R. C., Burnett, R. G., and Staley, N. A. (1957). Oogenesis in adult *Drosophila melanogaster*. IV. Hereditary ovarian tumors. *Growth* **21,** 239–261.

King, R. C., *et al.* (1968). The development of the female *Drosophila* reproductive system. *J. Morphol.* **124,** 143–166.

King, R. C., *et al.* (1978). Mutation that affects female and male germ-cells differentially in *Drosophila-melanogaster* meigen (diptera-drosophilidae). *Int. J. Insect Morphol. Embryol.* **7,** 359–375.

King, F. J., *et al.* (2001). Yb modulates the divisions of both germline and somatic stem cells through piwi- and hh-mediated mechanisms in the *Drosophila* ovary. *Mol. Cell* **7,** 497–508.

Koch, E. A., Smith, P. A., and King, R. C. (1967). Division and differentiation of *Drosophila* cystocytes. *J. Morphol.* **121,** 55–70.

Kolodkin, A. L., Pickup, A. T., Lin, D. M., Goodman, C. S., and Banerjee, U. (1994). Characterization of Star and its interactions with sevenless and EGF receptor during photoreceptor cell development in *Drosophila*. *Development* **120,** 1731–1745.

Laugé, G. (1966). Étude histologique des gonades à morphologie femelle des intersexués triploïdes de *Drosophila melanogaster* Meig. *C. R. Acad. Sci. Fr.* **265,** 767–770.

Laugé, G. (1969a). Étude des gonades des intersexués triploïdes de *Drosophila melanogaster*. Description morphologique ontogenese des structures histologiques. *Ann. Soc. Entomol. Fr.* **5,** 253–314.

Laugé, G. (1969b). [Origin and growth of the genital disk in triploid intersexual *Drosophila melanogaster* Meig]. *C. R. Seances Soc. Biol. Fil.* **163,** 1073–1078.

Laugé, G., and King, R. C. (1979). The ultrastructure of the ovaries of 2X 3A intersexes of *Drosophila melanogaster* meigen (Diptera: Drosophilidae). *Int. J. Insect Morphol. Embryol.* **8,** 277–288.

Lee, S. H., Kim, Y., and Kim-Ha, J. (2000). Requirement of *Rbp9* in the maintenance of *Drosophila* germline sexual identity. *FEBS Lett.* **465,** 165–168.

Lin, H., and Spradling, A. C. (1995). Fusome asymmetry and oocyte determination in *Drosophila*. *Dev. Genet.* **16**, 6–12.

Lin, H., Yue, L., and Spradling, A. C. (1994). The *Drosophila* fusome, a germline-specific organelle, contains membrane skeletal proteins and functions in cyst formation. *Development* **120**, 947–956.

Lu, J., Andrews, J., Pauli, D., and Oliver, B. (1998). *Drosophila* OVO zinc-finger protein regulates ovo and ovarian tumor target promoters. *Dev. Genes Evol.* **208**, 213–222.

Mangan, S., and Alon, U. (2003). Structure and function of the feed-forward loop network motif. *Proc. Natl. Acad. Sci. USA* **100**, 11980–11985.

Marsh, J. L., and Wieschaus, E. (1978). Is sex determination in germ line and soma controlled by separate genetic mechanisms? *Nature* **272**, 249–251.

McKearin, D., and Ohlstein, B. (1995). A role for the *Drosophila* bag-of-marbles protein in the differentiation of cystoblasts from germline stem cells. *Development* **121**, 2937–2947.

McKeown, M., Belote, J. M., and Baker, B. S. (1987). A molecular analysis of transformer, a gene in *Drosophila melanogaster* that controls female sexual differentiation. *Cell* **48**, 489–499.

Meyer, G. F., Hess, O., and Beermann, W. (1961). [Phase specific function structure in spermatocyte nuclei of *Drosophila melanogaster* and their dependence of Y chromosomes.]. *Chromosoma* **12**, 676–716.

Nagoshi, R. N., McKeown, M., Burtis, K. C., Belote, J. M., and Baker, B. S. (1988). The control of alternative splicing at genes regulating sexual differentiation in *D. melanogaster*. *Cell* **53**, 229–236.

Nagoshi, R. N., Patton, J. S., Bae, E., and Geyer, P. K. (1995). The somatic sex determines the requirement for ovarian tumor gene activity in the proliferation of the *Drosophila* germline. *Development* **121**, 579–587.

Narbonne-Reveau, K., Besse, F., Lamour-Isnard, C., Busson, D., and Pret, A. M. (2006). Fused regulates germline cyst mitosis and differentiation during *Drosophila* oogenesis. *Mech. Dev.* **123**, 197–209.

Nöthiger, R., Jonglez, M., Leuthold, M., Meier-Gerschwiler, P., and Weber, T. (1989). Sex determination in the germ line of *Drosophila* depends on genetic signals and inductive somatic factors. *Development* **107**, 505–518.

Nystul, T. G., and Spradling, A. C. (2007). An epithelial niche in the *Drosophila* ovary undergoes long-range stem cell replacement. *Cell Stem Cell* **1**, 277–285.

Ohlstein, B., *et al.* (2004). The stem cell niche: Theme and variations. *Curr. Opin. Cell Biol.* **16**, 693–699.

Oliver, B., and Pauli, D. (1998). Suppression of distinct ovo phenotypes in the *Drosophila* female germline by maleless and sex-lethal. *Dev. Genet.* **23**, 335–346.

Oliver, B., Pauli, D., and Mahowald, A. P. (1990). Genetic evidence that the ovo locus is involved in *Drosophila* germ line sex determination. *Genetics* **125**, 535–550.

Oliver, B., Kim, Y. J., and Baker, B. S. (1993). Sex-lethal, master and slave: a hierarchy of germ-line sex determination in *Drosophila*. *Development* **119**, 897–908.

Oliver, B. (2002). Genetic control of germline sexual dimorphism in *Drosophila*. *In* "International Review of Cytology - a Survey of Cell Biology" **219**, pp. 1–60.

Olivieri, G., and Olivieri, A. (1965). Autoradiographic study of nucleic acid synthesis during spermatogenesis in *Drosophila melanogaster*. *Mutat. Res.* **2**, 366–380.

Ortega, A. (2005). Localization of the *Drosophila* protein FL(2)D in somatic cells and female gonads. *Cell Tissue Res.* **320**, 361–367.

Panchin, Y. V. (2005). Evolution of gap junction proteins—the pannexin alternative. *J. Exp. Biol.* **208**, 1415–1419.

Pauli, D., Oliver, B., and Mahowald, A. P. (1993). The role of the ovarian tumor locus in *Drosophila melanogaster* germ line sex determination. *Development* **119**, 123–134.

Pennetta, G., and Pauli, D. (1997). Stand still, a *Drosophila* gene involved in the female germline for proper survival, sex determination, and differentiation. *Genetics* **145**, 975–987.

Petrella, L. N., Smith-Leiker, T., and Cooley, L. (2007). The Ovhts polyprotein is cleaved to produce fusome and ring canal proteins required for *Drosophila* oogenesis. *Development* **134**, 703–712.

Raftery, L. A., and Sutherland, D. J. (1999). TGF-beta family signal transduction in *Drosophila* development: From Mad to Smads. *Dev. Biol.* **210**, 251–268.

Rasmussen, S. W. (1973). Ultrastructural studies of spermatogenesis in *Drosophila melanogaster* Meigen. *Z. Zellforsch. Mikrosk. Anat.* **140**, 125–144.

Rawlings, J. S., Rosler, K. M., and Harrison, D. A. (2004).The JAK/STAT signaling pathway*J. Cell Sci.* **117**, 1281–1283.

Renkawitz-Pohl, R., Hempel, L., Hollmann, M., and Schafer, M. A. (2005). Spermatogenesis. *In* "Comprehensive insect physiology, biochemistry, pharmacology and molecular biology" (G. L.I., I. K., and G. S, eds.) pp. 157–178. Elsevier, Oxford.

Rodesch, C., Pettus, J., and Nagoshi, R. N. (1997).The *Drosophila* ovarian tumor gene is required for the organization of actin filaments during multiple stages in oogenesis. *Dev. Biol.* **190**, 153–164.

Sahut-Barnola, I., and Pauli, D. (1999). The *Drosophila* gene *stand still* encodes a germline chromatin-associated protein that controls the transcription of the *ovarian tumor* gene. *Development* **126**, 1917–1926.

Salz, H. K., Maine, E. M., Keyes, L. N., Samuels, M. E., Cline, T. W., and Schedl, P. (1989). The *Drosophila* female-specific sex-determination gene, Sex-lethal, has stage-, tissue-, and sex-specific rnas suggesting multiple modes of regulation. *Genes Dev.* **3**, 708–719.

Sarkar, A., Parikh, N., Hearn, S. A., Fuller, M. T., Tazuke, S. I., and Schulz, C. (2007). Antagonistic roles of rac and rho in organizing the germ cell microenvironment. *Curr. Biol.* **17**, 1253–1258.

Sass, G. L., Comer, A. R., and Searles, L. L. (1995). The *ovarian tumor* protein isoforms of *Drosophila melanogaster* exhibit differences in function, expression, and localization. *Dev. Biol.* **167**, 201–212.

Schulz, C., Wood, C. G., Jones, D. L., Tazuke, S. I., and Fuller, M. T. (2002). Signaling from germ cells mediated by the rhomboid homolog stet organizes encapsulation by somatic support cells. *Development* **129**, 4523–4534.

Schulz, C., Kiger, A. A., Tazuke, S. I., Yamashita, Y. M., Pantalena-Filho, L. C., Jones, D. L., Wood, C. G., and Fuller, M. T. (2004). A misexpression screen reveals effects of bag-of-marbles and TGF beta class signaling on the *Drosophila* male germ-line stem cell lineage. *Genetics* **167**, 707–723.

Schüpbach, T. (1985). Normal female germ cell differentiation requires the female X chromosome to autosome ratio and expression of sex-lethal in *Drosophila melanogaster*. *Genetics* **109**, 529–548.

Seidel, S. (1963). [Experimental studies on the principles of sterility of transformer (Tra) males in *Drosophila* melanogaster]. *Z. Vererbungsl.* **94**, 217–241.

Shivdasani, A. A., and Ingham, P. W. (2003). Regulation of stem cell maintenance and transit amplifying cell proliferation by tgf-beta signaling in *Drosophila* spermatogenesis. *Curr. Biol.* **13**, 2065–2072.

Smith, P. A., Bodenstein, D., and King, R. C. (1965). Autonomy of fu and fu-59 ovarian implants with respect to rate of tumor production. *J. Exp. Zool.* **159**, 333–335.

Smith, J. E., Cummings, C. A., and Cronmiller, C. (2002). Daughterless coordinates somatic cell proliferation, differentiation and germline cyst survival during follicle formation in *Drosophila*. *Development* **129**, 3255–3267.

Snapp, E. L., Iida, T., Frescas, D., Lippincott-Schwartz, J., and Lilly, M. A. (2004). The fusome mediates intercellular endoplasmic reticulum connectivity in *Drosophila* ovarian cysts. *Mol. Biol. Cell* **15**, 4512–4521.
Song, X., and Xie, T. (2002). DE-cadherin-mediated cell adhesion is essential for maintaining somatic stem cells in the *Drosophila* ovary. *Proc. Natl. Acad. Sci. USA* **99**, 14813–14818.
Song, X., Wong, M. D., Kawase, E., Xi, R., Ding, B. C., McCarthy, J. J., and Xie, T. (2004). Bmp signals from niche cells directly repress transcription of a differentiation-promoting gene, bag of marbles, in germline stem cells in the *Drosophila* ovary. *Development* **131**, 1353–1364.
Sosnowski, B. A., Belote, J. M., and McKeown, M. (1989). Sex-specific alternative splicing of RNA from the transformer gene results from sequence-dependent splice site blockage. *Cell* **58**, 449–459.
Spradling, A. C. (1993). Developmental Genetics of Oogenesis. In "The Development of *Drosophila melanogaster*" (M. Bate and A. Martinez Arias, eds.). Cold Spring Harbor Laboratory Press, New York.
Stebbings, L. A., Todman, M. G., Phelan, P., Bacon, J. P., and Davies, J. A. (2000). Two *Drosophila* innexins are expressed in overlapping domains and cooperate to form gap-junction channels. *Mol. Biol. Cell* **11**, 2459–2470.
Stebbings, L. A., Todman, M. G., Phillips, R., Greer, C. E., Tam, J., Phelan, P., Jacobs, K., Bacon, J. P., and Davies, J. A. (2002). Gap junctions in *Drosophila*: Developmental expression of the entire innexin gene family. *Mech. Dev.* **113**, 197–205.
Steinhauer, W. R., and Kalfayan, L. J. (1992). A specific ovarian tumor protein isoform is required for efficient differentiation of germ cells in *Drosophila* oogenesis. *Genes Dev.* **6**, 233–243.
Steinmann-Zwicky, M., Schmid, H., and Nothiger, R. (1989). Cell-autonomous and inductive signals can determine the sex of the germ line of *drosophila* by regulating the gene Sxl. *Cell* **57**, 157–166.
Stevens, N. R., *et al.* (2007). From stem cell to embryo without centrioles. *Curr. Biol.* **17**, 1498–1503.
Storto, P. D., and King, R. C. (1989). The role of polyfusomes in generating branched chains of cystocytes during *Drosophila* oogenesis. *Dev. Genet.* **10**, 70–86.
Tazuke, S. I., Schulz, C., Gilboa, L., Fogarty, M., Mahowald, A. P., Guichet, A., Ephrussi, A., Wood, C. G., Lehmann, R., and Fuller, M. T. (2002). A germline-specific gap junction protein required for survival of differentiating early germ cells. *Development* **129**, 2529–2539.
Tran, J., Brenner, T. J., and DiNardo, S. (2000). Somatic control over the germline stem cell lineage during *Drosophila* spermatogenesis. *Nature* **407**, 754–757.
Tulina, N., and Matunis, E. (2001). Control of stem cell self-renewal in *Drosophila* spermatogenesis by JAK-STAT signaling. *Science* **294**, 2546–2549.
Van Buskirk, C., and Schüpbach, T. (2002). Half pint regulates alternative splice site selection in *Drosophila*. *Dev. Cell* **2**, 343–353.
Van Deusen, E. B. (1977). Sex determination in germ line chimeras of *Drosophila melanogaster*. *J. Embryol. Exp. Morphol.* **37**, 173–185.
Wang, H., Singh, S. R., Zheng, Z., Oh, S. W., Chen, X., Edwards, K., and Hou, S. X. (2006). Rap-GEF signaling controls stem cell anchoring to their niche through regulating DE-cadherin-mediated cell adhesion in the *Drosophila* testis. *Dev. Cell* **10**, 117–126.
Wasserman, J. D., and Freeman, M. (1998). An autoregulatory cascade of EGF receptor signaling patterns the *Drosophila* egg. *Cell* **95**, 355–364.
Wasserman, J. D., Urban, S., and Freeman, M. (2000). A family of rhomboid-like genes: *Drosophila* rhomboid-1 and roughoid/rhomboid-3 cooperate to activate EGF receptor signaling. *Genes Dev.* **14**, 1651–1663.

Waterbury, J. A., Horabin, J. I., Bopp, D., and Schedl, P. (2000). Sex determination in the *Drosophila* germline is dictated by the sexual identity of the surrounding soma. *Genetics* **155,** 1741–1756.

Wieschaus, E., and Szabad, J. (1979). The development and function of the female germ line in *Drosophila melanogaster*: A cell lineage study. *Dev. Biol.* **68,** 29–46.

Wilson, P. G. (2005). Centrosome inheritance in the male germ line of *Drosophila* requires hu-li tai-shao function. *Cell Biol. Int.* **29,** 360–369.

Xie, T., and Spradling, A. C. (1998). Decapentaplegic is essential for the maintenance and division of germline stem cells in the *Drosophila* ovary. *Cell* **94,** 251–260.

Xie, T., and Spradling, A. C. (2000). A niche maintaining germ line stem cells in the *Drosophila* ovary. *Science* **290,** 328–330.

Yamashita, Y. M., Jones, D. L., and Fuller, M. T. (2003). Orientation of asymmetric stem cell division by the APC tumor suppressor and centrosome. *Science* **301,** 1547–1550.

Yamashita, Y. M., Fuller, M. T., and Jones, D. L. (2005). Signaling in stem cell niches: lessons from the *Drosophila* germline. *J. Cell Sci.* **118,** 665–672.

Yamashita, Y. M., Mahowald, A. P., Perlin, J. R., and Fuller, M. T. (2007). Asymmetric inheritance of mother versus daughter centrosome in stem cell division. *Science* **315,** 518–521.

CHAPTER FIVE

Sexual Development of the Soma in the Mouse

Danielle M. Maatouk *and* Blanche Capel

Contents

1. History of Sex Determination in Mammals 152
   1.1. Genetic control of sex determination in mammals 152
   1.2. How does the Y chromosome control sex determination in mammals? 153
   1.3. Two theoretical models predicting an active ovarian pathway 155
   1.4. The bipotential gonad is balanced between two fates 156
2. Molecular Control of Sex Determination 159
   2.1. Molecular specification of the gonad 159
   2.2. Choosing the testis pathway 160
   2.3. Testis morphogenesis 162
   2.4. Choosing the ovarian pathway 165
   2.5. Ovarian morphogenesis 168
3. Secondary Sex Differentiation 170
   3.1. Sex-specific development of the sex ducts 170
   3.2. Differentiation of the external genitalia 173
   3.3. Sex-specific differentiation of the brain 174
4. Summary 176
References 177

Abstract

Sex determination in mammals results in two discrete sexes, male and female. The sexes are genetically distinct at fertilization (XY = male and XX = female). However, there is little evidence for differences in their development until mid-gestation when the gonadal primordium forms. Recent research suggests that signals within this tissue maintain the gonad in a bipotential state, balanced between two alternative fates, testis or ovary. At mid-gestation, expression of the Y-linked gene *Sry* in gonadal cells triggers the divergence of gonad development toward the testis pathway. The critical role of *Sry* may be simply to promote the stable expression of SOX9 in the key cell population in the gonad.

Department of Cell Biology, Duke University Medical Center, Durham, North Carolina 27710

Establishment of SOX9 in this lineage is opposed by female promoting factors that push the gonad toward an ovarian fate. Both the male and female sides of these antagonistic pathways are augmented by feedback loops and reinforcing signals that canalize development, once the initial choice is made. Hormones and growth factors produced by the developing testis regulate the male differentiation of the sex ducts and the external genitalia during fetal life. In contrast, the ovary is not required for the fetal development of female genital ducts or female external genitalia, as these organs develop in the absence of a gonad. At puberty, hormones produced by the testis or the ovary control the sex-specific differentiation of the musculature, mammary tissue, and body hair. Male or female development of the brain was previously thought to depend on hormones produced by the testis or ovary. However, recent evidence reveals expression differences between XX and XY brains prior to the time that hormones are circulating, suggesting that some influences on brain development may be autonomous to the cells of the brain.

1. History of Sex Determination in Mammals

1.1. Genetic control of sex determination in mammals

In 1923, Painter showed that human males carry both an X and a Y chromosome whereas females carry two X's (Painter, 1923). At the time, it was known from earlier work that *Drosophila* males also have one X and one Y chromosome. However, Bridges had shown that the *Drosophila* Y is dispensable for male development, and sex determination in flies actually depends on the X to autosome ratio (Bridges, 1914). For this reason, it was not clear if the presence of the Y or the lack of two X's was responsible for male development in mammals. This question was resolved in 1959 by the work of two groups, Welshons and Russell, and Jacobs and Strong (Jacobs and Strong, 1959; Welshons and Russell, 1959). These two groups examined the human conditions known as Turner's Syndrome and Klinefelter's Syndrome. They showed that individuals with Turner's Syndrome have only one X chromosome (XO), but nonetheless, develop as phenotypic females; whereas, individuals with Klinefelter's Syndrome are XXY and develop as phenotypic males, despite the fact that they have two X chromosomes. These studies clearly indicated that it was not the X to autosome ratio that determines sex in mammals, but instead the presence or the absence of the Y chromosome. Therefore, although genetic sex (XX or XY) is determined at fertilization, during the first half of embryonic development in mice, sex determination is not initiated (Fig. 5.1).

In mammals, the mechanism for determining sex is unlinked from the mechanism for regulating dosage of the heterogametic sex chromosomes. In both flies and worms, the X to autosome ratio is calibrated in embryonic

Mammalian Somatic Sex Determination 153

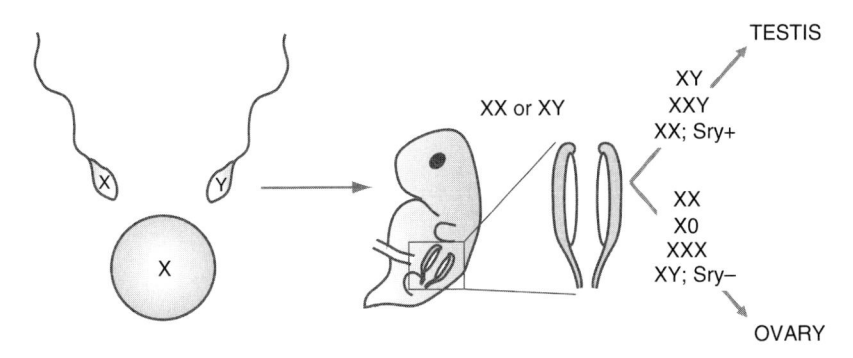

Figure 5.1 Mammalian sex determination is genetically determined. At fertilization, the mother's egg carries a single X chromosome while the father's sperm can carry either an X or a Y chromosome. When fertilization occurs the embryo will either be XX or XY and be genetically determined to follow the female (XX) or male (XY) pathway. Following fertilization, embryonic development is identical in male and female embryos, and it is not until the fetal gonad undergoes differentiation, that male and female embryos initiate separate developmental pathways. The fetal gonad is initially bipotential and the presence of a Y chromosome diverts the gonad to a testicular fate, whereas the absence of a Y chromosome diverts the gonad to an ovarian fate. Various chromosomal abnormalities can cause the bipotential gonad to switch fates. For example, XX embryos that inherit all or part of the Y chromosome (containing *Sry*) develop as males, while XY embryos, carrying Y chromosome deletions encompassing *Sry*, develop as females.

cells at the outset of embryogenesis (reviewed in Cline and Meyer, 1996). One arm of this process initiates the sex determination pathway while the other appropriately adjusts the dosage of genes on the X. In mammals, the dosage system works by inactivating one X in female cells at early embryonic stages (Lyon, 1961, 1962). If supernumery X chromosomes are present, only one X remains active, and all others are silenced by X-inactivation. This process occurs in the early embryo and adjusts dosage of most genes on the X to one-X in all cells in male and female embryos (for review see Panning and Jaenisch, 1998). The process of sex determination is distinct in mammals and occurs much later, specifically in cells of the gonad.

1.2. How does the Y chromosome control sex determination in mammals?

Working at the end of World War II in France, Alfred Jost performed a critical embryological experiment that revealed the connection between the gonad and sexual development of the embryo (Fig. 5.2). Jost removed undifferentiated gonads from fetal rabbits *in utero* and found that all operated embryos, regardless of their X/Y genotype, developed female anatomy including the development of an oviduct, a uterus, a vagina, and female

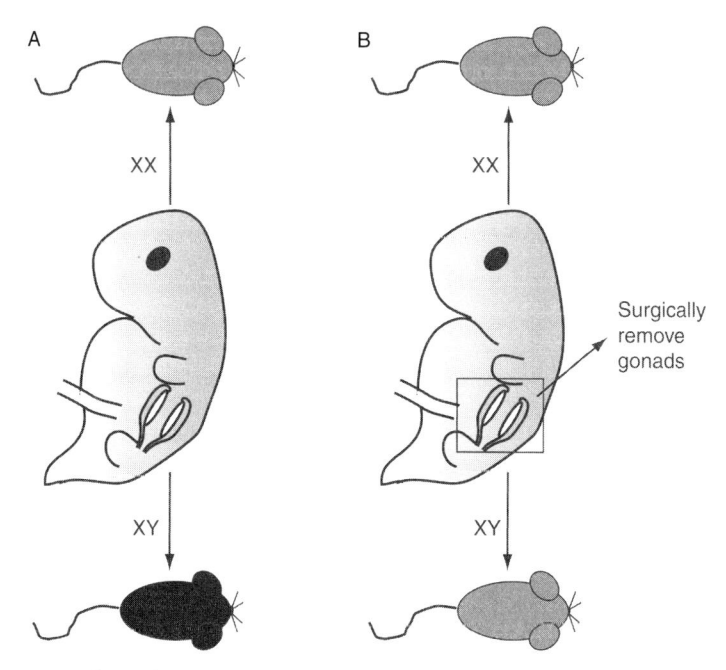

Figure 5.2 The influence of the gonad over sexual development. XX embryos normally develop as female, while XY embryos develop as male (A). Alfred Jost investigated the influence of the gonad over sexual development by surgically removing the gonads from rabbit embryos *in utero* (B). Removing the gonads from XX embryos had no effect on female sexual development. However, when the gonads were removed from XY embryos, the embryos developed as female. Based on the Jost experiments, it was determined that factors present in the XY gonad are required for male development, while female development does not require the influence of the gonad.

external genitalia (Jost, 1952). From these experiments, he concluded that male sexual development requires a gonad (a testis) whereas female sexual development can occur in the absence of a gonad.

These critical experiments established the testis as a central regulator of male sexual development in mammals, regardless of the influence of the rest of the embryo. Size differences exist during the first half of development of XX versus XY embryos that result from their different sex chromosome constitution (Scott and Holson, 1977; Seller and Perkins-Cole, 1987). However, Burgoyne and colleagues showed that these size differences do not control the sexual differentiation of the embryo (Burgoyne *et al.*, 1995). Consistent with these findings, various XY mutants in which the testis does not form develop female ductal anatomy and female external genitalia despite the fact that a Y chromosome and a male X-chromosome dose are present in all somatic cells. On the other hand, XO mice develop female sex ducts and female external genitalia, despite the fact that somatic cells have a male-like dose of the X chromosome. Thus, regardless of the potential for expression differences between XX and XY somatic cells that may influence their sex-specific development, the development of a testis or ovary is the over-riding determinant of sexual differentiation of the soma.

Based on these experiments, investigators recognized that male development follows from differentiation of a testis and predicted the existence of a male sex-determining gene on the Y chromosome whose primary role is to regulate differentiation of a testis from the bipotential gonad. Taking advantage of naturally occurring deletions and translocations in humans and mice, investigators began to pare down the Y-chromosome, looking for the testis determining region (*TDY* in humans), or testis determining factor (*Tdf* in mice). This 30-year process culminated in 1990 with the identification of a single gene, *SRY* (Sex-determining region of the Y; *Sry* in mice), which holds the secret of the switch between male and female development (Gubbay et al., 1990; Sinclair et al., 1990).

Consistent with these predictions, *Sry* was found to be expressed in the XY mouse gonad at the earliest stages of its development (Bullejos and Koopman, 2001; Hacker et al., 1995; Koopman et al., 1990). Both loss- and gain-of-function evidence indicated that *SRY/Sry* is the male sex-determining gene. Human XY male to female sex-reversed patients were discovered who carried mutations in the *SRY* gene (Hawkins et al., 1991). Similarly, a line of XY mice that develop an ovary and follow female sexual development pathways were shown to harbor a 10-kb Y-chromosome deletion including the *Sry* gene (Gubbay et al., 1992). Finally, XX mice carrying a transgenic autosomal copy of *Sry* were produced (Koopman et al., 1991). Although these mice are sterile in the absence of a Y chromosome (which is required for spermatogenesis), they developed a testis and all male secondary sex characteristics. By 1991, the relationship between the Y chromosome, the gonad, and mammalian sex determination was firmly established.

1.3. Two theoretical models predicting an active ovarian pathway

The pervasive idea that female is the "default" pathway in sex determination grew out of Jost's finding that in embryos where gonads were completely removed, the remaining reproductive organs developed as female, regardless of the embryo's genetic sex. Although this view has percolated into the field of gonad development, these ideas originally applied only to secondary sexual development and not to the fate decision in the gonad. The language that arose to describe the role of the Y chromosome as "a dominant determinant of sex" also fueled this general view by implying that the male pathway is active, whereas the female pathway is passive. However, based on observations of sex-reversing mouse models and human sex-reversal cases, several hypotheses were developed that predicted an actively competing ovarian pathway. In both of the primary theories that emerged, male and female pathways were seen as antagonistic.

Based on their work on XY sex-reversing mice (and prior to the discovery of *Sry*), Eicher and Washburn proposed that, similar to the Y-linked testis determining gene, an ovary determining gene (*Od*), located either on the X chromosome or an autosome, drives female development (Eicher and Washburn, 1983). In this model, the testis determining factor was predicted to function earlier than *Od*, possibly by inactivating the *Od* gene.

About a decade later, following the discovery of the *Sry* gene, McElreavey *et al.* put forth a similar model based on analysis of sex-reversed XX humans (McElreavey *et al.*, 1993). In a group of XX males, approximately 90% were found to carry a piece of the Y chromosome including the *Sry* gene. The remaining 10% lacked any detectable Y chromosomal DNA, suggesting that XX individuals could differentiate as males without the sex-determining gene *Sry*. To relieve the problem of activating the entire downstream testis pathway in the absence of a Y chromosome or *Sry*, McElreavey and colleagues proposed that a gene in XX individuals, the *Z* gene, functions to repress the expression of male-specific genes. This could explain the occurrence of human XX individuals that develop as males in the absence of a Y chromosome (and *SRY*). Mutation of the *Z* gene would allow for male-specific gene expression in XX individuals, and subsequent differentiation along the testis pathway. In this model, *Sry* would function in XY individuals to repress *Z*, and thus de-repress downstream genes in the testis pathway. The *Z* hypothesis differs from the *Od* hypothesis proposed by Eicher and Washburn in that, instead of functioning to promote ovarian development, *Z* was proposed to block genes in the male pathway (Fig. 5.3).

1.4. The bipotential gonad is balanced between two fates

The gonad is comprised of germ cells and somatic cells. Germ cells migrate to the gonad between embryonic day 8.0 and 11.0 (E8.0 and E11.0) (Chiquoine, 1954). They arrive in a field of somatic cells that arise by a thickening of the coelomic epithelium in the region of the intermediate mesoderm that overlies the mesonephric duct (for review see Capel, 1996). This likely occurs primarily by proliferation of cells in this domain, although additional contribution of cells from the mesonephric region to this initial population may also occur. Both the germ cells and somatic cells in this field are bipotential and can give rise to male or female cell types (McLaren, 1988). This unusual feature of cells in the gonadal primordium is clear from

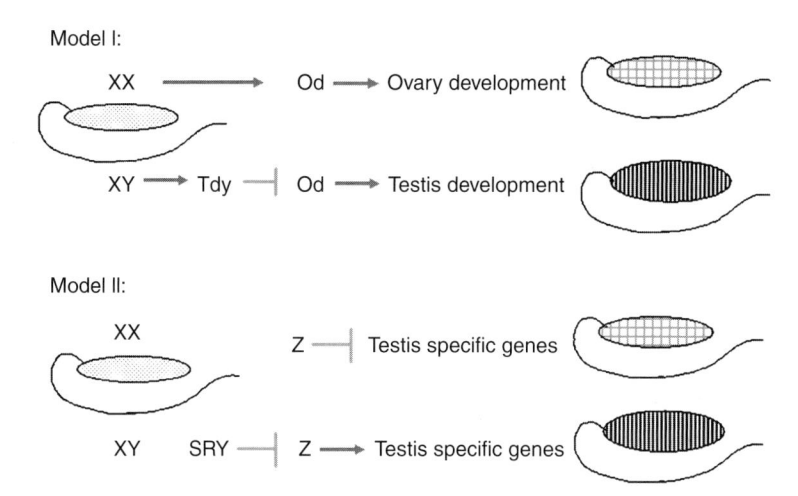

Figure 5.3 The hypothetical ovary-determining gene. Male sex determination depends on the presence of a single gene, *Tdy/Sry*, to initiate the male-specific differentiation events. No such gene has been identified in females, however an active sex-determining program in XX individuals was hypothesized. Model 1, proposed by Eicher and Washburn, hypothesized that an ovary-determining gene (*Od*) acts in XX individuals to promote the ovarian pathway. In this model, *Tdy* would act to block Od in XY gonads. *Od* would therefore promote ovarian development and block the testis pathway. Model 2, after the discovery of *Sry*, McElreavey, *et al.* proposed that the Z gene functions to antagonize the expression of testis-specific genes in XX individuals. In XY individuals, the Z gene would be inhibited by *Sry*, allowing for male development to occur.

sex reversal cases where an XX gonad develops as a testis, or an XY gonad develops as an ovary.

Some bipotential primordia (e.g., the *Drosophila* genital disc) are composed of two distinct regions, only one of which normally develops in any individual. Witschi suggested that the vertebrate gonad harbors a testis primordium in the medullary region and an ovary primordium in the cortical region (Witschi, 1951). It is true that the sperm-producing region of the testis develops within the central domain in the gonad, whereas, the oocyte producing region of the ovary develops cortically. However, more recent cell lineage experiments have indicated that the bipotential fate of the mammalian gonad results not from a regional specification, but from a bipotential fate decision that occurs in each cell (Albrecht and Eicher, 2001; Karl and Capel, 1998). The choice between these two fates is the primary sex-determining decision in vertebrates. Ovotestes are rare and when they occur in mice, do not partition along cortical and medullary lines, but instead typically partition longitudinally with a central testicular domain and polar ovarian regions (Fig. 5.4; Eicher *et al.*, 1980, 1982; Nagamine *et al.*, 1998). An exception to this rule is the Talpid mole that

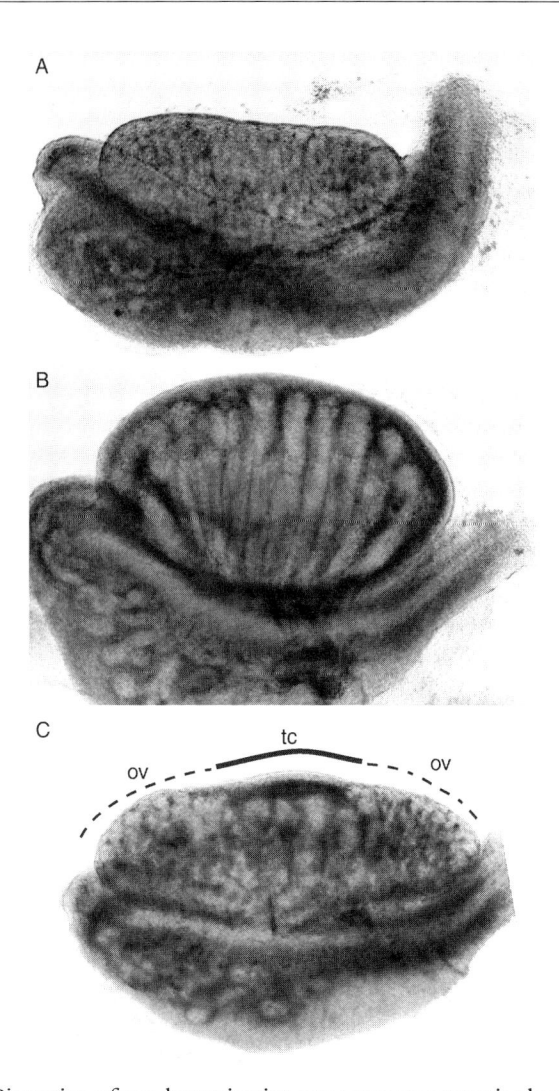

Figure 5.4 Disruption of sex determination can cause an ovotestis phenotype. Bright-field images of wild type female (A) and male (B) gonads at E14.5. On some backgrounds male development is incomplete, and leads to an ovotestis phenotype (C), where the gonad consists of ovarian and testicular regions. The testicular region is always located in the middle, with ovarian regions at each end of the gonad. Dotted lines mark the ovarian regions (ov) and a solid line marks the testicular region (tc). Image contributed by E. Eicher and L. Washburn.

has an ovotestis-like gonad in which the cortical domain produces oocytes during the breeding season, and the medullary region expands and produces high levels of testosterone when the animal is not breeding (Barrionuevo et al., 2004).

2. Molecular Control of Sex Determination

2.1. Molecular specification of the gonad

Between E10.0–11.0 in the mouse gonad, all somatic cells of XX and XY gonads are positive for several known markers, Steroidogenic factor 1 (*Sf1*), Lim-like homeodomain protein (*Lhx9*), the paired-like homeobox gene (*Emx2*), and the Wilms tumor gene (*Wt1*). Null mutations in any one of these genes leads to regression of the gonad in both sexes soon after E11.5 (for review see Brennan and Capel, 2004). Experiments suggest that the undifferentiated gonad passes through a transient phase between E10.5–12.0 when opposing signals hold gonadal cells in a bipotential state (Fig. 5.5; Kim and Capel, 2006). Prior to E11.5, expression patterns are very similar in XX and XY gonads: The male-promoting signal, *Fgf9* is expressed in the coelomic domain while the female-promoting signal *Wnt4* is expressed in a reciprocal domain overlapping the gonad/mesonephros border. *Sox9* (Sry-like HMG-box protein 9) is expressed at low levels near the gonad/mesonephric border in XX and XY gonads (Morais da Silva *et al.*, 1996). The gonad appears to be poised between two pathways of development.

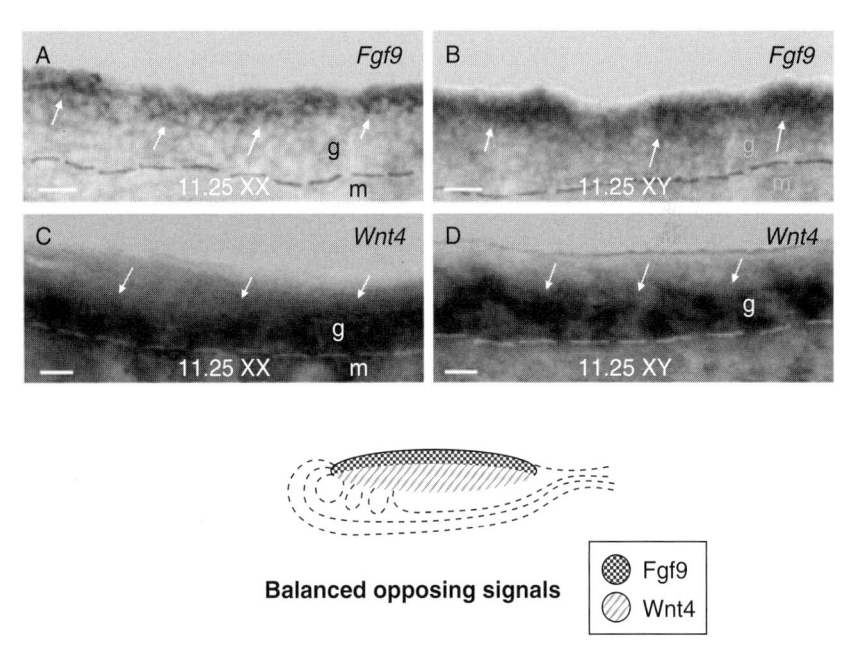

Figure 5.5 The bipotential gonad is balanced between two fates. Prior to E11.5, male and female gonads express genes which later become sex-specific. For example, the male-specific gene *Fgf9* is expressed along the surface of the gonad, while the female-specific gene *Wnt4*, is expressed along the border of the gonad and mesonephros. Figure modified from Kim *et al.*, 2006. (See Color Insert.)

2.2. Choosing the testis pathway

Several years before *Sry* was discovered, it was predicted that the male sex-determining gene would act in a cell lineage in the gonad referred to as the supporting cell lineage because of its key role in supporting development of germ cells during early organogenesis and throughout adult life. Burgoyne and co-workers produced chimeric mice composed of XX and XY cells and allowed them to develop into adults (Palmer and Burgoyne, 1991). When the XX versus XY genotype of the major cell types of the adult testis was analyzed, it became clear that only the Sertoli lineage showed a strong bias for the presence of a Y chromosome (and the *Sry* gene); all other lineages in the adult testis could be XX or XY with equal probability. Consistent with this finding, subsequent transgenic mice expressing *Sry* reporters (Albrecht and Eicher, 2001; Sekido et al., 2004), and antibodies detecting SRY (Wilhelm et al., 2005), confirmed that expression is restricted to cells that differentiate as Sertoli cells. These findings are all consistent with the idea that the critical role of SRY is to trigger a Sertoli cell (vs. follicle cell) fate decision in supporting cell precursors in the early gonad.

The fate decision in this lineage revolves around whether or not *Sox9* expression is established or repressed. If *Sox9* expression is established in a supporting cell, it will develop as a Sertoli cell in the testis. If *Sox9* expression is not established, this lineage will develop as a follicle cell in the ovary. SRY likely works by boosting expression of *Sox9*. Both gain and loss of function experiments suggest that *Sox9* lies at the center of the sex-determination pathway (Lovell-Badge et al., 2002). If *Sox9* is upregulated in XX gonads, either as a transgene (Vidal et al., 2001), or as the result of an activating insertion in the *Sox9* locus (Qin et al., 2004), testis development is initiated. In contrast, mutations in *SOX9* in humans (Foster et al., 1994; Wagner et al., 1994), or deletion of the gene in mice (Chaboissier et al., 2004), cause male to female sex-reversal. These experiments imply that *Sry* works primarily (if not exclusively) through the up-regulation of *Sox9*. Sox protein binding sites have been identified in the *Sox9* regulatory region. These sites respond to SRY binding and also to SOX9 itself in an auto-regulatory loop (Sekido and Lovell-Badge, 2008). Substitution of another Sox gene under the control of the *Sry* promoter leads to the establishment of SOX9 expression and Sertoli cell differentiation (Bergstrom et al., 2000). These experiments suggest that all that is needed to break the stand-off in sex determination is to elevate Sox levels in the supporting cell lineage.

Both the level and timing of *Sry* expression appears to be critical to establish the male pathway. Mutants that have reduced and/or delayed expression of *Sry* produce ovotestes or ovaries. For example, when the Y^{POS} chromosome is carried on the C57BL/6 inbred strain (Albrecht and Eicher, 1997; Nagamine et al., 1998), in some transgenic lines (R. Lovell-Badge, personal communication), and in cases where regions 5′ of *Sry* are deleted (Capel et al., 1993), the *Sry* allele is expressed late and/or at lower levels, leading to partial or complete sex reversal (Burgoyne and Palmer, 1991). It is likely that *Sox9* expression is not established at high enough levels in individual cells or that not enough cells stabilize SOX9 to propagate the male pathway.

Several lines of evidence indicate that this decision is influenced not only by the cell-autonomous activity of *Sry*, but also by non-cell-autonomous signals in the gonadal field. The finding that XX⇔XY chimeric gonads develop as a testis, if >30% of their cells are XY strongly implies that secreted signals can play a significant role. Furthermore, in these chimeras, only an average of 90% of the adult Sertoli population was XY; 10% of Sertoli cells were found to be XX indicating that paracrine signals can induce a Sertoli fate in XX cells (Burgoyne and Palmer, 1992; Palmer and Burgoyne, 1991). Two early characterized signals have an antagonistic effect on the establishment of *Sox9* expression (Fig. 5.6; Kim et al., 2006). *Fgf9*, promotes SOX9 expression while *Wnt4* represses expression of SOX9. *Sox9* works in a feed-forward loop with *Fgf9*: *Sox9* is required for *Fgf9* expression, and *Fgf9* and its receptor, *Fgfr2*, are required to maintain SOX9 expression. When this feed-forward loop is established, *Wnt4* is silenced. In contrast, in

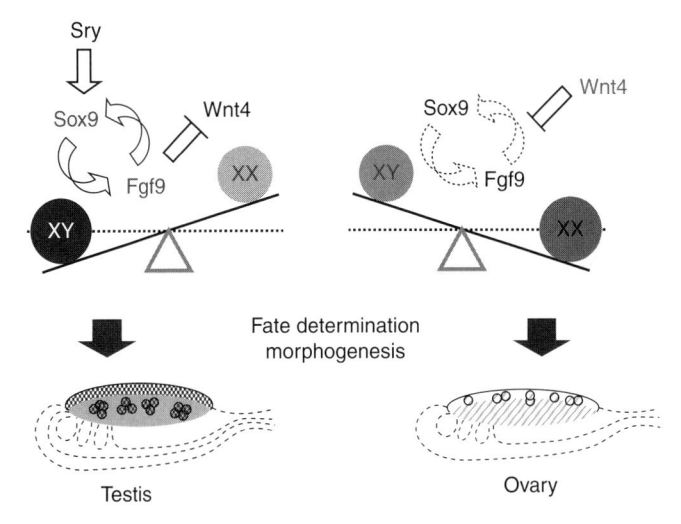

Figure 5.6 Antagonistic signals control sex determination. Prior to sex-determination, male and female factors are balanced within the bipotential gonad. In mammalian XY embryos, the switch (*Sry*), tips the balance towards the male pathway by upregulating *Sox9* expression. *Sox9* upregulates *Fgf9*, which is required to maintain *Sox9* expression, creating a feed-forward loop that drives male development. This feed-forward loop also blocks the expression of *Wnt4*, inhibiting the female pathway. Similarly, in XX embryos, *Wnt4* tips the balance towards the female pathway and blocks aspects of the male pathway by antagonizing the expression of *Fgf9* and *Sox9*. Altering the levels of these factors early in sex determination can influence the fate of the gonad. Figure reproduced from Kim et al., 2006.

XX gonads, where SOX9 expression is not boosted by *Sry*, *Fgf9* is silenced and *Wnt4* expression is elevated.

Wnt4 normally acts to downregulate SOX9 and FGF9 to establish the female pathway. In *Wnt4* loss of function experiments, SOX9 and FGF9 are transiently elevated in XX gonads (even though *Sry* is not present), but complete sex reversal to testis development does not occur. This is likely due to additional signals that reinforce the female pathway (discussed below).

The cell autonomous decision in Sertoli progenitors is propagated to other lineages in the gonad by extracellular signals like Fgfs, Hedgehogs, Pdgfs, and prostaglandin D2 produced by Sertoli cells. Both *Fgf9* and *Pgd2* can recruit more Sertoli cells to reinforce the testis pathway (Kim and Capel, 2006; Wilhelm *et al.*, 2005). One of the earliest known changes downstream of *Sry* is an upregulation of proliferation in the cells of the coelomic epithelium (Schmahl *et al.*, 2000). BrdU pulse-chase experiments, and DiI lineage tracing indicate that cells in this epithelium give rise to both Sertoli and interstitial cells prior to E11.5, but after this time point, only to interstitial cells (Karl and Capel, 1998; Schmahl *et al.*, 2000). Blocking proliferation prior to E11.5 in the XY gonad, either chemically, by administering 5-fluorouracil or methyltrexate, or genetically by eliminating *Fgf9* signaling (and perhaps also Igf signaling), led to a disruption of the male pathway and sex reversal to female (Schmahl and Capel, 2003; Schmahl *et al.*, 2004). These results suggest that an increase in the number of Sertoli cells is also important to counterbalance female signals, and push the gonad toward a male fate.

2.3. Testis morphogenesis

In mice, the primary steps in morphogenesis of the testis occur rapidly after *Sry* and SOX9 expression is initiated. Within 36 h after the initiation of *Sry* expression, Sertoli cells begin to aggregate and surround germ cells, partitioning the testis into two compartments: the testis cords and the interstitium (Fig. 5.7). Although loss of germ cells from the testis slightly delays the formation of testis cords, it does not affect testis cord structure, Sertoli differentiation, or the steroidogenic function of the testis (McLaren, 1991). However, there is evidence that the introduction of meiotic germ cells into an XY gonadal environment can interfere with testis development (Yao *et al.*, 2003).

Signals downstream of *Sry* induce cells from the adjacent mesonephros to migrate into the gonad. When migration is blocked, testis cord structures do not form, strongly suggesting that migrating cells induce changes required for the *de novo* formation of testis cords (Martineau *et al.*, 1997; Tilmann and Capel, 1999). Peritubular myoid cells, which surround testis cords, have been shown *in vitro* to support Sertoli cell differentiation and to cooperate in the deposition of the basal lamina (Skinner *et al.*, 1985; Tung *et al.*, 1984). Based on this literature, the spindle shape of migrating cells, and

Mammalian Somatic Sex Determination   163

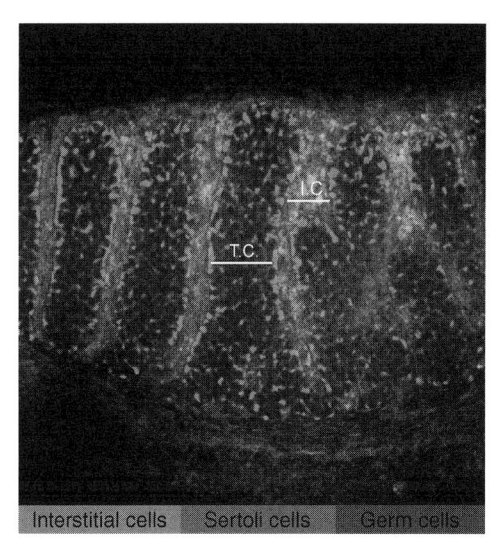

Figure 5.7 Distinct morphology of the XY gonad following sex determination. Shortly after *Sry* is expressed, the cells of the XY gonad reorganize into testis cords (comprised of Sertoli cells and germ cells) or interstitial cell domains. The confocal microscopy image was taken of a mouse gonad at E13.5 after immunostaining for germ cells (blue), Sertoli cells (red), and interstitial cells (green). The gonad is oriented with the coelomic epithelium at the top and the mesonephric border at the bottom. T.C., testis cords; I.C., interstitial cells. (See Color Insert.)

their location proximal to cord boundaries, it was proposed that peritubular myoid cells were the critical cell in the migrating population, required for the induction of testis cord formation. However, as cell-specific markers became available, it became clear that the majority (if not all) migrating cells are endothelial cells (Cool *et al.*, 2008). This finding suggests that it is not peritubular myoid cells, but endothelial cells that are critical for cord formation.

Live imaging, using a GFP labeled population of endothelial cells, indicates that these cells arise from a vascular plexus at the mesonephric border and enter the gonad in a directed pattern of migration toward the coelomic domain. As they migrate into the gonad, their paths partition the gonad into approximately 10 domains where testis cords begin to condense. It is not yet clear whether their migratory paths reflect domain boundaries already established in the gonadal field, or whether endothelial cells play an active, inductive role in partitioning the gonad into cord-forming units. By E12.5, a new arterial system is established in the coelomic domain of the testis, with branches extending back toward the mesonephros, closely apposed to each testis cord boundary (Fig. 5.8; Coveney *et al.*, 2008). As the vasculature in the adult testis has been associated with the regulation of spermatogonial stem cells (Yoshida *et al.*, 2007), the establishment of the

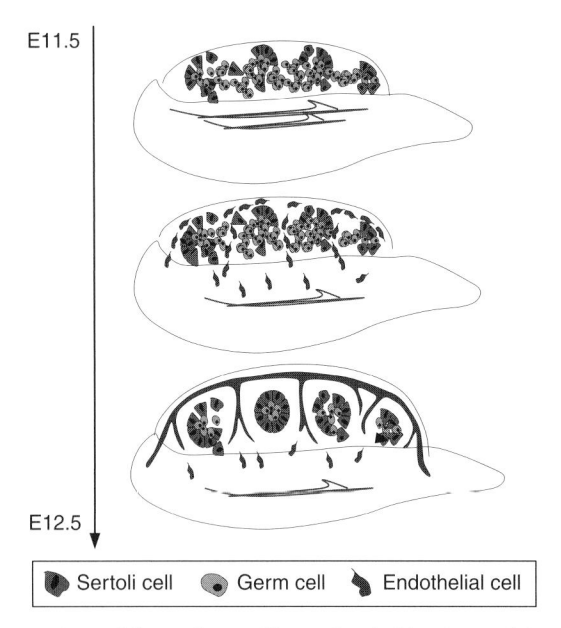

Figure 5.8 Formation of the male specific coelomic blood vessel in XY gonads. At E11.5, vasculature is apparent at the gonad/mesonephric border of both XX and XY gonads. Shortly after *Sry* is expressed in XY gonads, this vascular bed dissociates and endothelial cells migrate into the gonad. By E12.5 a newly formed blood vessel has formed along the coelomic surface of the gonad. Formation of the coelomic blood vessel occurs concurrent with the formation of testis cords. (See Color Insert.)

arterial system of the testis may have multiple functions. The signals that regulate the integration between the vasculature and testis cord structures are the subject of intense investigation.

The fact that there are no cases where *Sox9* expression persists in the absence of testis cord structural organization suggests that the formation of testis cords is critical to reinforce commitment to Sertoli fate and the male pathway. This may reflect a relationship between cell structure and the activation of Sertoli differentiation pathways. In addition to structural reinforcement of the testis pathway, cells in the gonad secrete many reinforcing signals that canalize the different lineages along one of the two competing pathways and coordinate their differentiation (DiNapoli and Capel, 2008; Wilhelm *et al.*, 2007). These findings illustrate an important concept in sex determination: despite being very plastic, the system is resistant to perturbation. This is likely the result of feedback loops and parallel pathways at both intra- and extra-cellular levels that reinforce the decision to embark on one pathway or the other. Discordant development of different lineages within the gonad is very rare. Even when an ovotestis forms, there is a regional distribution of testicular and ovarian tissue, but within a testis or ovarian region, the cell lineages are concordant.

Leydig cells, the steroidogenic lineage of the testis, are detected at E12.5 using probes for enzymes in the testosterone pathway. While the differentiation of these cells may begin earlier, P450 side chain cleavage and 3βHSD appear about 24–30 h after SOX9 is detected in Sertoli cells. Some studies have suggested that the steroidogenic cell population is complex. Steroidogenic cells have been shown to migrate into the gonad from the mesonephros (Jeays-Ward *et al.*, 2003) and also to derive from the initial somatic population produced from divisions in the coelomic epithelium (Brennan *et al.*, 2003). Several lines of evidence suggest that Leydig cells are also involved in feedback loops that stabilize and reinforce testis morphogenesis and play an important role in regulating both Sertoli and germ cells in the early testis (Tang *et al.*, in press).

2.4. Choosing the ovarian pathway

Many investigators have searched for the theoretical *Od* or *Z* gene, predicted to play a role in ovary development parallel to *Sry's* role in testis development. *DAX1* was an exciting *Z* candidate based on evidence that duplication of this X-linked gene in humans leads to female development in XY patients (Bardoni *et al.*, 1994). At one time, it was predicted that *DAX1* might be a remnant of a dosage-based sex-determining system. Swain and colleagues showed that over-expression of *Dax1* in XY mice led to sex-reversal in situations where *Sry* function was attenuated (Swain *et al.*, 1998). However, results from loss-of-function mutations in *Dax1* have confounded the picture (Bouma *et al.*, 2005; Meeks *et al.*, 2003). The gene is not required for ovary development, but instead is required for the testis pathway. Mutations in *Dax1* cause complete male-to-female sex-reversal on some genetic backgrounds. The explanation for the disparate results in gain and loss of function situations in humans and mice remains a mystery.

Another candidate *Z* gene was identified initially in goats homozygous for the *Polled* mutation. XX goats that carried the mutation were sex-reversed to male, although, no Y chromosomal DNA was detected, suggesting sex-reversal occurred independent of *Sry*. Mapping of the mutation identified the PIS locus (Polled/Intersex Syndrome) in goats, a 100 kb region which is syntenic to a human locus where blepharophimosis ptosis epicanthus inversus syndrome (BPES) maps (Pailhoux *et al.*, 2001). This syndrome is associated with premature ovarian failure. Of the two genes associated with this locus, *Foxl2* emerged as a potential *Z* candidate (Crisponi *et al.*, 2001). *Foxl2*, a basic helix-loop-helix transcription factor, initiates expression in the murine XX gonad at E12.5, and marks the commitment of the supporting cell lineage to the follicle cell pathway (Braw Tal *et al.*, 1993; Crisponi *et al.*, 2001; Ottolenghi *et al.*, 2005; Pailhoux *et al.*, 2001). Disappointingly, in the absence of *Foxl2*, initial development of the ovary is unaffected (Ottolenghi *et al.*, 2005). One week after birth, when follicles are beginning to form around oocytes in wild type ovaries, defects in folliculogenesis are apparent in *Foxl2* mutant ovaries. However, this effect appeared too late to implicate this gene either as a *Z* gene or an *Od* gene as the ovary initially forms in its absence. The effect of *Foxl2* gain-of-function has not been investigated. However, no primary sex-reversal has been observed due to a *Foxl2* defect in mice or in human BPES patients (Uhlenhaut and Treier, 2006) suggesting that, unlike the situation in goats, *Foxl2* is not required for the initial decision leading to female sex determination.

While the existence of a single ovarian determining gene is still debatable, two genes have been identified that are required for initial development of the ovary. The most well studied ovarian-determining gene is *Wnt4*. In mice, *Wnt4* is expressed in both male and female gonads during the bipotential stage, and becomes female-specific by E12.5. This is consistent with what would be expected of a gene involved in female sex-determination. Mutation of *Wnt4* in mice, however, exhibited only a partial female-to-male sex-reversal (Vainio *et al.*, 1999). XX gonads lacking *Wnt4* developed vasculature along the coelomic surface of the gonad, resembling the pattern normally seen in developing testes. Male-specific steroidogenic cells derived from adrenal precursors were also present, *Sox9* expression was initially upregulated, but not maintained, and no testis cords were formed (Jeays-Ward *et al.*, 2003; Kim *et al.*, 2006). In addition, germ cells in the XX *Wnt4* mutant gonads entered meiosis similar to normal XX gonads, suggesting that an ovarian environment persisted (Yao *et al.*, 2004). This suggests that *Wnt4* does not act alone as the primary *Od* or *Z* gene since ovarian differentiation was not completely blocked, and testis development was only partially initiated.

Incomplete sex reversal in XX *Wnt4* mutants suggested that other genes might cooperate with *Wnt4* to oppose the testis pathway in XX gonads. A second candidate emerged from studies in human XX patients who were sex-reversed to male. In two independent cases where testis development occurred in genetic females, *R-SPONDIN1* (*RSPO1*) was found to harbor disrupting mutations (Parma *et al.*, 2006). This data strongly implied that this gene plays an important role in blocking male development and/or promoting female development. Disrupting mutations in *Rspo1* in mice lead to a phenotype similar to *Wnt4* mutants, where ovarian development is disrupted, but complete sex reversal of XX mutants does not occur (Chassot *et al.*, 2008; Tomizuka *et al.*, 2008). A human XY patient exhibiting male-to-female sex-reversal was found to carry a duplication of a distal portion of chromosome 1 which includes both the *WNT4* and *RSPO1* loci, suggesting that these two genes may operate together to establish the ovarian pathway (Elejalde *et al.*, 1984; Jordan *et al.*, 2001).

The identification of *RSPO1* focused attention on β-catenin signaling as a mediator of the female pathway. *Wnt4* can function through canonical or non-canonical Wnt pathways, however RSPO1 has been strongly implicated as a player in the canonical β-catenin pathway. While loss of *Rspo1* or *Wnt4* alone does not lead to complete sex-reversal, overexpression of β-catenin causes male-to-female sex-reversal of XY mice (Maatouk *et al.*, 2008). This occurs through a dramatic downregulation of SOX9 in XY gonads around the time of sex-determination, blocking the formation of testis cords and diverting the fate of the gonad towards the ovarian pathway. Intracellular antagonism between SOX9 and β-catenin during the fate decision that occurs in chondroblasts has been well documented (Akiyama *et al.*, 2004). By analogy, it is possible that the fate decision in supporting cell precursors in the gonad occurs through a similar mechanism where either SOX9 or β-catenin gains control of cell fate and triggers the differentiation of Sertoli versus follicle cells (Fig. 5.9). If this model is correct, the outcome may depend on the relative levels of SOX9 and β-catenin. Each of the two genes has its booster team of extracellular signals. In addition to *Fgf9*, *Pgd2* also promotes SOX9 expression (Kim *et al.*, 2006; Wilhelm *et al.*, 2007). *Wnt4* and *Rspo1* may be joined in promoting the female fate by other Wnts such as *Wnt9a* that also shows early female-specific expression (Nef *et al.*, 2005). Much work remains to be done to test this hypothesis. However, sex reversal as a result of loss of function mutations (both male to female, and female to male), strongly argues for two active and opposing pathways that the gonad can switch between during development.

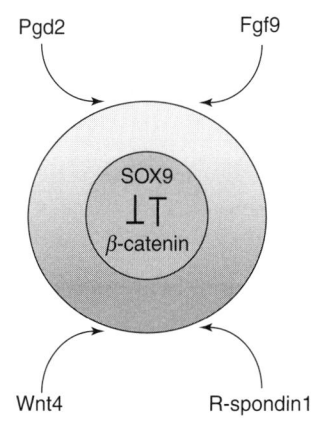

Figure 5.9 Model for intracellular antagonism between SOX9 and β-catenin. Within the bipotential gonad, male and female factors are coexpressed, and the gonad is balanced between two fates. Somatic cells of the gonad are influenced by extracellular signaling molecules including *Pgd2*, *Fgf9*, *Wnt4*, and *Rspo1*. Within the cell, an intracellular antagonism between SOX9 and β-catenin is proposed to exist. In XY gonads, *Sry* expression leads to an upregulation of *Sox9*, which allows *Sox9* to out-compete β-catenin and promote the male pathway. Conversely, in XX gonads, the lack of *Sry* allows for *Wnt4* and *Rspo1* signaling to influence the somatic cells, leading to an upregulation of β-catenin, and differentiation towards the female pathway.

2.5. Ovarian morphogenesis

Unlike the rapid changes that occur in XY gonads following sex determination, XX gonads do not undergo striking morphological changes until close to birth. In the mouse, the ovary can be divided into two main regions: the cortex and the medulla (Fig. 5.10). Germ cells accumulate in the cortex, while those in the medulla eventually die by apoptosis (Yao *et al.*, 2004). Little is known about the pathways that regulate ovarian morphogenesis. However, several genes have been identified that function downstream of *Wnt4*. Gonads lacking *Follistatin* (*Fst*) have some characteristics similar to those lacking *Wnt4*. Loss of *Fst* in XX gonads results in formation of a male-specific coelomic blood vessel. However, unlike *Wnt4* mutants, steroidogenic cells are not present (Yao *et al.*, 2004). Epistasis studies reveal that *Fst* acts downstream of *Wnt4* and carries out a subset of *Wnt4* functions, specifically, antagonizing formation of the coelomic vessel. *Inhibin beta B* (*Inhbb*), which promotes formation of the coelomic vessel, is normally repressed by *Wnt4* and *Fst* in the ovary (Yao *et al.*, 2006).

The *bone morphogenetic protein*, *Bmp2*, was found to localize to the coelomic domain of the XX gonad in a female specific expression pattern that is also dependent on *Wnt4* (Yao *et al.*, 2004). Although there is little morphological evidence of a cortical/medullary structure in the ovary at E12.5, this expression pattern suggests that molecularly distinct domains are already established. However, *Bmp2* expression is maintained in XX *Fst* mutant gonads, suggesting that *Bmp2* and *Fst* act in separate pathways

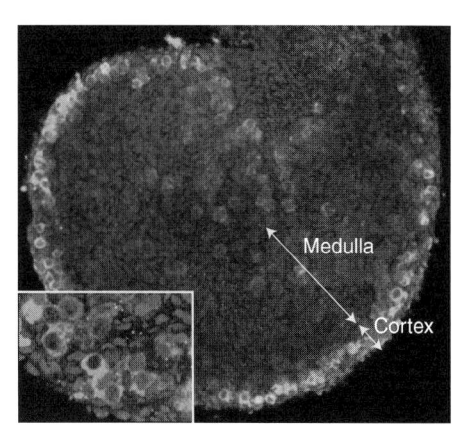

Figure 5.10 A distinct ovarian morphology is observed just prior to birth. While expression of *Sry* leads to a dramatic reorganization of the testis, changes in ovarian structure are subtle and are not apparent until near birth. A confocal microscopy image of an E18.5 mouse ovary shows germ cells (green) accumulating in the cortex, while those in the medulla die by apoptosis. The inset shows an enlargement of germ cells in the cortex. Blue staining indicates the nuclei of all cells. (See Color Insert.)

downstream of *Wnt4*. *Bmp2* null mice are lethal prior to sex determination, therefore a conditional mutation will be necessary to determine what role it plays during ovarian development.

Although the effect of a germ cell loss has not been thoroughly investigated during fetal stages, morphological evidence suggests that the cortical/medullary structure of the ovary at the end of fetal life does not depend on the presence of germ cells (Merchant-Larios and Centeno, 1981). More recent studies indicate that the known molecular elements of the fetal somatic pathway are not disrupted when germ cells are eliminated (Guigon and Magre, 2006). However, germ cells are required for the morphological development of the ovary after birth. At this stage of ovary development, beds of germ cells are surrounded by follicle cells, which gradually extend cytoplasmic processes to pinch off individual germ cells and enclose them in a single layer of squamous follicle cells (Fig. 5.11; Pepling and Spradling, 2001). A number of genes are known to be required within germ cells during folliculogenesis, including *Figla*, *Nobox*, *Kit*, and *Ntrk2* (for review see Choi and Rajkovic, 2006). In agametic XX gonads, follicles never form and the cortical/medullary structure of the ovary collapses (Merchant Larios and Coello, 1979). In cases of germ cell loss after birth, follicle cells often lose identity and express markers characteristic of Sertoli cells (McLaren, 1991). Post-menopausal ovaries, or ovaries depleted of germ cells following radiation or chemotherapy, also show evidence of follicle cell transdifferentiation. Double mutants in estrogen receptors ($Er\alpha^{-/-}$; $Er\beta^{-/-}$) or mutants in aromatase ($Cyp19a1^{-/-}$, the enzyme that produces estrogen) loose all germ cells from the ovary soon after birth (Britt and Findlay, 2003; Couse *et al.*, 1999). While these mutants also loose ovarian structure and express Sertoli cell markers, it has been

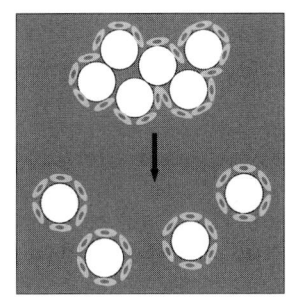

Figure 5.11 Follicle formation occurs in the ovary near the time of birth. At the end of gestation, oocytes (white cells) are arrested at the end of the first prophase of meiosis, arranged in clusters surrounded by follicle cells (pink cells). A high rate of germ cell atresia eliminates many germ cells at this stage. Follicle cells extend cytoplasmic processes and completely surround surviving oocytes forming a squamous cell layer around individual oocytes.

impossible to determine whether the loss of follicle cell fate is a primary effect or a secondary consequence of germ cell loss. These reported instances of "transdifferentiation" of follicle cells provide evidence of the plasticity of the female supporting cell lineage to switch between follicle and Sertoli cell fate.

3. Secondary Sex Differentiation

3.1. Sex-specific development of the sex ducts

The sex ducts in mammals arise within the tissue adjacent to the gonad, the mesonephros. This tissue is also bipotential in the sense that it contains the primordia for both the male and female ductal systems. The primordium for the male ductal system is the mesonephric duct (or Wolffian duct), while the primordium for the female ductal system is the Müllerian duct. Only one of the two ductal primordia will develop and the other will undergo atresia.

The mesonephric duct, the antecedent of the male ductal system, forms by E10.0 by anterior to posterior extension of the pronephric duct within the intermediate mesoderm (Orvis and Behringer, 2007). By E10.5, mesonephric tubules condense in the anterior third of the mesonephros, attach to the mesonephric duct, and coil toward the coelomic surface (for further references, see Capel, 1996). In other vertebrates, this system is very elaborate, and functions as the excretory system during development and afterwards (Grobstein, 1967). In mammals, there is little evidence to support a nephric function for this duct during fetal life. Instead, in mammals, this function is taken over by the kidney, which is induced by a branch from the mesonephric duct into the mesenchyme at the caudal end of the mesonephros. The mesonephric duct, called the Wolffian duct later in male development, differentiates into the epididymis and vas deferens, whereas the mesonephric tubules are remodeled to form the efferent ductules connecting the testis cords to the epididymis (Barsoum and Yao, 2006).

The second duct within the mesonephros is the Müllerian duct, which forms slightly later than the mesonephric duct. The formation of the Müllerian duct depends on signals from *Wnt4* and *β*-catenin that induce the invagination of the surface epithelium of the mesonephros (Deutscher and Hung-Chang Yao, 2007; Vainio *et al.*, 1999). After this step, extension of the Müllerian duct in a caudal direction requires the presence of the mesonephric duct, which appears to act as a guide as cells within the ductal tissue proliferate (Fig. 5.12; Guioli *et al.*, 2007; Orvis and Behringer, 2007). At E12.5, both the mesonephric and the Müllerian ducts are visible in the mesonephros. Thereafter, following differentiation of the gonad, one of the two ductal systems regresses and the other is elaborated.

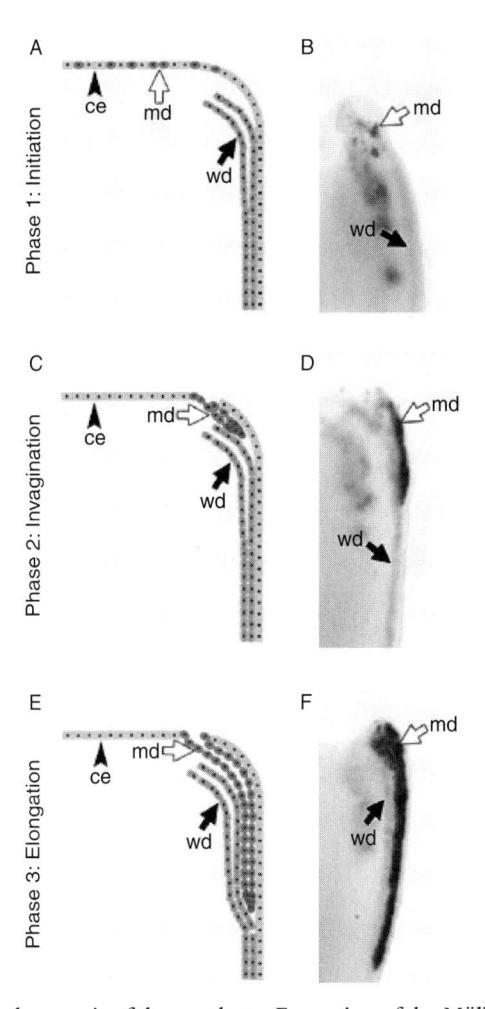

Figure 5.12 Morphogenesis of the sex ducts. Formation of the Müllerian duct can be divided into three phases. (A, B) During the initiation phase, Müllerian duct cells become specified. (C, D) In the second phase, these cells invaginate towards the Wolffian duct, with proliferation of the precursor cells at the leading edge of the duct. (E, F) In the third phase, the Müllerian duct elongates along the Wolffian duct. Müllerian duct cells (blue), proliferating Müllerian duct precursor cells (red), coelomic epithelial cells (brown), Wolffian epithelial cells (yellow). ce; coelomic epithelium, md; Müllerian duct, wd, Wolffian duct. Reproduced with permission (Orvis and Behringer, 2007). (See Color Insert.)

The pioneering work on sex-specific development of the ducts was done by Jost (Jost, 1970). Following on experiments showing that the testis was required for the development of secondary sex characteristics, Jost showed that there are at least two active factors produced by the fetal testis that lead to male sexual development (Fig. 5.13). By implanting a crystal of testosterone in a rabbit fetus in which the testis had been removed, Jost showed that he could recover development of the epididymis, the vas deferens, and male external genitalia. However, this experiment did not cause the regression of the Müllerian duct, indicating that testosterone produced by the testis was sufficient to promote differentiation of the male reproductive tract, but that there was a second factor produced by the testis that was required to repress the development of the female reproductive tract in males. This factor was later identified and named Anti-Müllerian hormone (AMH; Josso et al., 1993). AMH is responsible for inducing regression of the Müllerian duct, the antecedent of the oviduct, uterus, and vagina (Behringer et al., 1994).

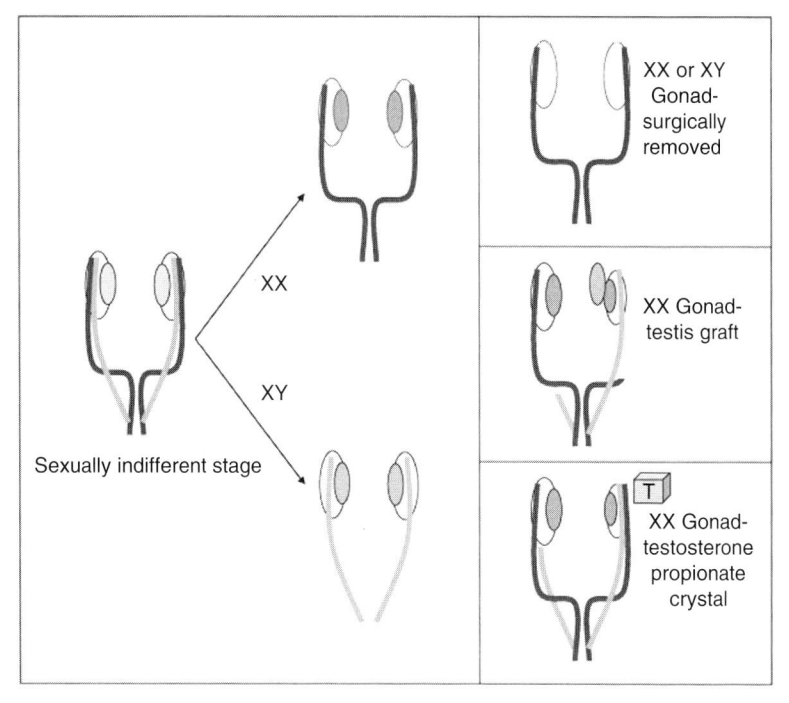

Figure 5.13 The XY gonad produces two factors required for male ductal differentiation. Alfred Jost performed experiments on fetal rabbits to remove the fetal gonad in utero. He found that when the gonad is surgically removed from XX or XY embryos only a female ductal system develops. When a testis is grafted unilaterally next to an XX gonad, factors from the testis cause both regression of the Müllerian duct and development of the mesonephric (Wollfian) duct. However, when a testosterone crystal is placed next to an XX gonad, both male and female ducts are present. These results suggested that testosterone is sufficient to support development of the mesonephric duct, but another factor produced by the testis is required to cause regression of the Müllerian duct.

3.2. Differentiation of the external genitalia

In mice, the appearance of the primordia for the external genitalia coincides with differentiation of the fetal gonads, however, no sex-specific differences are apparent until after E16.5, four days after sexually divergent changes in the gonads occur (Yamada *et al.*, 2006). The external genitalia primordium is also bipotential, and arises between E10.5 and E11.5 with the formation of the genital tubercle (GT). Differentiation of the GT occurs in two phases. First, a hormone-independent phase occurs between E10.5 and E16.5, followed by a second, hormone-dependent phase (Fig. 5.14).

The first phase of external genitalia development involves the outgrowth and patterning of the GT. Interestingly, many similarities have been identified between the mechanisms controlling initial differentiation of the GT, and the more extensively studied appendages, the limb buds (Cohn, 2004). Initial outgrowth of the limb buds and the GT is controlled by an epithelial component in both systems; the apical ectodermal ridge in the limb, and the distal urethral epithelium in the GT. *Fgf8* is expressed in these epithelial cells and is required for outgrowth of both the GT and limb buds. Patterning of the limb buds is controlled, in part, by a region of the posterior limb mesenchyme termed the zone of polarizing activity (ZPA). *Shh*, the effector of the ZPA, sets up a concentration gradient which leads to patterning and

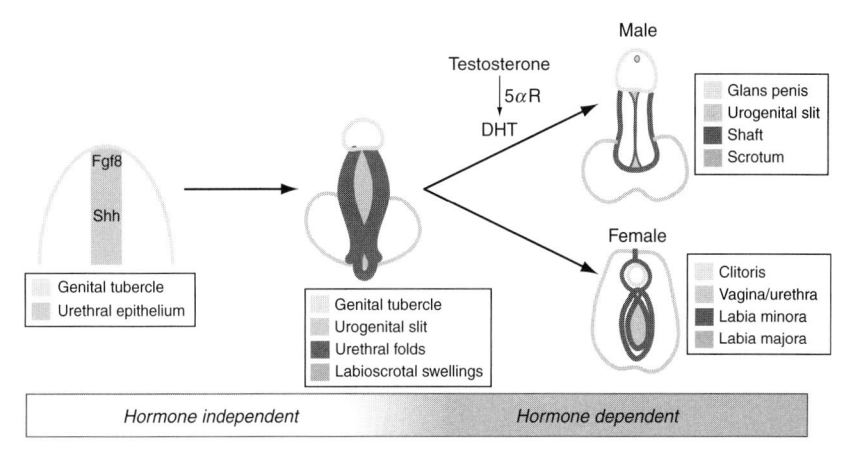

Figure 5.14 Morphogenesis of the external genitalia. Development of the external genitalia occurs in two phases; a hormone independent phase, followed by a hormone dependent phase. During the hormone independent phase, outgrowth of the genital tubercle is controlled by *Fgf8*, expressed in the urethral epithelium, and outgrowth is maintained by *Shh*. This process occurs in an identical fashion in both male and female embryos. After E16.5, sex-specific differentiation of the external genitalia begins and is dependent on the presence or absence of testosterone. Testosterone is converted to the active derivative, dihydrotestosterone (DHT) by the 5α-reductase enzyme (5αR). (See Color Insert.)

specification of the digits. In the GT, *Shh* is expressed in the urethral epithelium and in the absence of *Shh*, outgrowth is initiated, but arrests before a GT is formed (Perriton *et al.*, 2002). Similar to its function in limb development, Shh is necessary to maintain the expression of several signaling molecules in the GT including Bmp, Wnt, and Fgf ligands.

After E16.5, external genitalia differentiation switches from hormone-independent to hormone-dependent and the first sex-specific changes become apparent (Yamada *et al.*, 2003). In both XX and XY embryos, the androgen receptor is expressed throughout the GT. In the fetal testis, testosterone is produced and subsequently converted to DHT by the enzyme 5α-reductase (Russell and Wilson, 1994). DHT then binds to the androgen receptor, activating the signaling pathway that masculinizes the male external genitalia. Since XX embryos express both 5 α-reductase and the androgen receptor in the GT, it is the absence of testosterone that ultimately leads to development of the female genitalia.

External genitalia development is hormonally controlled in lower vertebrate species as well, including aquatic species such as fish and alligators. The effects of many environmental contaminants that resemble androgens or estrogens can influence and override the normal developmental pathway. For example, metabolites from the insecticide DDT act as anti-androgens and have been reported to cause reduced penis size in juvenile alligator populations (Guillette *et al.*, 1996). Such observations highlight the control androgens exert on the latter stages of external genitalia development.

3.3. Sex-specific differentiation of the brain

The concept that exposure, or lack or exposure, to hormones secreted by the testis was responsible for sex-specific differences in the brain was founded on Jost's work. These ideas were reinforced when experiments on female guinea pigs showed that prenatal exposure to testosterone led to male-typical behavior patterns (Phoenix *et al.*, 1959). However, this concept has been challenged by more recent data which suggests that in addition to hormones, a genetic component may influence sexual differentiation of the brain.

Cells of XX and XY embryos are genetically different, with females carrying two copies of the X chromosome, while males have a Y chromosome and only one X. As a general rule, X inactivation in females equalizes X chromosome gene dosage between males and females, however, some genes escape inactivation, allowing for an imbalance of some X-linked gene products between male and female cells (Heard and Disteche, 2006). There is evidence for differences in expression between XX and XY brains based on the cohort of genes that escape X-inactivation (and thus, are expressed in a double dose in females) and on the presence of Y-linked genes only in males (Dewing *et al.*, 2003). In humans, the effects of X-linked gene dosage on neuronal development are exemplified in patients with Turner's

Syndrome, where individuals lack part or all of one X-chromosome. Compared to karyotypically normal XX females, Turner's Syndrome patients exhibit neurodevelopmental deficiencies and are more prone to disorders such as attention deficit disorder (ADHD), autism, and possibly schizophrenia. In the general population, these disorders are more common, and more severe in men than in women, suggesting a correlation with the lower dose of X-linked gene products (Davies and Wilkinson, 2006). Mice carrying only one X chromosome also exhibit behavioral changes similar to human Turner's Syndrome patients, making the XO mouse model an ideal tool for understanding the genetic basis of the disease (Davies et al., 2007).

It has been very difficult to distinguish the role of hormones versus a possible role of the sex chromosomes in brain development and behavior. In mice, the influence of sex chromosomes versus hormones on sexual differentiation of the brain was investigated using mouse models developed to study sex-reversal. XY mice lacking *Sry* are phenotypically female, but carry a male chromosome complement, and XX mice carrying an *Sry* transgene (XX *Sry*) are phenotypically male, but carry a female chromosome complement. Studies comparing such mice to their wild type male and female siblings identified a genetic influence over aspects of the brain (Fig. 5.15; De Vries et al., 2002; Markham et al., 2003). For example, the density of vasopressin innervation of the lateral septum was found to be higher (more masculine) in mice carrying a Y chromosome with or without a testis and male hormones. Additionally, sex-chromosome dependent variations in the number of dopamine neurons were found in cultures of

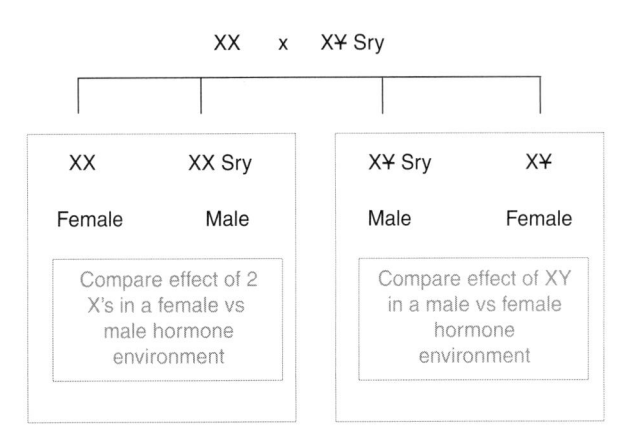

XX x X¥ Sry

XX	XX Sry	X¥ Sry	X¥
Female	Male	Male	Female
Compare effect of 2 X's in a female vs male hormone environment		Compare effect of XY in a male vs female hormone environment	

¥ = Y chromosome carrying a deletion of Sry

Figure 5.15 Sex-specific differentiation of the brain. Mice carrying both a Y chromosome lacking the *Sry* gene, and an autosomal *Sry* transgene, were crossed to wild type females. Offspring of this cross were used to isolate hormonal versus genetic effects on brain differentiation.

embryonic mesencephalic cells from these mice. XY cultures had consistently more dopamine neurons, whether or not *Sry* was present. This difference was also observed when XX and XY brain cells were harvested from E14.5 mouse embryos, before circulating hormone levels vary between the two sexes (Carruth *et al.*, 2002). Therefore, some cells of the brain have genetic differences that can influence their differentiation and function, independent of the differentiation of the gonads.

Direct evidence of a genetic component to sexual differentiation of the brain comes from recent work identifying regions of *Sry* expression in the brain. In humans, *SRY* is expressed in the hypothalamus and the frontal and temporal cortex (Mayer *et al.*, 1998). In rodent male brains, *Sry* is expressed in the substantia nigra, specifically in the TH-positive neurons (Dewing *et al.*, 2006; Mayer *et al.*, 2000). The numbers of TH neurons is sexually dimorphic, with approximately 20% fewer found in females as compared to males. The potential role of *Sry* in these cells, or in any region of the brain in which it has been detected, is not yet understood.

4. Summary

The discovery of the male sex determining gene, *SRY*, opened up a new era of investigation in the classic field of sex determination. A little less than two decades later, we are beginning to get a clearer picture of the molecular changes that guide the initial divergence of testis and ovary development. Many questions still remain about the early differentiation events that occur shortly after *SRY/Sry* expression. The process that leads to the rapid reorganization of Sertoli cells and germ cells into testis cords is not yet understood, nor have the interstitial cells that remain outside of the cords been well characterized. This includes the Leydig cell population, which synthesizes steroidogenic enzymes critical to influencing sexual differentiation throughout the embryo. Progress has been made in integrating hormone influences into the molecular pathways regulating the differentiation of the sex ducts and external genitalia. However, little is known about how these two influences lead to sexually dimorphic development of the brain. Human sex-reversing mutations have been key to unraveling some of the genes involved in the male pathway, however, the rare occurrence of human mutations that disrupt the female pathway (that do not involve *SRY* translocations), has slowed progress in understanding the mechanisms driving the ovarian pathway. The recent identification of two genes, *RSPO1* and *WNT4*, that are involved in the transition of the bipotential gonad to an ovary are an important starting point to uncover the female sex-determining mechanisms. Whether or not a single gene or multiple genes control female sex-determination still remains a question. Downstream of the initial sex-determining event in females, it is not clear how the ovary organizes into two morphologically distinct domains: the medulla and cortex. Further, the mechanisms that promote the accumulation of germ cells in the cortex, and the influence of germ cells over the somatic components of the ovary, remain to be investigated.

REFERENCES

Akiyama, H., Lyons, J. P., Mori-Akiyama, Y., Yang, X., Zhang, R., Zhang, Z., Deng, J. M., Taketo, M. M., Nakamura, T., Behringer, R. R., McCrea, P. D., and de Crombrugghe, B. (2004). Interactions between Sox9 and beta-catenin control chondrocyte differentiation. *Genes Dev.* **18,** 1072–1087.

Albrecht, K. H., and Eicher, E. M. (1997). DNA sequence analysis of Sry alleles (subgenus Mus) implicates misregulation as the cause of C57BL/6J-Y(POS) sex reversal and defines the SRY functional unit. *Genetics* **147,** 1267–1277.

Albrecht, K. H., and Eicher, E. M. (2001). Evidence that Sry is expressed in pre-Sertoli cells and Sertoli and granulosa cells have a common precursor. *Dev. Biol.* **240,** 92–107.

Bardoni, B., Zanaria, E., Guioli, S., Floridia, G., Worley, K. C., Tonini, G., Ferrante, E., Chiumello, G., McCabe, E. R., Fraccaro, M., Zuffardi, O., and Camerino, G. (1994). A dosage sensitive locus at chromosome Xp21 is involved in male to female sex reversal. *Nat. Genet.* **7,** 497–501.

Barrionuevo, F. J., Zurita, F., Burgos, M., and Jimenez, R. (2004). Testis-like development of gonads in female moles. New insights on mammalian gonad organogenesis. *Dev. Biol.* **268,** 39–52.

Barsoum, I., and Yao, H. H. (2006). The road to maleness: From testis to Wolffian duct. *Trends Endocrinol. Metab.* **17,** 223–228.

Behringer, R. R., Finegold, M. J., and Cate, R. L. (1994). Mullerian-inhibiting substance function during mammalian sexual development. *Cell* **79,** 415–425.

Bergstrom, D. E., Young, M., Albrecht, K. H., and Eicher, E. M. (2000). Related function of mouse SOX3, SOX9, and SRY HMG domains assayed by male sex determination. *Genesis* **28,** 111–124.

Bouma, G. J., Albrecht, K. H., Washburn, L. L., Recknagel, A. K., Churchill, G. A., and Eicher, E. M. (2005). Gonadal sex reversal in mutant Dax1 XY mice: A failure to upregulate Sox9 in pre-Sertoli cells. *Development* **132,** 3045–3054.

Braw Tal, R., McNatty, K. P., Smith, P., Heath, D. A., Hudson, N. L., Phillips, D. J., McLeod, B. J., and Davis, G. H. (1993). Ovaries of ewes homozygous for the X-linked Inverdale gene (FecXI) are devoid of secondary and tertiary follicles but contain many abnormal structures. *Biol. Reprod.* **49,** 895–907.

Brennan, J., and Capel, B. (2004). One tissue, two fates: Molecular genetic events that underlie testis versus ovary development. *Nat. Rev. Genet.* **5,** 509–521.

Brennan, J., Tilmann, C., and Capel, B. (2003). Pdgfr-alpha mediates testis cord organization and fetal Leydig cell development in the XY gonad. *Genes Dev.* **17,** 800–810.

Bridges, C. B. (1914). Direct proof through non-disjunction that the sex-linked genes of *Drosophila* are borne by the X-chromosome. *Science* **40,** 107–109.

Britt, K. L., and Findlay, J. K. (2003). Regulation of the phenotype of ovarian somatic cells by estrogen. *Mol. Cell Endocrinol.* **202,** 11–17.

Bullejos, M., and Koopman, P. (2001). Spatially dynamic expression of Sry in mouse genital ridges. *Dev. Dyn.* **221,** 201–205.

Burgoyne, P., and Palmer, S. (1991). The genetics of XY sex reversal in the mouse and other mammals. *Semin. Dev. Biol.* **2,** 277–284.

Burgoyne, P., and Palmer, S. (1992). "Cellular Basis of Sex Determination and Sex Reversal in Mammals from Gonadal Development and Function." Raven Press, NY.

Burgoyne, P. S., Thornhill, A. R., Boudrean, S. K., Darling, S. M., Bishop, C. E., and Evans, E. P. (1995). The genetic basis of XX-XY differences present before gonadal sex differentiation in the mouse. *Philos. Trans. R. Soc. Lond. B Biol. Sci.* **350,** 253–260 discussion 260–261.

Capel, B. (1996). The role of Sry in cellular events underlying mammalian sex determination. *Curr. Top Dev. Biol.* **32,** 1–37.

Capel, B., Rasberry, C., Dyson, J., Bishop, C. E., Simpson, E., Vivian, N., Lovell-Badge, R., Rastan, S., and Cattanach, B. M. (1993). Deletion of Y chromosome sequences located outside the testis determining region can cause XY female sex reversal. *Nat. Genet.* **5,** 301–307.

Carruth, L. L., Reisert, I., and Arnold, A. P. (2002). Sex chromosome genes directly affect brain sexual differentiation. *Nat. Neurosci.* **5,** 933–934.

Chaboissier, M. C., Kobayashi, A., Vidal, V. I., Lutzkendorf, S., van de Kant, H. J., Wegner, M., de Rooij, D. G., Behringer, R. R., and Schedl, A. (2004). Functional analysis of Sox8 and Sox9 during sex determination in the mouse. *Development* **131,** 1891–1901.

Chassot, A. A., Ranc, F., Gregoire, E. P., Roepers-Gajadien, H. L., Taketo, M. M., Camerino, G., de Rooij, D. G., Schedl, A., and Chaboissier, M. C. (2008). Activation of beta-catenin signaling by Rspo1 controls differentiation of the mammalian ovary. *Hum. Mol. Genet.* **17,** 1264–1277.

Chiquoine, A. D. (1954). The identification, origin, and migration of primordial germ cells in the mouse embryo. *Anat. Rec.* **118,** 135–146.

Choi, Y., and Rajkovic, A. (2006). Genetics of early mammalian folliculogenesis. *Cell Mol. Life Sci.* **63,** 579–590.

Cline, T. W., and Meyer, B. J. (1996). Vive la difference: Males vs females in flies vs worms. *Annu. Rev. Genet.* **30,** 637–702.

Cohn, M. J. (2004). Developmental genetics of the external genitalia. *Adv. Exp. Med. Biol.* **545,** 149–157.

Cool, J., Carmona, F. D., Szucsik, J. C., and Capel, B. (2008). Peritubular myoid cells are not the migrating population required for testis cord formation in the XY gonad. *Sex Dev.* **2,** 128–133.

Couse, J. F., Hewitt, S. C., Bunch, D. O., Sar, M., Walker, V. R., Davis, B. J., and Korach, K. S. (1999). Postnatal sex reversal of the ovaries in mice lacking estrogen receptors alpha and beta. *Science* **286,** 2328–2331.

Coveney, D., Cool, J., Oliver, T., and Capel, B. (2008). Four-dimensional analysis of vascularization during primary development of an organ, the gonad. *Proc. Natl. Acad Sci. USA*, **105,** 7212–7217.

Crisponi, L., Deiana, M., Loi, A., Chiappe, F., Uda, M., Amati, P., Bisceglia, L., Zelante, L., Nagaraja, R., Porcu, S., Ristaldi, M. S., Marzella, R., et al. (2001). The putative forkhead transcription factor FOXL2 is mutated in blepharophimosis/ptosis/epicanthus inversus syndrome. *Nat. Genet.* **27,** 159–166.

Davies, W., and Wilkinson, L. S. (2006). It is not all hormones: Alternative explanations for sexual differentiation of the brain. *Brain Res.* **1126,** 36–45.

Davies, W., Humby, T., Isles, A. R., Burgoyne, P. S., and Wilkinson, L. S. (2007). X-monosomy effects on visuospatial attention in mice: A candidate gene and implications for Turner syndrome and attention deficit hyperactivity disorder. *Biol. Psychiatry* **61,** 1351–1360.

De Vries, G. J., Rissman, E. F., Simerly, R. B., Yang, L. Y., Scordalakes, E. M., Auger, C. J., Swain, A., Lovell-Badge, R., Burgoyne, P. S., and Arnold, A. P. (2002). A model system for study of sex chromosome effects on sexually dimorphic neural and behavioral traits. *J. Neurosci.* **22,** 9005–9014.

Deutscher, E., and Hung-Chang Yao, H. (2007). Essential roles of mesenchyme-derived beta-catenin in mouse Mullerian duct morphogenesis. *Dev. Biol.* **307,** 227–236.

Dewing, P., Shi, T., Horvath, S., and Vilain, E. (2003). Sexually dimorphic gene expression in mouse brain precedes gonadal differentiation. *Brain Res. Mol. Brain Res.* **118,** 82–90.

Dewing, P., Chiang, C. W., Sinchak, K., Sim, H., Fernagut, P. O., Kelly, S., Chesselet, M. F., Micevych, P. E., Albrecht, K. H., Harley, V. R., and Vilain, E. (2006). Direct regulation of adult brain function by the male-specific factor SRY. *Curr. Biol.* **16,** 415–420.

DiNapoli, L., and Capel, B. (2008). SRY and the standoff in sex determination. *Mol. Endocrinol.* **22,** 1–9.

Eicher, E. M., and Washburn, L. L. (1983). Inherited sex reversal in mice: Identification of a new primary sex-determining gene. *J. Exp. Zool.* **228,** 297–304.

Eicher, E. M., Beamer, W. G., Washburn, L. L., and Whitten, W. K. (1980). A cytogenetic investigation of inherited true hermaphroditism in BALB/cWt mice. *Cytogenet. Cell Genet.* **28,** 104–115.

Eicher, E. M., Washburn, L. L., Whitney, J. B., III, and Morrow, K. E. (1982). Mus poschiavinus Y chromosome in the C57BL/6J murine genome causes sex reversal. *Science* **217,** 535–537.

Elejalde, B. R., Opitz, J. M., de Elejalde, M. M., Gilbert, E. F., Abellera, M., Meisner, L., Lebel, R. R., and Hartigan, J. M. (1984). Tandem dup (1p) within the short arm of chromosome 1 in a child with ambiguous genitalia and multiple congenital anomalies. *Am. J. Med. Genet.* **17,** 723–730.

Foster, J. W., Dominguez-Steglich, M. A., Guioli, S., Kowk, G., Weller, P. A., Stevanovic, M., Weissenbach, J., Mansour, S., Young, I. D., Goodfellow, P. N., Brook, J. D., and Schafer, A. J. (1994). Campomelic dysplasia and autosomal sex reversal caused by mutations in an SRY-related gene. *Nature* **372,** 525–530.

Grobstein, C. (1967). Mechanisms of organogenetic tissue interaction. *Natl. Cancer Inst. Monogr.* **26,** 279–299.

Gubbay, J., Collignon, J., Koopman, P., Capel, B., Economou, A., Munsterberg, A., Vivian, N., Goodfellow, P., and Lovell-Badge, R. (1990). A gene mapping to the sex-determining region of the mouse Y chromosome is a member of a novel family of embryonically expressed genes. *Nature* **346,** 245–250.

Gubbay, J., Vivian, N., Economou, A., Jackson, D., Goodfellow, P., and Lovell Badge, R. (1992). Inverted repeat structure of the Sry locus in mice. *Proc. Natl. Acad. Sci. USA* **89,** 7953–7957.

Guigon, C. J., and Magre, S. (2006). Contribution of germ cells to the differentiation and maturation of the ovary: Insights from models of germ cell depletion. *Biol. Reprod.* **74,** 450–458.

Guillette, L. J., Jr., Pickford, D. B., Crain, D. A., Rooney, A. A., and Percival, H. F. (1996). Reduction in penis size and plasma testosterone concentrations in juvenile alligators living in a contaminated environment. *Gen. Comp. Endocrinol.* **101,** 32–42.

Guioli, S., Sekido, R., and Lovell-Badge, R. (2007). The origin of the Mullerian duct in chick and mouse. *Dev. Biol.* **302,** 389–398.

Hacker, A., Capel, B., Goodfellow, P., and Lovell-Badge, R. (1995). Expression of Sry, the mouse sex determining gene. *Development* **121,** 1603–1614.

Hawkins, J. R., Koopman, P., and Berta, P. (1991). Testis-determining factor and Y-linked sex reversal. *Curr. Opin. Genet. Dev.* **1,** 30–33.

Heard, E., and Disteche, C. M. (2006). Dosage compensation in mammals: Fine-tuning the expression of the X chromosome. *Genes Dev.* **20,** 1848–1867.

Jacobs, P. A., and Strong, J. A. (1959). A case of human intersexuality having a possible XXY sex-determining mechanism. *Nature* **183,** 302–303.

Jeays-Ward, K., Hoyle, C., Brennan, J., Dandonneau, M., Alldus, G., Capel, B., and Swain, A. (2003). Endothelial and steroidogenic cell migration are regulated by WNT4 in the developing mammalian gonad. *Development* **130,** 3663–3670.

Jordan, B. K., Mohammed, M., Ching, S. T., Delot, E., Chen, X. N., Dewing, P., Swain, A., Rao, P. N., Elejalde, B. R., and Vilain, E. (2001). Up-regulation of WNT-4 signaling and dosage-sensitive sex reversal in humans. *Am. J. Hum. Genet.* **68,** 1102–1109.

Josso, N., Cate, R. L., Picard, J. Y., Vigier, B., di Clemente, N., Wilson, C., Imbeaud, S., Pepinsky, R. B., Guerrier, D., Boussin, L., Legeai, L., and Carre-Eusebe, D. (1993). Anti-mullerian hormone: The Jost factor. *Recent Prog. Horm. Res.* **48,** 1–59.

Jost, A. (1952). Investigation of hormonal control of genesis of the sex organs in rabbit and notes on certain malformations of the genital apparatus in man. *Ginecol. Obstet. Mex.* **7,** 477–492.

Jost, A. (1970). Hormonal factors in the sex differentiation of the mammalian foetus. *Philos. Trans. R. Soc. Lond. B Biol. Sci.* **259,** 119–130.

Karl, J., and Capel, B. (1998). Sertoli cells of the mouse testis originate from the coelomic epithelium. *Dev. Biol.* **203,** 323–333.

Kim, Y., and Capel, B. (2006). Balancing the bipotential gonad between alternative organ fates: A new perspective on an old problem. *Dev. Dyn.* **235,** 2292–2300.

Kim, Y., Kobayashi, A., Sekido, R., DiNapoli, L., Brennan, J., Chaboissier, M. C., Poulat, F., Behringer, R. R., Lovell-Badge, R., and Capel, B. (2006). Fgf9 and Wnt4 act as antagonistic signals to regulate mammalian sex determination. *PLoS Biol.* **4,** e187.

Koopman, P., Munsterberg, A., Capel, B., Vivian, N., and Lovell-Badge, R. (1990). Expression of a candidate sex-determining gene during mouse testis differentiation. *Nature* **348,** 450–452.

Koopman, P., Gubbay, J., Vivian, N., Goodfellow, P., and Lovell-Badge, R. (1991). Male development of chromosomally female mice transgenic for Sry. *Nature* **351,** 117–121.

Lovell-Badge, R., Canning, C., and Sekido, R. (2002). Sex-determining genes in mice: Building pathways. *Novartis Found. Symp.* **244,** 4–18; discussion 18–22, 35–42, 253–257.

Lyon, M. F. (1961). Gene action in the X-chromosome of the mouse (Mus musculus L.). *Nature* **190,** 372–373.

Lyon, M. F. (1962). Sex chromatin and gene action in the mammalian X-chromosome. *Am. J. Hum. Genet.* **14,** 135–148.

Maatouk, D. M., DiNapoli, L., Alvers, A., Parker, K. L., Taketo, M. M., and Capel, B. (2008). Stabilization of beta-catenin in XY gonads causes male-to-female sex-reversal. *Hum. Mol. Genet.* **17,** 2949–2955.

Markham, J. A., Jurgens, H. A., Auger, C. J., De Vries, G. J., Arnold, A. P., and Juraska, J. M. (2003). Sex differences in mouse cortical thickness are independent of the complement of sex chromosomes. *Neuroscience* **116,** 71–75.

Martineau, J., Nordqvist, K., Tilmann, C., Lovell-Badge, R., and Capel, B. (1997). Male-specific cell migration into the developing gonad. *Curr. Biol.* **7,** 958–968.

Mayer, A., Lahr, G., Swaab, D. F., Pilgrim, C., and Reisert, I. (1998). The Y-chromosomal genes SRY and ZFY are transcribed in adult human brain. *Neurogenetics* **1,** 281–288.

Mayer, A., Mosler, G., Just, W., Pilgrim, C., and Reisert, I. (2000). Developmental profile of Sry transcripts in mouse brain. *Neurogenetics* **3,** 25–30.

McElreavey, K., Vilain, E., Abbas, N., Herskowitz, I., and Fellous, M. (1993). A regulatory cascade hypothesis for mammalian sex determination: SRY represses a negative regulator of male development. *Proc. Natl. Acad. Sci. USA* **90,** 3368–3372.

McLaren, A. (1988). Somatic and germ-cell sex in mammals. *Philos. Trans. R. Soc. Lond. B Biol. Sci.* **322,** 3–9.

McLaren, A. (1991). Development of the mammalian gonad: The fate of the supporting cell lineage. *Bioessays* **13,** 151–156.

Meeks, J. J., Weiss, J., and Jameson, J. L. (2003). Dax1 is required for testis determination. *Nat. Genet.* **34**, 32–33.

Merchant Larios, H., and Coello, J. (1979). The effect of busulfan on rat primordial germ cells at the ultrastructural level. *Cell Differ.* **8**, 145–155.

Merchant-Larios, H., and Centeno, B. (1981). Morphogenesis of the ovary from the sterile W/Wv mouse. *Prog. Clin. Biol. Res.* **59B**, 383–392.

Morais da Silva, S., Hacker, A., Harley, V., Goodfellow, P., Swain, A., and Lovell-Badge, R. (1996). Sox9 expression during gonadal development implies a conserved role for the gene in testis differentiation in mammals and birds. *Nat. Genet.* **14**, 62–68.

Nagamine, C. M., Capehart, J., Carlisle, C., and Chang, D. (1998). Ovotestes in B6-XXSxr sex-reversed mice. *Dev. Biol.* **196**, 24–32.

Nef, S., Schaad, O., Stallings, N. R., Cederroth, C. R., Pitetti, J. L., Schaer, G., Malki, S., Dubois-Dauphin, M., Boizet-Bonhoure, B., Descombes, P., Parker, K. L., and Vassalli, J. D. (2005). Gene expression during sex determination reveals a robust female genetic program at the onset of ovarian development. *Dev. Biol.* **287**, 361–377.

Orvis, G. D., and Behringer, R. R. (2007). Cellular mechanisms of Mullerian duct formation in the mouse. *Dev. Biol.* **306**, 493–504.

Ottolenghi, C., Omari, S., Garcia-Ortiz, J. E., Uda, M., Crisponi, L., Forabosco, A., Pilia, G., and Schlessinger, D. (2005). Foxl2 is required for commitment to ovary differentiation. *Hum. Mol. Genet.* **14**, 2053–2062.

Pailhoux, E., Vigier, B., Chaffaux, S., Servel, N., Taourit, S., Furet, J. P., Fellous, M., Grosclaude, F., Cribiu, E. P., Cotinot, C., and Vaiman, D. (2001). A 11.7-kb deletion triggers intersexuality and polledness in goats. *Nat. Genet.* **29**, 453–458.

Painter, T. S. (1923). Further observations on the sex chromosomes of mammals. *Science* **58**, 247–248.

Palmer, S. J., and Burgoyne, P. S. (1991). In situ analysis of fetal, prepuberal and adult XX—XY chimaeric mouse testes: Sertoli cells are predominantly, but not exclusively, XY. *Development* **112**, 265–268.

Panning, B., and Jaenisch, R. (1998). RNA and the epigenetic regulation of X chromosome inactivation. *Cell* **93**, 305–308.

Parma, P., Radi, O., Vidal, V., Chaboissier, M. C., Dellambra, E., Valentini, S., Guerra, L., Schedl, A., and Camerino, G. (2006). R-spondin1 is essential in sex determination, skin differentiation and malignancy. *Nat. Genet.* **38**, 1304–1309.

Pepling, M. E., and Spradling, A. C. (2001). Mouse ovarian germ cell cysts undergo programmed breakdown to form primordial follicles. *Dev. Biol.* **234**, 339–351.

Perriton, C. L., Powles, N., Chiang, C., Maconochie, M. K., and Cohn, M. J. (2002). Sonic hedgehog signaling from the urethral epithelium controls external genital development. *Dev. Biol.* **247**, 26–46.

Phoenix, C. H., Goy, R. W., Gerall, A. A., and Young, W. C. (1959). Organizing action of prenatally administered testosterone propionate on the tissues mediating mating behavior in the female guinea pig. *Endocrinology* **65**, 369–382.

Qin, Y., Kong, L. K., Poirier, C., Truong, C., Overbeek, P. A., and Bishop, C. E. (2004). Long-range activation of Sox9 in Odd Sex (Ods) mice. *Hum. Mol. Genet.* **13**, 1213–1218.

Russell, D. W., and Wilson, J. D. (1994). Steroid 5 alpha-reductase: Two genes/two enzymes. *Annu. Rev. Biochem.* **63**, 25–61.

Schmahl, J., and Capel, B. (2003). Cell proliferation is necessary for the determination of male fate in the gonad. *Dev. Biol.* **258**, 264–276.

Schmahl, J., Eicher, E. M., Washburn, L. L., and Capel, B. (2000). Sry induces cell proliferation in the mouse gonad. *Development* **127**, 65–73.

Schmahl, J., Kim, Y., Colvin, J. S., Ornitz, D. M., and Capel, B. (2004). Fgf9 induces proliferation and nuclear localization of FGFR2 in Sertoli precursors during male sex determination. *Development* **131,** 3627–3636.

Scott, W. J., and Holson, J. F. (1977). Weight differences in rat embryos prior to sexual differentiation. *J. Embryol. Exp. Morphol.* **40,** 259–263.

Sekido, R., Bar, I., Narvaez, V., Penny, G., and Lovell-Badge, R. (2004). SOX9 is upregulated by the transient expression of SRY specifically in Sertoli cell precursors. *Dev. Biol.* **274,** 271–279.

Sekido, R., Lovell-Badge, R. (2008). Sex determination involves synergistic action of SRY and SF1 on a specific Sox9 enhancer. *Nature* **453,** 930–934.

Seller, M. J., and Perkins-Cole, K. J. (1987). Sex difference in mouse embryonic development at neurulation. *J. Reprod. Fertil.* **79,** 159–161.

Sinclair, A. H., Berta, P., Palmer, M. S., Hawkins, J. R., Griffiths, B. L., Smith, M. J., Foster, J. W., Frischauf, A. M., Lovell Badge, R., and Goodfellow, P. N. (1990). A gene from the human sex-determining region encodes a protein with homology to a conserved DNA-binding motif [see comments]. *Nature* **346,** 240–244.

Skinner, M. K., Tung, P. S., and Fritz, I. B. (1985). Cooperativity between Sertoli cells and testicular peritubular cells in the production and deposition of extracellular matrix components. *J. Cell Biol.* **100,** 1941–1947.

Swain, A., Narvaez, V., Burgoyne, P., Camerino, G., and Lovell-Badge, R. (1998). Dax1 antagonizes Sry action in mammalian sex determination. *Nature* **391,** 761–767.

Tilmann, C., and Capel, B. (1999). Mesonephric cell migration induces testis cord formation and Sertoli cell differentiation in the mammalian gonad. *Development* **126,** 2883–2890.

Tomizuka, K., Horikoshi, K., Kitada, R., Sugawara, Y., Iba, Y., Kojima, A., Yoshitome, A., Yamawaki, K., Amagai, M., Inoue, A., Oshima, T., and Kakitani, M. (2008). R-spondin1 plays an essential role in ovarian development through positively regulating Wnt-4 signaling. *Hum. Mol. Genet.* **17,** 1278–1291.

Tung, P. S., Skinner, M. K., and Fritz, I. B. (1984). Cooperativity between Sertoli cells and peritubular myoid cells in the formation of the basal lamina in the seminiferous tubule. *Ann. N. Y. Acad. Sci.* **438,** 435–446.

Uhlenhaut, N. H., and Treier, M. (2006). Foxl2 function in ovarian development. *Mol. Genet. Metab.* **88,** 225–234.

Vainio, S., Heikkila, M., Kispert, A., Chin, N., and McMahon, A. P. (1999). Female development in mammals is regulated by Wnt-4 signalling. *Nature* **397,** 405–409.

Vidal, V. P., Chaboissier, M. C., de Rooij, D. G., and Schedl, A. (2001). Sox9 induces testis development in XX transgenic mice. *Nat. Genet.* **28,** 216–217.

Wagner, T., Wirth, J., Meyer, J., Zabel, B., Held, M., Zimmer, J., Pasantes, J., Bricarelli, F. D., Keutel, J., Hustert, E., Wolf, U., Tommerup, N., et al. (1994). Autosomal sex reversal and campomelic dysplasia are caused by mutations in and around the SRY-related gene SOX9. *Cell* **79,** 1111–1120.

Welshons, W. J., and Russell, L. B. (1959). The Y-chromosome as the bearer of male determining factors in the mouse. *Proc. Natl. Acad. Sci. USA* **45,** 560–566.

Wilhelm, D., Martinson, F., Bradford, S., Wilson, M. J., Combes, A. N., Beverdam, A., Bowles, J., Mizusaki, H., and Koopman, P. (2005). Sertoli cell differentiation is induced both cell-autonomously and through prostaglandin signaling during mammalian sex determination. *Dev. Biol.* **287,** 111–124.

Wilhelm, D., Palmer, S., and Koopman, P. (2007). Sex determination and gonadal development in mammals. *Physiol. Rev.* **87,** 1–28.

Witschi, E. (1951). I. Gonad development and function. *Recent Prog. Horm. Res.* **6,** 1–27.

Yamada, G., Satoh, Y., Baskin, L. S., and Cunha, G. R. (2003). Cellular and molecular mechanisms of development of the external genitalia. *Differentiation* **71,** 445–460.

Yamada, G., Suzuki, K., Haraguchi, R., Miyagawa, S., Satoh, Y., Kamimura, M., Nakagata, N., Kataoka, H., Kuroiwa, A., and Chen, Y. (2006). Molecular genetic cascades for external genitalia formation: An emerging organogenesis program. *Dev. Dyn.* **235,** 1738–1752.

Yao, H. H., DiNapoli, L., and Capel, B. (2003). Meiotic germ cells antagonize mesonephric cell migration and testis cord formation in mouse gonads. *Development* **130,** 5895–5902.

Yao, H. H., Matzuk, M. M., Jorgez, C. J., Menke, D. B., Page, D. C., Swain, A., and Capel, B. (2004). Follistatin operates downstream of Wnt4 in mammalian ovary organogenesis. *Dev. Dyn.* **230,** 210–215.

Yao, H. H., Aardema, J., and Holthusen, K. (2006). Sexually dimorphic regulation of inhibin beta B in establishing gonadal vasculature in mice. *Biol. Reprod.* **74,** 978–983.

Yoshida, S., Sukeno, M., and Nabeshima, Y. (2007). A vasculature-associated niche for undifferentiated spermatogonia in the mouse testis. *Science* **317,** 1722–1726.

CHAPTER SIX

DEVELOPMENT OF GERM CELLS IN THE MOUSE

Gabriela Durcova-Hills* and Blanche Capel[†]

Contents

1. Origin of the Germ Cell Lineage — 185
2. Specification of Primordial Germ Cells in Mice — 186
3. Migration and Population of the Urogenital Ridge — 190
4. Maintaining Pluripotency in the Germ Cell Lineage — 192
5. Sex-Specific Development of Germ Cells in XX and XY Gonads — 193
6. The Sexually Dimorphic Cell Cycle — 195
7. Epigenetic Regulation in the Mouse Germ Line — 198
8. Sexually Dimorphic Epigenetic Regulation — 200
9. Germ Cells to Stem Cells and Back — 202
   9.1. Generation of pluripotent stem cells from PGCs — 202
   9.2. Generation of pluripotent stem cells from neonatal testis cells — 204
   9.3. From pluripotent ES cells to germ cells — 205
10. Conclusions — 205
Acknowledgments — 206
References — 206

1. Origin of the Germ Cell Lineage

Germ cells in mammals have a fundamentally different origin than in *Drosophila* or *Caenorhabditis elegans*. During the first cleavage stages in many organisms, cytoplasmic determinants of the germ plasm are allocated to regions of the embryo that give rise to the germ cell lineage. Referred to as the "nuage" in the older literature, the germ plasm is known to contain a cloud of mitochondria, specific RNAs, and RNA binding proteins that regulate the translation of

---

* Wellcome Trust/Cancer Research UK Gurdon Institute of Cancer and Developmental Biology, University of Cambridge, Cambridge CB2 1QN, United Kingdom
[†] Department of Cell Biology, Duke University Medical Center, Durham, North Carolina 27710

cytoplasmic messages. As cellularization occurs in these organisms, cells inheriting germ plasm determinants develop as germ cells. This is referred to as the allocation or preformation model of germline formation. This model suggests the continuity of the germline by means of a substance present in the zygote that is passed from one germline generation to the next. The alternative model, referred to as regulative specification of germ cells, proposes that germ cells are re-specified at each generation by signals from the soma (Fig. 6.2).

2. Specification of Primordial Germ Cells in Mice

Investigators working in the mouse failed to identify a nuage or any other indication of the allocation of the germline during early cleavage stages. The finding that a normal fertile mouse could develop from a single cell of the 4-cell embryo or from random fractions of morula stage embryos, strongly suggested that germ cells were not allocated during early cleavage stages in the mouse embryo (Gardner, 1977, Eddy and Hahnel, 1983). The first marker for the germ cell lineage in the mouse was tissue nonspecific alkaline phosphatase (TNAP) which identified early germ cells called primordial germ cells (PGCs) at embryonic day E7.5–E8.5 (Chiquoine, 1954, Ginsburg et al., 1990; Ozdzenski, 1967; Fig. 6.1). Lawson and colleagues labeled single cells in the proximal epiblast at E6.0–E6.5 and, using the TNAP marker, showed that some of the labeled cells gave rise to PGCs. No labeled clones were identified in this study that contained only germ cells and no somatic cells, suggesting that the germ cell lineage was not segregated by E6.0–E6.5. From her experiments, Lawson calculated the size of the founding germ cell population to be approximately 45 cells and their time of specification to be between E7.2 and E7.4. (Lawson and Hage, 1994).

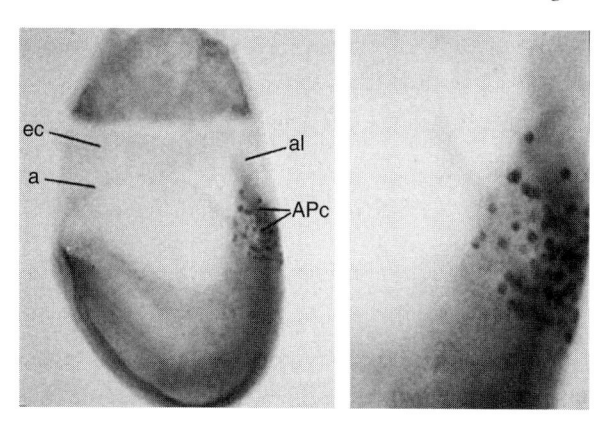

Figure 6.1 Alkaline phosphatase is strongly expressed in a cluster of cells at the base of the allantois. Alkaline Positive cells (APc); allantois (al); extraembryonic ectoderm (ec); anterior (a). Used by permission (Ginsburg et al., 1990). (See Color Insert.)

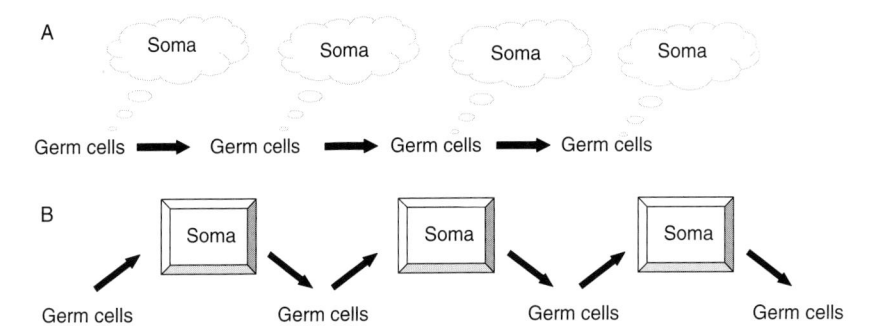

Figure 6.2 Classic models of the relationship between germ cells and the soma. (A) The germ line is continuous via a determinant passed from generation to generation. (B) The soma is required to induce formation of the germ line in each generation.

The definitive experiment indicating that the mouse embryo relies on an induction or epigenetic mechanism to establish the germ cell lineage, was performed by Patrick Tam and colleagues. These investigators performed reciprocal transplants between the distal and proximal regions of the epiblast at E6.5, and showed that cells transplanted to the proximal epiblast domain gave rise to PGCs, whereas cells transferred away from the proximal epiblast domain gave rise to other somatic lineages (Tam and Zhou, 1996). Shortly thereafter, *Bmp4, Bmp2,* and *Bmp8b* were identified as signals in the extraembryonic ectoderm that are required to predispose the adjacent proximal epiblast cells to give rise to PGCs among their descendants (Lawson *et al.*, 1999, Ying and Zhao, 2001, Ying *et al.*, 2001). The extraembryonic ectoderm was shown to be capable of inducing the formation of germ cells in ectopic tissue juxtaposed to this region *in vitro* (Yoshimizu *et al.*, 2001; Fig. 6.3).

The regulative mechanism of germ line induction is not unique to mammals, but is an ancient mechanism proposed by P.D. Nieuwkoop to regulate formation of the germ cell lineage in the *Axolotl*, an amphibian urodele (Nieuwkoop, 1947). More recent experiments support this mechanism in the *Axolotl* as well as echinoderms, and planaria (Extavour and Akam, 2003; Juliano *et al.*, 2006; Wang *et al.*, 2007). The idea that signals from the soma can specify the germline stem (GS) cell population *de novo* at each generation (or during regeneration in the case of planaria) has reoriented our thinking about the functional importance of the soma and the stem cell niche.

To gain more detailed insight into the genetic program of PGC specification, cDNA libraries were made from individual cells in the region of the epiblast where the founder PGCs arise (Saitou *et al.*, 2002; Fig. 6.4). A variety of markers were used to distinguish between cDNA libraries from PGCs and somatic cells. PGC libraries were identified by selecting for expression of TNAP (a positive marker for germ cells), against *Bmp4* (a marker of extraembryonic somatic cells), and against *Hoxb1* (a marker

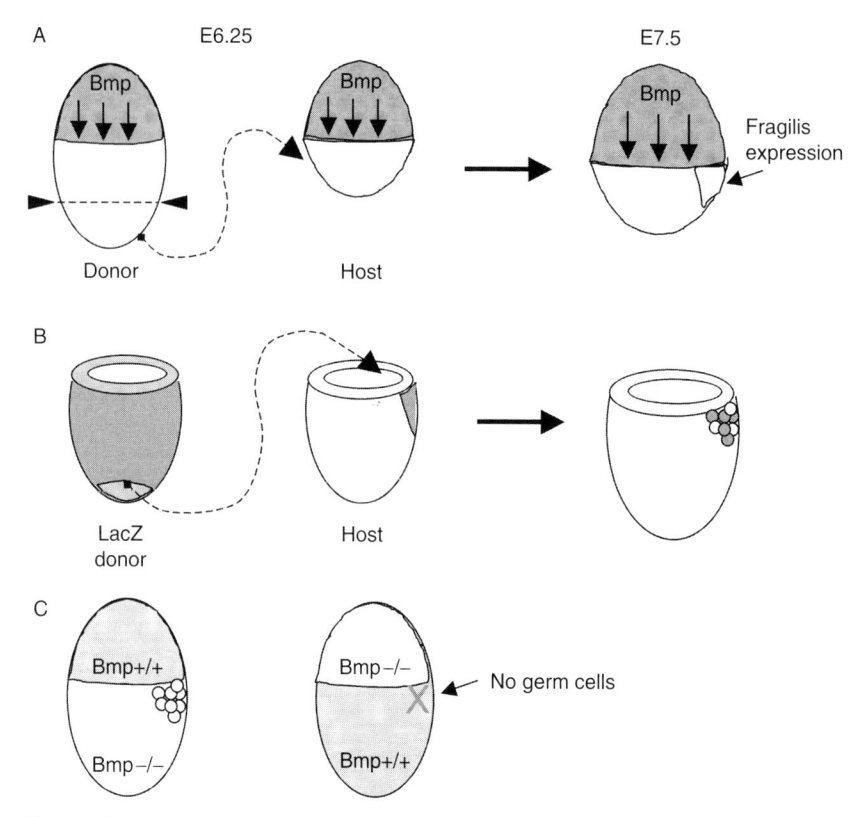

Figure 6.3 PGCs are induced by Bmp signals from the extraembryonic ectoderm. (A) The epiblast was removed from a host embryo, and the distal third of a donor epiblast was cultured proximal to the extraembryonic ectoderm. After culture, *Fragilis* positive germ cells were found in the transplanted fragment (after Saitou et al., 2002). (B) A small fragment from the tip of the epiblast of a donor embryo expressing LacZ was transplanted proximal to the extraembryonic ecoderm of a wild type host. After culture, germ cells in the host embryo were LacZ positive (after Tam and Zhou, 1996). (C) Using tetraploid chimeras, it was shown that when the extraembryonic tissue is mutant for *Bmp4*, no germ cells form, even though the embryonic tissue is wild type. Conversely, when only the embryonic tissue is mutant for *Bmp4*, but the extraembryonic tissue is wild type, germ cells still form (after Lawson and Hogan, 1999).

of somatic cells in the epiblast). Differential screening of single-cell PGC cDNA libraries led to identification of two genes: *Fragilis* and *Stella*.

Fragilis (*Ifitm/Mil*) is a novel member of the interferon inducible transmembrane protein family. It is expressed in many different embryonic and adult tissues where it is associated with homotypic cell adhesion. Detailed mRNA *in situ* hybridization analysis of the fragilis genes showed that PGCs express both *Ifitm1* and *Ifitm3*, and their expression is highly dynamic during gastrulation (Lange et al., 2003; Saitou et al., 2002; Tanaka and Matsui, 2002; Tanaka et al., 2004). Based on these patterns and RNAi disruption, it

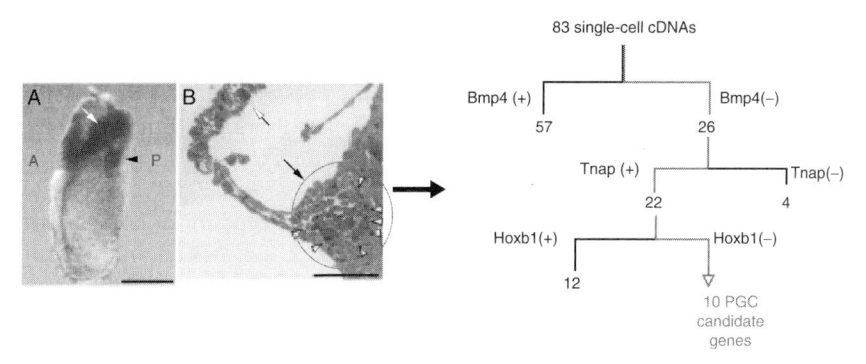

Figure 6.4 To screen for new genes expressed during the specification of PGCs, 83 single cells were picked from the TNAP-positive region of the epiblast proximal to the extraembryonic ectoderm. A cDNA library was made from each cell, and 10 candidate PGC libraries were selected based on the observation that they did not express *Bmp4* (26), did express TNAP (22), and did not express *Hoxb1* (10). Differential expression analysis revealed two new PGC genes common to these libraries and not present in others: *Fragilis* and *Stella*. Used by permission (Saitou *et al.*, 2002). (See Color Insert.)

was proposed that *Ifitm1* and *Ifitm3* regulate adhesion changes that lead to expulsion of PGCs from their site of origin at the base of the allantois into the endoderm, where they travel to the site of the gonads by active and passive transport processes (Tanaka *et al.*, 2005). To investigate Ifitm function during germ cell development, Lange and colleagues used targeted chromosome engineering to generate mutants that either lack the entire Ifitm locus, or carry a disrupted *Ifitm3* gene only (Lange *et al.*, 2008). Contrary to expectations, disruption of these genes resulted in no detectable effects on development of the germline, on migration of PGCs, or on the generation of live young, suggesting that the Ifitm family may be functionally compensated by other pathways during development.

The other gene identified by the screen was the novel gene *Stella* (*Dppa3*, *Pgc7*). *Stella* expression begins at E7.2 in the PGC cluster region, in the centre of the *Fragilis* expression domain, and is maintained in female germ cells until E13.5 and in male germ cells until E15.5. In the adult ovary, STELLA is deposited in maturing oocytes as a maternally inherited protein. The gene is subsequently zygotically expressed in preimplantation embryos from the two-cell stage (E1.5) until the blastocyst stage (E3.5) (Payer *et al.*, 2003; Sato *et al.*, 2002). After E3.5, Stella expression declines and is exclusively reexpressed by nascent PGCs as they are specified, closing its expression-circle in the germline. *Stella*-deficient females displayed severely reduced fertility due to a lack of maternally inherited STELLA-protein in their oocytes, and $Stella^{-/-}$ embryos rarely reached the blastocyst stage (Payer *et al.*, 2003).

Based on the knowledge that repression of the somatic transcriptional program is a key feature of PGC specification in *Drosophila* and *C. elegans*,

Ancelin and colleagues preformed a single-cell cDNA screen to identify candidate genes that regulate gene repression in mouse PGCs (Ancelin et al., 2006). They analyzed 25 candidate SET-PR-domain-containing genes for expression in E7.5 PGCs and found that *Blimp1*, *G9a*, *Set1*, *Ezh2*, and *Pfm1* are expressed in the embryonic region where PGCs arise. However, only *Blimp1* (B-lymphocyte induced maturation protein 1) was expressed in PGCs and not neighboring somatic cells. *Blimp1* has DNA-binding activity and the potential to recruit diverse chromatin-modifying proteins (e.g., G9a, Hdac1, Hdac2, and the transcriptional co-repressor Groucho) predicted to regulate complexes for gene silencing during cell differentiation. *Blimp1* expression is first seen in the E6.25 epiplast in 6–7 cells adjacent to the extraembryonic ectoderm. Based on *Blimp1-Cre* lineage tracing experiments, all of these cells are committed to the PGC lineage. Loss of function of *Blimp1* showed that it is a key determinant of PGC specification in mice (Ohinata et al., 2005). At E7.5, *Blimp1*-deficient embryos contain a small, aberrant cluster of 20 PGC-like cells, and this number declines further by E8.5. To reconcile this data with earlier experiments of Tam and Lawson, an accretion model of PGC specification has been proposed (McLaren and Lawson, 2005). While the 6–7 *Blimp1* positive cells at E6.25 are likely the earliest inhabitants of the PGC lineage, additional cells are proposed to activate *Blimp1* and join the PGC lineage by E7.25. *Blimp1* is highly expressed throughout the phases of PGC specification and migration (Chang et al., 2002).

Given the importance of transcriptional repressors in the *Drosophila* and *C. elegans* germ cell lineages, expression of *Blimp1* in mammalian PGCs implies highly conserved mechanisms of PGC regulation. Although BLIMP1 is known to act as a transcriptional repressor in other contexts, how BLIMP1 regulates germ-cell specification and suppresses the somatic program is currently obscure. Although *Blimp1* has a histone-methyltransferase motif, activity has not been detected. However, as discussed below, BLIMP1 binds to a histone-arginine methyltransferase, PRMT5 (protein arginine N-methyltransferase 5), to repress premature expression of some germ-cell specific genes in more advanced E11.5 PGCs (Ancelin et al., 2006). It is possible that similar epigenetic modifications contribute to the somatic repression role of *Blimp1* in early PGCs.

3. MIGRATION AND POPULATION OF THE UROGENITAL RIDGE

After their expulsion from the embryonic mesoderm into the endoderm, PGCs move toward the gonads by active and passive processes. As gastrulation occurs, PGCs are incorporated into the hindgut where they move randomly among the cells of the gut epithelium. During this period, germ cells are connected by long filapodia (Gomperts *et al.*, 1994). Between E9.0 and E9.5, they emerge in the region of the intermediate mesoderm, just as the urogenital ridge is forming. From here they traverse the bilateral mesonephroi and populate the gonads between E10 and E11.5. *Kit* (a tyrosine kinase receptor expressed in germ cells), and its secreted ligand (*KitL*, or stem cell factor (SCF) expressed in somatic cells), were among the earliest mutations characterized in the mouse, and both are required for germ cell survival and migration (Bennett, 1956; Coulombre and Russell, 1954). A number of other mutations interfere with germ-cell migration, including deletion of *β1-integrin* (Anderson *et al.*, 1999), *Fgf8* (Sun *et al.*, 1999), and the insertion mutation *germ cell deficient* (Pellas *et al.*, 1991).

Study of PGC behavior has been significantly advanced by the development of a live marker, Oct4-GFP (Molyneaux *et al.*, 2001). Using mice expressing this marker, PGCs have been tracked during migratory stages. Genetic evidence in zebrafish indicates that the interaction between stromal cell-derived factor 1 (SDF1) and its G-protein-coupled receptor CXCR4, already known to control many types of normal and pathological cell migrations, is also required for the normal migration of PGCs (Doitsidou *et al.*, 2002; Knaut *et al.*, 2003). In the mouse, germ cell migration and survival requires the SDF1/CXCR4 interaction (Molyneaux *et al.*, 2003). PGCs are actively attracted to the gonad by expression of SDF1/CXCL12, which acts as a chemo-attractant at the stage when germ cells leave the gut and, responding through the receptor CXCR4, move toward the gonad (Doitsidou *et al.*, 2002; Molyneaux *et al.*, 2003). Fgfs affect both motility and germ cell numbers during PGC migration stages (Takeuchi *et al.*, 2005). These effects are mediated by two Fgf receptors expressed in germ cells, FGFR1IIIc and FGFR2IIIb. FGF2, a ligand for FGFR1IIIc, affects motility, whereas, FGF7, a ligand for FGFR2IIIb, affects germ cell numbers, likely by promoting their survival.

It has also been proposed that Bmp signaling mediates aspects of PGC survival and motility. Members of the bone morphogenetic protein (BMP) family have been identified as key regulators of germ cells in the *Drosophila* gonad. Dudley and colleagues used an organ culture system to study the role of BMPs during PGC migration. Results showed that BMP signaling regulates PGC migration by controlling gene expression in both the somatic cells along the migration route, and within the genital ridges (Dudley *et al.*, 2007). In mice, *Bmp7* is expressed in gonads of both sexes and is required for germ cell proliferation during a narrow window of development between E10.5 and E11.5. *Bmp7* mutants show a drop in the numbers of germ cells just as they enter the genital ridge (Ross *et al.*, 2007).

Nanos is implicated in migration and maintenance of PGCs in *Drosophila* (Kobayashi *et al.*, 1996), *C. elegans* (Subramaniam and Seydoux, 1999) and zebrafish (Köprunner *et al.*, 2001). Of the two orthologs characterized in mice, *Nanos3* is found in migrating PGCs. Elimination of this factor results in the complete loss of germ cells in both sexes. Hence, although mice and flies differ in their mechanisms for germ cell specification, there seems to be a conserved function for Nanos proteins among invertebrates and vertebrates (Tsuda et al., 2003).

4. Maintaining Pluripotency in the Germ Cell Lineage

In mice, PGCs are derived from pluripotent proximal epiblast cells and they maintain expression of the triad of pluripotency genes *Oct4*, *Sox2*, and *Nanog* during their development (Chambers et al., 2007; Kehler et al., 2004; Pesce et al., 1998; Scholer et al., 1990; Yamaguchi et al., 2005). Single-cell transcriptional profiling was employed to examine gene expression dynamics of pluripotency genes in PGC isolated from E6.75–E8.25 embryos (Yabuta et al., 2006). Expression levels of *Oct4* were high in PGC precursors and PGCs. In contrast, *Sox2* expression was barely detectable in the neighboring somatic cells, and low or absent in PGCs at E6.75. However, it was progressively upregulated between E6.75 and E8.5. These results indicate that the OCT4-SOX2 complex exists only in PGC precursors after E6.75. *Nanog* was expressed continuously in PGC precursors, but sharply downregulated in somatic neighbors after E6.75. In contrast, the expression of another early PGC marker, *Dppa5* (*Esg1*), was detected at low levels in both the PGC precursors and somatic cells at E6.75, and was not upregulated until E8.25 in PGCs (Western et al., 2005). Taken together, this analysis suggests that the transitions from PGC precursors to more advanced PGCs involve a stage-dependent transcriptional program that begins with the regaining of the pluripotency associated gene network (Yabuta et al., 2006; Fig. 6.5).

In the same study the expression of other germline-associated genes was also examined. Expression of the tyrosine kinase receptor, *Kit*, was not detected between E6.75 and E7.75 in neighboring somatic cells, but was specifically upregulated in PGCs at E7.25 and maintained between E6.75 and E8.25. In contrast, expression of the ligand for the Kit receptor, *Kitl*, was detected transiently at high levels in E7.25 PGCs and decreased within 24 h. Expression dynamics of two genes encoding RNA binding proteins required for germ cell survival in mice and other species were also investigated, *Nanos3* and *Dnd1*. *Nanos3* expression was initiated specifically in PGCs at E7.25, thus designating *Nanos3* as an early germ cell-specific transcript similar to *Stella*. Loss of *Nanos3* leads to a germ cell-less phenotype in both sexes. Interestingly, this phenotype can be rescued by exogenous expression of *Nanos2* in early PGCs (Suzuki et al., 2007). *Dnd1* was expressed in PGC precursors, and was sharply upregulated in PGCs after E7.25. *Dnd1* has been recently identified as the gene responsible for the classic mouse mutation, *Ter* (Youngren et al., 2005). This mutation leads to

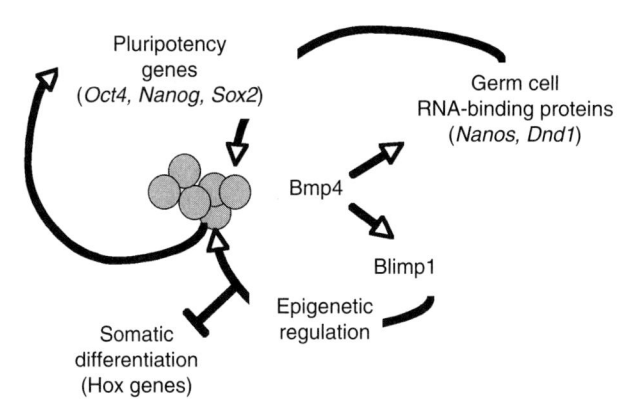

Figure 6.5 During early specification stages, PGCs are maintained by expression of the pluripotency genes, *Oct4, Nanog,* and *Sox2*, and by expression of RNA-binding proteins including *Nanos* and *Dnd1*, and the transcriptional repressor, *Blimp1*.

the loss of the majority of PGCs on all genetic backgrounds, and the development of a high incidence of testicular teratomas on the 129 Sv/J genetic background (Stevens, 1973).

5. SEX-SPECIFIC DEVELOPMENT OF GERM CELLS IN XX AND XY GONADS

Sex-specific development of germ cells depends on their gonadal environment (Fig. 6.6). XX germ cells that arrive in a developing testis, for example in an XX animal carrying an *Sry* transgene, or in an XX↔XY chimera developing a testis, are incorporated into testis cords and initiate a male pattern of development. Similarly, XY germ cells that arrive in a developing ovary, for example in an XY animal carrying a deletion of *Sry*, remain interspersed with somatic cells and follow the female developmental pathway (Fig. 6.7). Soon after germ cells reach the gonads, their sex specific differentiation begins. Between E10.5 and E13.5, they continue to divide mitotically at similar levels in XX and XY gonads. However, both loss of *Fgf9* and loss of *Vasa* affect germ cells in the testis and not the ovary by E12.5. The earliest of these is *Fgf9*, which can promote survival of XY but not XX germ cells isolated from their respective environments as early as E11.5 (DiNapoli et al., 2006). *Vasa* is a second dead end protein characteristic of early specification of germ cells in many species including *Drosophila*, zebrafish, and turtles. The mouse ortholog of this gene (*Mvh*) is activated relatively late in mouse PGCs, just as they arrive in the gonad. In *Mvh* mutants, germ cells are specifically affected in the testis, where they show proliferation defects, and fail to localize within testis cords (Tanaka et al., 2000). Similarly, loss of *Bmp7* shows a more severe loss of germ cells in the

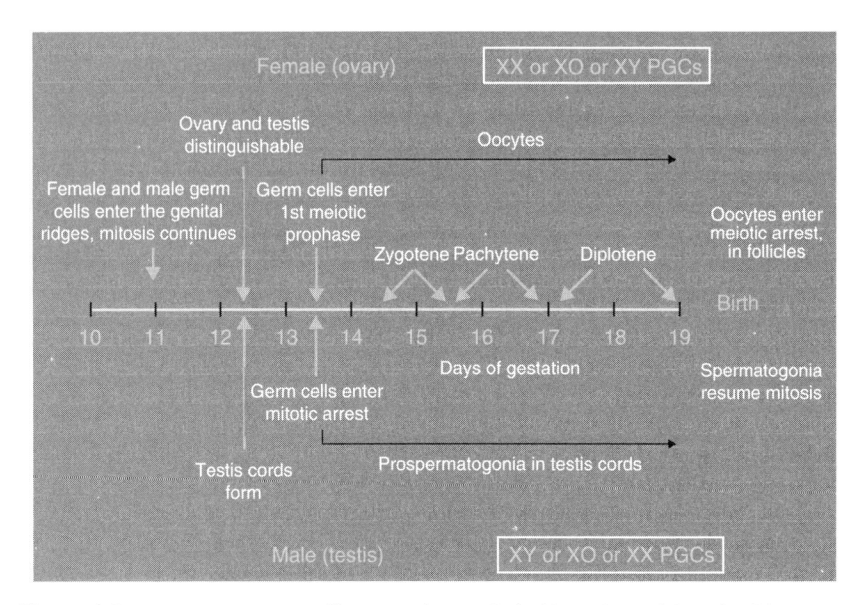

Figure 6.6 At E11.0, germ cells enter the genital ridges (gonads) and initiate sex-specific differentiation. In the ovary germ cells are interspersed with somatic cells, and enter meiosis between E13.5 and E14.5. They pass through zygotene, pachytene and arrest in diplotene around the time of birth when individual follicles form. In the testis, germ cells are enclosed in testis cords by E12.5 where they undergo mitotic arrest between E13.5 and E14.5. They transition to pro-spermatogoina, and remain arrested until after birth, when mitosis resumes, a spermatogonial stem cell population takes up residence at the periphery of the testis cords, and the first wave of meiosis is initiated. (*Drawing from A. McLaren*).

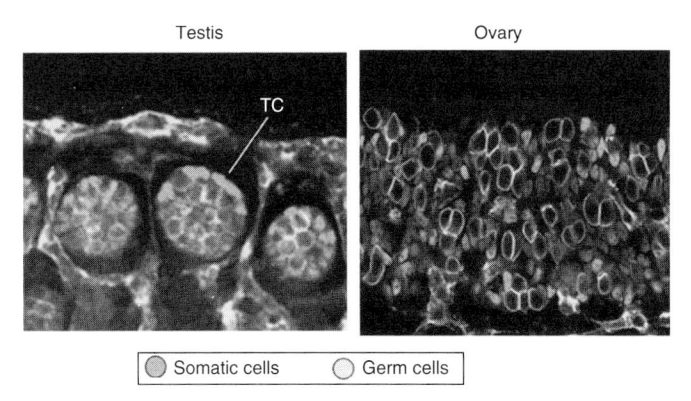

Figure 6.7 In the testis, germ cells (round, green) are enclosed by Sertoli cells (red) inside testis cords (TC) by E12.5. In the ovary, germ cells (round, green) remain scattered between somatic cells (red) until follicles form soon after birth. The marker for germ cells is PECAM1, which also labels vasculature, most evident in the interstitial space between cords in the testis. (See Color Insert.)

testis than in the ovary during this period, suggesting that there are sexually dimorphic compensatory pathways (Ross *et al.*, 2007). These findings indicate that germ cell survival pathways are different in the ovary and the testis at least by E11.5, and imply that sex-specific pathways are initiated between E10.5 and E11.5, coincident with expression of the male pathway genes, *Sry* and *Sox9* in XY somatic cells. After this stage, many genes are expressed in sexually dimorphic patterns.

6. THE SEXUALLY DIMORPHIC CELL CYCLE

The most conspicuous marker of the divergence of germ cell pathways in the testis and ovary is the regulation of the cell cycle. The net result of this difference is the establishment of spermatogonial stem cells in the adult testis, that act as a renewing stem cell population to produce sperm throughout reproductive life. In contrast, germ cells in the ovary synchronously arrest in prophase of meiosis, establishing a finite population of oocytes by birth. These differences are reflected in diverging expression patterns in germ cells (Fig. 6.8).

In the XY gonad, germ cells undergo mitotic arrest between E13.5 and E14.5 and undergo a differentiation transition to pro-spermatogonia. Characterization of cell cycle differences between germ cells in the testis versus the ovary revealed the upregulation of the cell cycle regulators $p27^{Kip1}$ and $p15^{INK4b}$, both of which regulate the hypophosphorylation of pRB (Western *et al.*, 2008). Hypophosphorylation of pRB is associated with a strong down-regulation of CyclinE which blocks the G1-S phase transition. DND1, an RNA binding protein, binds to the p27 mRNA and protects it from microRNA-mediated degradation (Kedde *et al.*, 2007). A requirement for p27 activity to arrest the cell cycle in male germ cells may explain why loss of function of DND1 in $Dnd1^{Ter/Ter}$ mutants leads to the formation of germ cell tumors in the testis.

In contrast, germ cells in the XX gonad enter meiotic prophase at this stage and progress to diplotene by birth. As this difference is evident using classic histological stains, this dimorphic pattern has been investigated for many years. Two mechanisms have been proposed to account for this divergent pattern: (1) the ovary produces a factor that promotes germ cell entry into meiosis or (2) germ cells follow an intrinsic clock and enter meiosis unless blocked by a factor produced by the fetal testis.

With respect to the first theory, there was compelling evidence for the existence of a somatically derived meiosis-inducing factor. Meiosis could be induced by culturing fetal XY gonads with conditioned media from ovaries or adult testes actively engaged in meiosis (Byskov, 1974). However, Zamboni had observed that germ cells occasionally fail to reach the gonad, and wind up in the mesonephros or the adrenal gland (Zamboni and

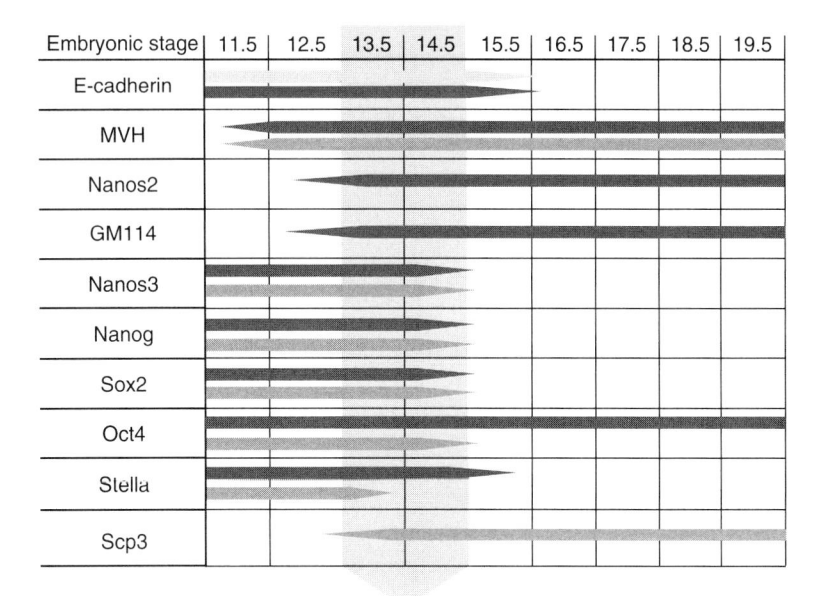

Figure 6.8 Expression timeline for germ cells in the testis (black) or ovary (grey). E-cadherin and MVH are activated in XX and XY germ cells as they enter the gonads. While, E-cadherin declines, MVH expression is maintained in both sexes. The triad of pluripotentcy genes (*Oct4*, *Sox2*, and *Nanog*), and *Nanos3* are expressed in both sexes. *Nanos3*, *Sox2*, and *Nanog* decline by E14.5–E15.5. *Oct4* expression is maintained in the testis, but declines in the ovary at E13.5. Several male-specific genes are activated just prior to mitotic arrest (*Nanos2*, *Gm114*). Stella expression declines in the ovary at E13.5, and in the testis at E15.5. The activation of markers of meiosis (including synaptonemal comples proteins, e.g., *Scp3*) is specific to germ cells in the ovary.

Upadhyay, 1983). When germ cells are found in these locations, whether in an XX or XY embryo, they invariably enter meiosis. Based on these experiments, McLaren proposed that entry into meiosis is a cell autonomous property of germ cells that is inhibited by testis-specific pathways. Using tissue recombination experiments, McLaren showed that germ cells from the testis enter meiosis if removed from the testis environment prior to E12.5, and cultured either with somatic cells from fetal lung tissue or from the XX gonad (Fig. 6.9). The fate of XX germ cells is not fixed until after E12.5, when their culture with somatic cells from the XY gonad no longer interferes with their entry into meiosis (McLaren and Southee, 1997).

The resolution of this debate combines elements of both theories. The first toe-hold into a molecular explanation for the sexually dimorphic behavior of germ cells came with the discovery that *Stra8* is expressed in female germ cells between E12.5 and E15.5 in an anterior to posterior wave (Menke *et al.*, 2003), a pattern associated with the appearance of markers of

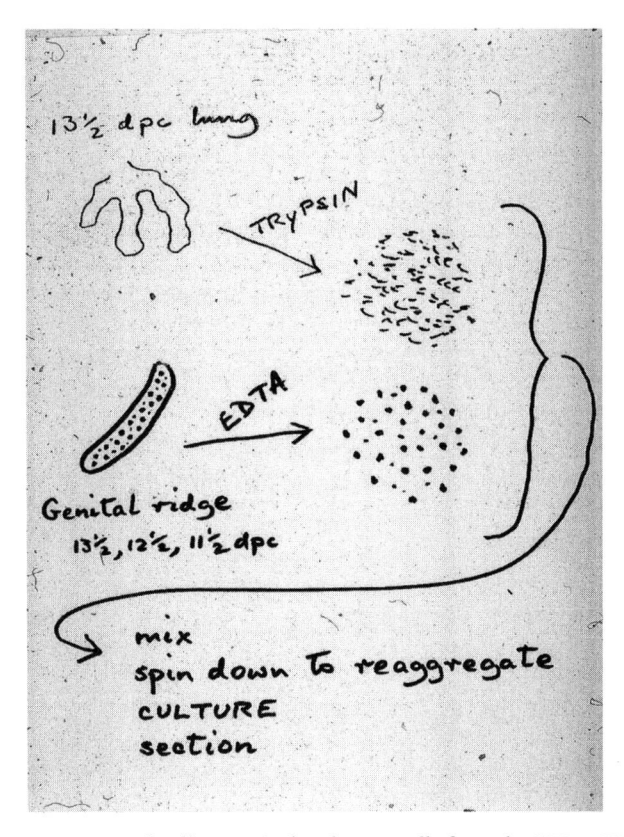

Figure 6.9 McLaren and colleagues isolated germ cells from the XX or XY gonad at E.11.5, E12.5, or E13.5 and cultured them with dissociated cells from the lung. They showed that germ cells from both sexes enter meiosis when isolated from the gonads at E11.5. However, after E12.5, only germ cells isolated from the XX gonad enter meiosis. From these experiments, she concluded that germ cells are committed to spermatogenesis in the XY gonad after E12.5. (*Hand drawing from A. McLaren.*)

meiosis (Yao *et al.*, 2003; Menke *et al.*, 2003). *Stra8*, stimulated by retinoic acid, is a gene originally identified in ES cells after retinoic acid treatment. Expression of this gene in a meiotic pattern in XX germ cells led to the idea that retinoic acid, which is present in the mesonephros during germ cell migration and population of the urogenital ridge (Bowles *et al.*, 2006), could be responsible for inducing meiosis through the activation of *Stra8*. Blocking retinoic acid by inhibitors or receptor antagonists led to the downregulation of *Stra8* in XX germ cells (Bowles and Koopman, 2007; Koubova *et al.*, 2006). Subsequently, it was shown that loss of function of *Stra8* results in a failure of germ cells to enter meiosis in the XX gonad (Baltus *et al.*, 2006). At about the same time, *Cyp26b1* was identified as a gene expressed in gonads of both sexes at E11.5, but downregulated in the XX and highly upregulated in the XY gonad by E12.5. As *Cyp26b1* encodes an enzyme that degrades retinoic acid, it was recognized that upregulation of this gene in the testis pathway could explain why retinoic acid is not effective in activating meiosis in germ cells in the testis (Bowles et al., 2006; Koubova et al., 2006; MacLean et al., 2007). Loss of function of *Cyp26b1* led to entry of germ cells into meiosis in XY gonads and their subsequent loss through apoptosis (Bowles et al., 2006; MacLean et al., 2007). This suggests a mechanism where high levels of retinoic acid and low levels of *Cyp26b1* lead to expression of *Stra8* and entry into meiosis in XX gonads, whereas, high levels of *Cyp26b1* lead to low levels of retinoic acid and a failure to enter meiosis in XY gonads (reviewed in Bowles and Koopman, 2007). Thus, the resolution of this long-standing debate involves both a meiosis inducing factor and a testis inhibitory factor.

Other genes are involved in the decision to undergo mitotic arrest or enter meiosis. For example, the RNA binding protein *Nanos2* is expressed specifically in XY germ cells just prior to mitotic arrest. In *Nanos2* mutants, when germ cell apoptosis is blocked by inactivation of the Bax pathway, XY germ cells enter meiosis (Suzuki and Saga, 2008), suggesting that RNA metabolism may play a role. Although the mouse ortholog of *Drosophila bam*, *Gm114*, is also specifically expressed during mitotic arrest and the transition to spermatogonia in the fetal testis, deletion of this gene did not disrupt the pattern of male germ cell development (Tang et al., 2008).

The anterior to posterior wave of meiosis in the ovary could be due to serial exposure of germ cells to an inducing factor expressed in an anterior to posterior pattern. Retinoic acid is expressed at higher levels in the anterior mesonephros and may diffuse into the gonad from this region where mesonephric tubules are attached to the gonad (Bowles and Koopman, 2007; Bowles et al., 2006; Karl and Capel, 1995). However, it is difficult to exclude the possibility that the wave pattern is due to a cell-autonomous timing factor, with entry into meiosis timed from germ cell specification in the niche at the base of the allantois and entry into the endoderm. As germ cells are carried along passively as the hindgut continues to invaginate and elongate, it has been proposed that the first PGCs to leave the specification niche colonize the most anterior region of the gonad, whereas the last to leave the niche are located in the most posterior position (McLaren, personal communication). One developmental factor that might affect the preparation of germ cells for meiosis is epigenetic programming.

7. Epigenetic Regulation in the Mouse Germ Line

Epigenetic information is encoded by DNA methylation, histone modifications, histone variants and non-histone chromatin proteins. In pluripotent cells, evidence suggests that epigenetic marks are plastic.

However, as cells begin to differentiate, epigenetic marks become more rigid, coincident with restrictions in cell potentcy.

Despite differences in PGC origins, a common theme has emerged from studies in *Drosophila*, *C. elegans* and mice: PGC specification depends on mechanisms that inhibit the expression of somatic genes. In *Drosophila* and *C. elegans* embryos, somatic cell nuclei initiate mRNA transcription soon after fertilization, but PGCs and their precursors do not accumulate zygotic mRNAs until after the onset of gastrulation. In *C. elegans*, this transcriptional block depends on PIE-1, a maternally inherited germ plasm component that accumulates in both cytoplasm and nuclei of PGCs. In *Drosophila* the germ cell components *gcl* (germ cell-less) and *pgc* (polar granule component) have been proposed to play similar roles (for review, see Blackwell, 2004).

PGCs in the mouse undergo unique global epigenetic reprogramming soon after their specification (Seki *et al.*, 2005, 2007; Fig. 6.10). Several methylation marks associated with transcriptional repression disappear between E7.5 and 8.75, while new repressive marks appear beginning at E8.25. During this transition, RNA PolII is transiently repressed (Seki *et al.*, 2007). The key changes observed are the erasure of H3K9-methylation at E8.0, together with a decrease in the levels of HP1α by E9.0 within the euchromatic and the pericentric heterochromatic regions. At the same time there is a decline in the global levels of DNA methylation in PGCs from E8.0 onwards. The loss of DNA methylation is coincident with the repression of the de novo DNA methyltransferases, *Dnmt3a* and *Dnmt3b*, and the transient decline in the maintenance methyltransferase, *Dnmt1*.

Hajkova and colleagues studied the changes in PGC chromatin and showed that the changes occur in two steps (Hajkova *et al.*, 2008). The first change in E8.5 PGCs is to establish a distinctive chromatin signature associated with the expression of pluripotency-specific genes: *Sox2*, *Oct4*, *Nanog*, and *Stella*. This process may reflect the programming of the PGC genome to an epigenetic profile similar to the one described for ES cells, where bivalent domains—regions that show both active and repressive methylation marks – are associated with the key transcriptional regulators (Bernstein *et al.*, 2006).

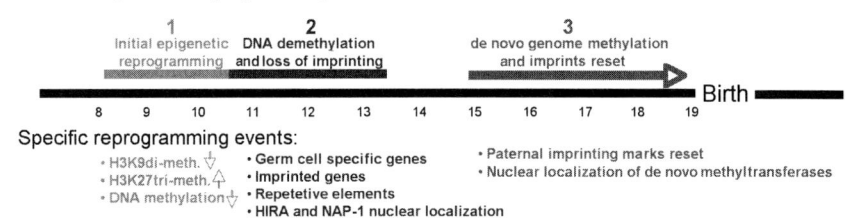

Figure 6.10 Phases of germline epigenetic reprogramming and specific reprogramming events associated with each stage.

When PGCs enter the gonads additional extensive epigenetic reprogramming events occur, including major changes in nuclear architecture accompanied by an extensive erasure of several histone modifications and the exchange of histone variants. Increases in H3K4-methylation and H3K9-acetylation (characteristic of permissive chromatin) are seen, along with decline of H3K9-methylation (a repressive mark). In addition, there is extensive genome-wide DNA demethylation. The promoter regions of several genes known to be transcribed once PGCs reach the urogenital ridge (e.g., *Mvh* and *Dazl*) are demethylated at this stage (Maatouk et al., 2006). Based on DAPI-stained nuclei of gonadal PGCs, Hajkova and colleagues showed that chromatin decondensation and restructuring are a transient phenomena. Hajkova and colleagues suggested that DNA demethylation may be connected with DNA repair, because DNA repair driven demethylation could directly induce chromatin changes and histone replacement (Hajkova et al., 2008). However, not all epigenetic marks are completely removed at this stage. For example, methylation of the IAP retrotransposon family is only partially reprogrammed (Lane et al., 2003).

Recently, a novel gene called *Meisetz* was identified that links epigenetic reprogramming with meiotic entry. *Meisetz* is characterized by a SET/PR domain with catalytic activity for H3K4-trimethylation (Hayashi and Matsui, 2006). *Meisetz* is expressed in XX germ cells at the time of their entry into meiotic prophase (E13.5) and in XY pre-meiotic spermatogonia in the postnatal testis. *Meisetz* deficient mice are sterile in both sexes, suggesting that epigenetic modifications play an essential role in meiotic entry.

8. Sexually Dimorphic Epigenetic Regulation

The majority of genes are expressed equally from the maternal and paternal allele. However a group of ~100 genes in mammals undergo parental-specific imprinting. These genes are somehow marked differently when inherited from the mother or the father and are expressed in the offspring from only one of the two parental alleles. A property of genomic imprinting is that the pattern must be erased in the germline at each generation and reset on both alleles in a sex-specific manner. Although the initial mark has not been clearly defined, loci that are imprinted show differential methylation of the two alleles. This pattern has been used to follow the process of imprint erasure and reestablishment in the germline.

The erasure of parental imprints and the reactivation of the inactive X chromosome in females appear to be an active processes that occur coincident with global demethylation. PGCs undergo imprint erasure beginning at E9.5, completing the process at all loci by E12.5 (Hajkova et al., 2002). Not all epigenetic marks are removed during this phase, which could lead to transmission of epigenetic information to the next generation. Parental imprints are reestablished in the germline in a sexually dimorphic manner (Fig. 6.11). In the ovary, maternal imprints are not reestablished until after birth, during folliculogenesis in growing oocytes. By contrast in the testis, imprints are re-established between E12.5 and E14.5, around the time that germ cells enter mitotic arrest and transition to prospermatogonia. Establishment of paternal imprints requires the DNA methyl transferases, DNMT3A and DNMT3B, and the interacting protein DNMT3L; whereas DNMT3B is dispensable for establishment of maternal imprints. These enzymes also play a role in methylation of retrotransposons (Kaneda *et al.*, 2004, 2007).

What controls the sex specific methylation patterns, the testis or ovary environment or the XX/XY sex of the germ cells? McLaren addressed this question by examining the pattern of imprinting in cases where XX germ cells were located in a testis, or XY germ cells were located in an ovary (Chuva de Sousa Lopes *et al.*, 2008). She found that XY females produce offspring with a typical maternal imprint, indicating that imprint methylation in the female germline is in response to oogenesis, and is not dependent on XX chromosome constitution. In contrast, XX germ cells developing in a testis show a reduction in the number of methylation marks at imprinted loci, suggesting that the sex of the germ cells may play a role in establishing the XY imprinting mark. Although there is clearly an important influence of the fetal testis environment on the paternal imprinting process, the extra X chromosome may exert an inhibitory influence, or the absence of the Y chromosome in this situation may impair the process.

X-chromosome inactivation occurs in XX PGCs by E6.5, synchronously with their somatic neighbors. Inactivation of Xm (the maternal X)

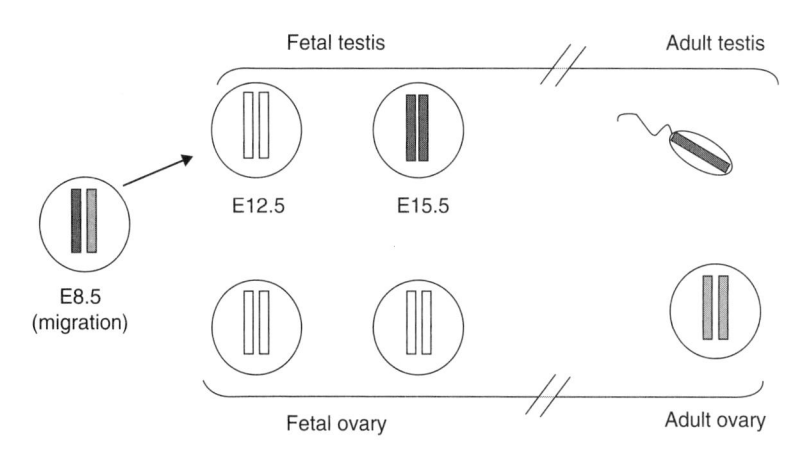

Figure 6.11 Genomic imprints (light grey = maternal; dark grey = paternal) are erased during germ cell migration and reestablished at different stages in the testis and ovary. In the testis, paternal imprints are established in the fetal testis about the time of mitotic arrest. In the ovary maternal imprints are established during ooctye growth in the adult ovary.

or Xp (the paternal X) is random, as in somatic cells. Between E7.5 and E10.5, the H3K27me3 spot associated with the inactive X gradually declines in XX germ cells (Chuva de Sousa Lopes et al., 2008). *Xist* expression from the inactive X progressively declines between E11.5 and E13.5 (Nesterova et al., 2002). Using XX^{GFP} as a reporter, all XX^{GFP} germ cells are GFP positive indicating that both X chromosomes are active by E13.5. XX germ cells isolated from the hindgut or from the E11.5 gonad and cultured on transwell filters undergo X-chromosome reactivation only in the presence of XX gonadal tissue, implying that there is a factor in XX gonads that triggers this process (Chuva de Sousa Lopes et al., 2008).

In contrast, the single X chromosome in XY germ cells remains active until adult spermatogenesis stages (Jamieson et al., 1997). During meiotic prophase the X and Y chromosomes associate through pairing of their pseudo-autosomal regions and form a structure referred to as the XY- or sex-body. Within this structure, both chromosomes are transcriptionally silent (Turner et al., 2005). The process of meiotic sex chromosome inactivation (MSCI) is mediated by epigenetic machinery. The histone variant, γH2AX, which is phosphorylated at the zygotene-pachytene transition is a central feature of MSCI. Formation of the sex body is dependent on BRACA1 (breast cancer related 1) and a histone trimethyltransferase, PRDM9 (Turner et al., 2004). Specific histone modifications within this structure mark meiotic progression. Some of these are mediated by SCMH1 (sex comb on midleg homolog 1), a component of the PRC1 (polycomb repressive complex1). Loss of function of *Scmh1* leads to meiotic arrest, however, the structure of the sex-body remains intact (Takada et al., 2007). Although the formation of the sex-body is clearly required for meiotic progression, it's function is not yet well characterized. It has been proposed to mask the unpaired regions of the X and Y chromosomes to bypass the meiotic checkpoint (Turner et al., 2005), or to be involved in the acquisition of a paternal imprint on the inactive X chromosome (Namekawa et al., 2006).

9. GERM CELLS TO STEM CELLS AND BACK

9.1. Generation of pluripotent stem cells from PGCs

PGCs are specified in the early mouse embryo at about the same stage that ES cells can be derived *in vitro*, leading to speculation that the two lineages have a common origin (Matsui and Okamura, 2005; Saitou et al., 2002). While germ cells normally give rise only to male and female gametes, they harbor the totipotent information necessary to generate a complete organism. Multipotent cells can be derived from germ cells at multiple stages. Conversely, it is also possible to derive germ cells from embryonic stem cells (Fig. 6.12).

Development of Germ Cells in the Mouse

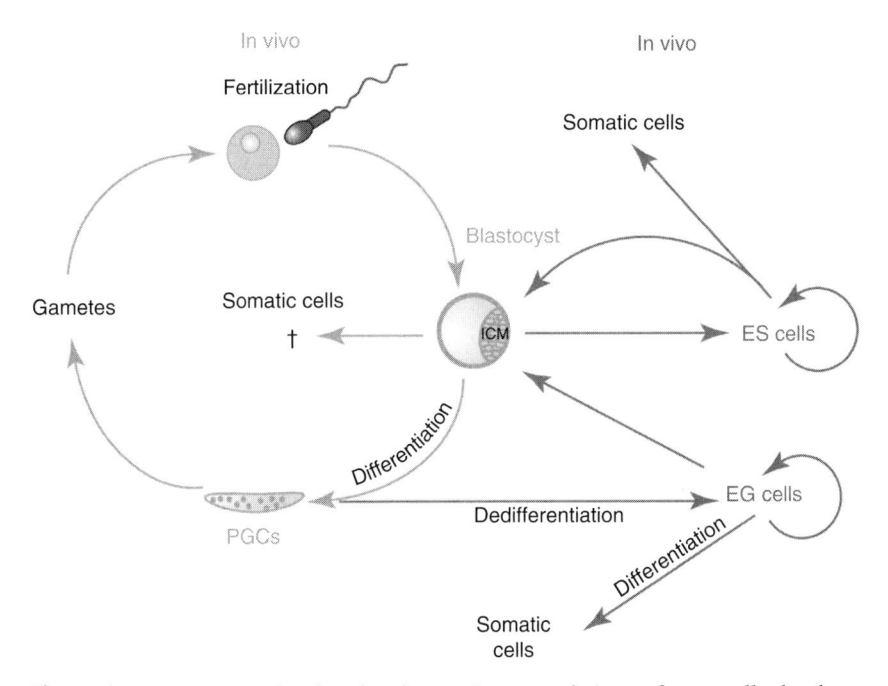

Figure 6.12 PGCs are closely related to various populations of stem cells that have been experimentally establisihed. *In vivo*, PGCs are specified from pluripotent epiblast cells at about the same stage when ES cells can be experimentally established. PGCs migrate and colonize genital ridges. Although they are destined to form specialized gametes *in vivo*, during migratory stages and until E12.5, they can be induced *in vitro* to form EG cells, a pluripotent stem cell population similar to ES cells. Under appropriate conditions, ES cells can also be differentiated into PGCs *in vitro*. The facility with which these transitions take place suggest that the genomic program in PGCs is closely related to that in various stem cell populations. (*Drawing from A. McLaren.*)

The first pluripotent stem cells identified were derived from teratocarcinomas, which are malignant gonadal tumors that arise from germ cells, and contain an admixture of differentiated cells representing all germ layers, and a core of undifferentiated cells. EC cells were derived from the core of the tumor, and could be propagated by serial transfer *in vivo*, or cultured and manipulated *in vitro*. EC cells are tumor cells and they are often aneuploid. Although they can contribute to chimeras when injected into a blastocyst (Brinster, 1974; Illmensee and Mintz, 1976), their contribution is usually low and they Matsui *et al*, 1992; Resnick *et al.*, 1992; Stewart *et al.*, 1994) rarely give rise to germ cells (Papaioannou *et al.*, 1975, 1978).

Early attempts to culture PGCs, however they were all unsuccessful until PGCs were cultured in the right cocktail of growth factors (Matsui *et al.*, 1992; Resnick *et al.*, 1992; Stewart *et al.*, 1994). Pluripotent embryonic germ (EG) EG cell lines were derived from PGCs cultured in the presence of leukaemia inhibitory factor (LIF), fibroblast growth factor 2 (FGF-2), and the membrane-bound form of SCF. Once established, EG cells require only LIF in their media, and exhibit full developmental capacity. They generate chimeras when injected into blastocysts, and can be induced to differentiate into three germ layers under appropriate culture conditions. When injected under the kidney capsule, they generate teratomas.

EG cells have been derived from PGCs harvested from E8.0 to 12.5 fetuses although efficiency declines after E11.5. Several mutations are known to increase the efficiency of generating EG cells including mutations in *Dnd1*, *Pten*, *Pgct1*, and *Akt* (Kanatsu-Shinohara et al., 2008). However, the presence of cytokines is essential for this process. Like ES cells, EG cells have the ability to reprogram somatic cell lineages after cell fusion, resulting in the upregulation of pluripotency genes and the downregulation of genes involved in differentiation (Shovlin et al., 2008; Tada et al., 1997, 2001).

In many respects EG cells are similar to ES cells, except that EG cells exhibit extensive erasure of parental imprints during their derivation (Tada et al., 1998). Mid and late EG cells do not reflect the methylation status of the cells from which they are derived. However, with rare exceptions, they are fully demethylated for all imprinted genes investigated (Durcova-Hills et al., 2001; Labosky et al., 1994; Tada et al., 1998). It was concluded that the methylation erasure process continued after PGCs were placed in culture. EG cell lines derived from E8.5 PGCs have strikingly heterogeneous DNA methylation patterns, in contrast to EG cells derived from E11.5 PGCs. Heterogeneity in DNA methylation patterns in EG cells established from E8.5 PGCs may reflect rapid changes in DNA methylation taking place in the germ cell lineage soon after its specification (Shovlin et al., 2008).

EG establishment and reprogramming takes place during the first 10 days of culture, and, moreover, the presence of exogenous FGF2 is critical only during the first 24 h (Durcova-Hills et al., 2006). More recently it has been established that LIF in combination with retinoic acid or forskolin can substitute for FGF-2 to generate EG cells (Koshimizu et al., 1996).

9.2. Generation of pluripotent stem cells from neonatal testis cells

Recently cell lines have been derived from much later stages of germ cell development. When neonatal testis cells were cultured in the presence of glial cell line-derived neurotrophic factor (GDNF), LIF, epidermal growth factor (EGF), and FGF2, two types of cells were generated. GS cells, which can differentiate only into germline cells and multipotent germline stem (mGS) cells, that act as pluripotent stem cells similar to ES or EG cells (Kanatsu-Shinohara et al., 2004, 2008). Pluripotent stem cells, which can differentiate into all three embryonic germ layers including germline, have even been derived from spermatogonia in the adult testis (Guan et al., 2006).

9.3. From pluripotent ES cells to germ cells

ES cells can differentiate into all types of somatic cells including germ cells in chimeric animals. When ES cells are and to differentiate in culture, they also generate somatic cell types and PGCs. However, the efficiency of deriving germ cells is not high. Recent studies have revealed that it is also possible to generate gametes from ES cells (for review see Ko and Scholer, 2006). Gametes or gamete-like cells were derived when mouse ES cells were cultured under various differentiation conditions including simple monolayer culture, embryoid body formation, embryoid body formation followed by treatment with retinoic acid, and retinoic acid induction alone. In the most successful case, ES-derived sperm cells were able to fertilize oocytes after intracytoplasmic injection, and support embryonic development to term. However, pups survived only up to five months, suggesting that reprogramming of the germ-cell genome was not properly accomplished (Nayernia *et al.*, 2006).

The facility with which these transitions can be induced suggests a close relationship between the genomic programming of a germ cell and the genomic programming of a multipotent stem cell. A comparison of the transcriptional profiles between GS cells (capable only of renewing the spermatogonial stem cell population) and mGS cells (capable of contributing to all cells of the embryo) revealed differences in the levels of pluripotent transcription factors and in DNA methylation patterns (Kanatsu-Shinohara *et al.*, 2008). While the potential to proliferate indefinitely and escape growth control is normally harnessed in the fetal and adult testis by the somatic cell niche within testis cords, and by mechanisms within the germ cells, the incidence of testicular germ cell tumors in humans is high, highlighting the importance of the regulatory controls that suppress the potential for germ cells to escape to a multipotent state capable of unregulated growth and differentiation.

10. CONCLUSIONS

Germ cells are induced in the proximal mouse epiblast by signals from the extraembryonic ectoderm. During their specification, PGCs establish a pluripotent genomic program characterized by the expression of *Oct4*, *Sox2*, and *Nanog*. As in other species, establishment of the germ cell population requires several RNA-binding proteins, including Nanos and Dnd proteins, and a transcriptional repressor, *Blimp1*. PGCs undergo dramatic epigenetic changes including histone modifications and genome-wide reprogramming of DNA methylation patterns. Once specified, germ cells migrate to the gonads where they initiate sex-specific patterns of differentiation including cell cycle arrest (in the case of the male) or entry into meiosis (in the case of the female). Male and female germ cells undergo differential programming of the X chromosome, and during the course of their maturation, acquire paternal- or maternal-specific imprints. The facility of transitions between germ cells and their derivative stem cells (EG, EC, or GS cells) suggests a close relationship among their genomic programs.

ACKNOWLEDGMENTS

We would like to thank Anne McLaren for her outstanding contributions to the field, and for leaving us several illustrations from her last presentation on Mouse Germ Cells at the Cold Spring Harbor Mouse Course in 2007 (Figs. 6.1, 6.6, 6.9, and 6.12). We would also like to thank Matthew Cook for his major contribution to Fig. 6.8, and Danielle Maatouk for her contribution of Fig. 6.10, and careful reading of the manuscript.

REFERENCES

Ancelin, K., Lange, U. C., Hajkova, P., Schneider, R., Bannister, A. J., Kouzarides, T., and Surani, M. A. (2006). BLIMP1 associates with PRMT5 and directs histone arginine methylation in mouse germ cells. *Nat. Cell Biol.* **6,** 623–630.

Anderson, R., Fässler, R., Georges-Labouesse, E., Hynes, R. O., Bader, B. L., Kreidberg, J. A., Schaible, K., Heasman, J., and Wylie, C. (1999). Mouse primordial germ cells lacking beta-1 integrins enter the germline but fail to migrate normally to the gonads. *Development* **126,** 1655–1664.

Baltus, A. E., Menke, D. B., Hu, Y. C., Goodheart, M. L., Carpenter, A. E., de Rooij, D. G., and Page, D. C. (2006). In germ cells of mouse embryonic ovaries, the decision to enter meiosis precedes premeiotic DNA replication. *Nat. Genet.* **38,** 1430–1434.

Bennett, D. (1956). Developmental analysis of a mutant with pleiotropic effects in the mouse. *J. Morphol.* **98,** 199–234.

Bernstein, B. E., Mikkelsen, T. S., Xie, X., Kamal, M., Huebert, D. J., Cuff, J., Fry, B., Meissner, A., Wernig, M., Plath, K., Jaenisch, R., Wagschal, A., Feil, R., Schreiber, S. L., and Lander, E. S. (2006). A bivalent chromatin structure marks key developmental genes in embryonic stem cells. *Cell* **125,** 315–326.

Blackwell, T. K. (2004). Germ cells: Finding programs of mass repression. *Curr. Biol.* **14,** R229–R230.

Bowles, J., Knight, D., Smith, C., Wilhelm, D., Richman, J., Mamiya, S., Yashiro, K., Chawengsaksophak, K., Wilson, M. J., Rossant, J., Hamada, H., and Koopman, P. (2006). Retinoid signaling determines germ cell fate in mice. *Science* **312,** 596–600.

Bowles, J., and Koopman, P. (2007). Retinoic acid, meiosis and germ cell fate in mammals. *Development* **134,** 3401–3411.

Brinster, R. L. (1974). The effect of cells transferred into the mouse blastocyst on subsequent development. *J. Exp. Med.* **140,** 1049–1056.

Byskov, A. G. (1974). Does the rete ovarii act as a trigger for the onset of meiosis? *Nature* **252,** 396–397.

Chambers, I., Silva, J., Colby, D., Nichols, J., Nijmeijer, B., Robertson, M., Vrana, J., Jones, K., Grotewold, L., and Smith, A. (2007). Nanog safeguards pluripotency and mediates germline development. *Nature* **450,** 1230–1234.

Chiquoine, A. D. (1954). The identification, origin, and migration of the primordial germ cells in the mouse embryo. *Anat. Rec.* **118,** 135–146.
Chang, D. H., Cattoretti, G., and Calame, K. L. (2002). The dynamic expression pattern of B lymphocyte induced maturation protein-1 (Blimp-1) during mouse embryonic development. *Mech. Dev.* **117,** 305–309.
Chuva de Sousa Lopes, S. M., Hayashi, K., Chovlin, T. C., Mifsud, W., Surani, M. A., and McLaren, A. (2008). X chromosome activity in mouse XX primordial germ cells. *PloS Genet.* **4,** e30.
Coulombre, J. L., and Russell, E. S. (1954). Analysis of the pleiotropism at the W-locus in the mouse. The effects of W and W^v substitution upon postnatal development of germ cells. *J. Exp. Zool.* **126,** 277–296.
DiNapoli, L., Batchvarov, J., and Capel, B. (2006). FGF9 promotes survival of germ cells in the fetal testis. *Development* **133,** 1519–1527.
Doitsidou, M., Reichman-Fried, M., Stebler, J., Koprunner, M., Dorries, J., Meyer, D., Esguerra, C. V., Leung, T., and Raz, E. (2002). Guidance of primordial germ cell migration by the chemokine SDF-1. *Cell* **111,** 647–659.
Dudley, B. M., Runyan, C., Takeuchi, Y., Schaible, K., and Molyneaux, K. (2007). BMP signaling regulates PGC numbers and motility in organ culture. *Mech. Dev.* **124,** 68–77.
Durcova-Hills, G., Ainscough, J. F. X., and McLaren, A. (2001). Pluripotent stem cells derived from migrating primordial germ cells. *Differentiation* **68,** 220–226.
Durcova-Hills, G., Wianny, F., Merriman, J., Zernicka-Goetz, M., and McLaren, A. (2003). Developmental fate of embryonic germ cells (egcs), *in vivo* and *in vitro*. *Differentiation* **71,** 135–141.
Durcova-Hills, G., Adams, I. R., Barton, S. C., Surani, M. A., and McLaren, A. (2006). The role of exogenous FGF-2 on the reprogramming of primordial germ cells into pluripotent stem cells. *Stem Cells* **24,** 1441–1449.
Eddy, E. M., and Hahnel, A. C. (1983). Establishment of the germ cell line in mammals. *In* "Current Problems in Germ Cell Differentiation" (A. McLaren and C. Wylie, Eds.), *Brit. Soc. Dev. Biol. Symp.* **7**: pp. 41–69.
Extavour, C. G., and Akam, M. (2003). Mechanisms of germ cell specification across the metazoans: Epigenesis and preformation. *Development* **130,** 5869–5884.
Gardner, R. L. (1977). Developmental potency of normal and neoplastic cells of the early mouse embryo. *In* "Birth Defects, Excerpta Medica International Congress, Ser. 432. Excerpta Medica" (J. W. Littlefield and J. de Grouchy, Eds.), Amsterdam, pp. 154–166.
Gomperts, M., Garcia-Castro, M., Wylie, C., and Heasman, J. (1994). Interactions between primordial germ cells play a role in their migration in mouse embryos. *Development* **120,** 135–141.
Ginsburg, M., Snow, M. H., and McLaren, A. (1990). Primordial germ cells in the mouse embryo during gastrulation. *Development* **110,** 521–528.
Guan, K., Nayernia, K., Maier, L. S., Wagner, S., Dressel, R., Lee, J. H., Nolte, J., Wolf, F., Li, M., Engel, W., , and Hasenfuss, G. (2006). Pluripotency of spermatogonial stem cells from adult mouse testis. *Nature* **440,** 1199–1203.
Hajkova, P., Erhardt, S., Lane, N., Haaf, T., El-Maarri, O., Reik, W., Walter, J., and Surani, M. (2002). Epigenetic reprogramming in primordial germ cells. *Mech. Dev.* **117,** 15.
Hajkova, P., Ancelin, K., Waldmann, T., Lacoste, N., Lange, U. C., Cesari, F., Lee, C., Almouzni, G., Schneider, R., and Surani, M. A. (2008). Chromatin dynamics during epigenetic reprogramming in the mouse germ line. *Nature* **452,** 877–881.
Hayashi, K., and Matsui, Y. (2006). *Meisetz*, a novel histone tri-methyltransferase, regulates meiosis-specific epigenesis. *Cell Cycle* **5,** 615–620.
Illmensee, K., and Mintz, B. (1976). Totipotency and normal differentiation of single teratocarcinoma cells cloned by injection into blastocysts. *Proc. Natl. Acad. Sci. USA* **73,** 549–553.

Jamieson, R. V., Zhou, S. X., Tan, S. S., and Tam, P. P. (1997). X-chromosome inactivation during the development of the male urogenital ridge of the mouse. *Int. J. Dev. Biol.* **41,** 49–55.

Juliano, C. E., Voronina, E., Stack, C., Aldrich, M., Cameron, A. R., and Wessel, G. M. (2006). Germ line determinants are not localized early in sea urchin development, but do accumulate in the small micromere lineage. *Dev. Biol.* **300,** 406–415.

Kanatsu-Shinohara, M., Inoue, K., Lee, J., Yoshimoto, M., Ogonuki, N., Miki, H., Baba, S., Kato, T., Kazuki, Y., Toyokuni, S., Toyoshima, M., Niwa, O., et al. (2004). Generation of pluripotent stem cells from neonatal mouse testis. *Cell* **119,** 1001–1012.

Kanatsu-Shinohara, M., Lee, J., Inoue, K., Ogonuki, N., Miki, H., Toyokuni, S., Ikawa, M., Nakamura, T., Ogura, A., and Shinohara, T. (2008). Pluripotency of a single spermatogonial stem cell in mice. *Biol. Reprod.* **78,** 681–687.

Kaneda, M., Okano, M., Hata, K., Sado, T., Tsujimoto, N., Li, E., and Sasaki, H. (2004). Essential role for *de novo* DNA methyltransferase *Dnmt3a* in paternal and maternal imprinting. *Nature* **429,** 900–903.

Karl, J., and Capel, B. (1995). Three-dimensional structure of the developing mouse genital ridge. *Philos. Trans. R. Soc. Lond.* **350,** 235–242.

Kato, Y., Kaneda, M., Hata, K., Kumaki, K., Hisano, M., Kohara, Y., Okano, M., Li, E., Nozaki, M., and Sasaki, H. (2007). Role of the Dnmt3 family in de novo methylation of imprinted and repetitive sequences during male germ cell development in the mouse. *Hum. Mol. Genet.* **16,** 2272–2280.

Kedde, M., Strasser, M. J., Boldajipour, B., Vrielink, J. A., Slanchev, K., le Sage, C., Nagel, R., Voorhoeve, P. M., van Duijse, J., Orom, U. A., Lund, A. H., Perrakis, A., et al. (2007). RNA-binding protein DND1 inhibits microrna access to target mrna. *Cell* **131,** 1273–1286.

Kehler, J., Tolkunova, E., Koschorz, B., Pesce, M., Gentile, L., Boiani, M., Lomelí, H., Nagy, A., McLaughlin, K. J., Schöler, H. R., and Tomilin, A. (2004). *Oct4* is required for primordial germ cell survival. *EMBO Rep.* **5,** 1078–1083.

Knaut, H., Werz, C., Geisler, R., and Nusslein-Volhard, C. (2003). A zebrafish homologue of the chemokine receptor *Cxcr4* is a germ-cell guidance receptor. *Nature* **421,** 279–282.

Ko, K., and Scholer, H. R. (2006). Embryonic stem cells as a potential source of gametes. *Semin. Reprod. Med.* **24,** 322–329.

Kobayashi, S., Yamada, M., Asaoka, M., and Kitamura, T. (1996). Essential role of the posterior morphogen nanos for germline development in *Drosophila*. *Nature* **380,** 708–711.

Köprunner, M., Thisse, C., Thisse, B., and Raz, E. (2001). A zebrafish nanos-related gene is essential for the development of primordial germ cells. *Genes Dev.* **15,** 2877–2885.

Koubova, J., Menke, D. B., Zhou, Q., Capel, B., Griswold, M. D., and Page, D. C. (2006). Retinoic acid regulates sex-specific timing of meiotic initiation in mice. *Proc. Natl. Acad. Sci. USA* **103,** 2474–2479.

Koshimizu, U., Taga, T., Watanabe, M., Saito, M., Shirayoshi, Y., Kishimoto, T., and Nakatsuji, N. (1996). Functional requirement of gp130-mediated signaling for growth and survival of mouse primordial germ cells in vitro and derivation of embryonic germ (EG) cells. *Development* **122,** 1235–1242.

Lane, N., Dean, W., Erhardt, S., Hajkova, P., Surani, A., Walter, J., and Reik, W. (2003). Resistance of iaps to methylation reprogramming may provide a mechanism for epigenetic inheritance in the mouse. *Genesis* **35,** 88–93.

Lange, U. C., Saitou, M., Western, P. S., Barton, S. C., and Surani, M. A. (2003). The fragilis interferon-inducible gene family of transmembrane proteins is associated with germ cell specification in mice. *BMC Dev. Biol.* **3,** 1.

Lange, U. C., Adams, D. J., Lee, C., Barton, S., Schneider, R., Bradley, A., and Surani, M. A. (2008). Normal germline establishment in mice carrying a deletion of the Ifitm/Fragilis gene family cluster. *Mol. Cell Biol.* 2008 doi:10.1128/MCB. 00272–08.

Labosky, P. A., Barlow, D. P., and Hogan, B. L. M. (1994). Mouse embryonic germ (EG) cell lines: Transmission through the germline and differences in the methylation imprint of insulin-like growth factor 2 receptor (*Igf2r*) gene compared with embryonic stem (ES) cell lines. *Development* **120,** 3197–3204.

Lawson, K. A., and Hage, W. J. (1994). Clonal analysis of the origin of primordial germ cells in the mouse. *Ciba Found. Symp.* **182,** 68–84; Discussion 84–91.

Lawson, K. A., Dunn, R. R., Roelen, B. A., Zeinstra, L. M., Davis, A. M., Wright, C. V., Korving, J. P., and Hogan, B. L. (1999). Bmp4 is required for the generation of primordial germ cells in the mouse embryo. *Genes Dev.* **13,** 424–436.

Maatouk, D. M., Kellam, L. D., Mann, M. R., Lei, H., Li, E., Bartolomei, M. S., and Resnick, J. L. (2006). DNA methylation is a primary mechanism for silencing postmigratory primordial germ cell genes in both germ cell and somatic cell lineages. *Development* **133,** 3411–3418.

MacLean, G., Li, H., Metzger, D., Chambon, P., and Petkovich, M. (2007). Apoptotic extinction of germ cells in testes of *Cyp26b1* knockout mice. *Endocrinology* **148,** 4560–4567.

Matsui, Y., and Okamura, D. (2005). Mechanisms of germ-cell specification in mouse embryos. *Bioessays* **27,** 136–143.

Matsui, Y., Zsebo, K., and Hogan, B. L. M. (1992). Derivation of pluripotential embryonic stem cells from murine primordial germ cells in culture. *Cell* **70,** 841–847.

McLaren, A., and Lawson, K. A. (2005). How is the mouse germ-cell lineage established? *Differentiation* **73,** 435–437.

McLaren, A., and Southee, D. (1997). Entry of mouse embryonic germ cells into meiosis. *Dev. Biol.* **187,** 107–113.

Menke, D. B., Koubova, J., and Page, D. C. (2003). Sexual differentiation of germ cells in XX mouse gonads occurs in an anterior-to-posterior wave. *Dev. Biol.* **262,** 303–312.

Molyneaux, K. A., Stallock, J., Schaible, K., and Wylie, C. (2001). Timelapse analysis of living mouse germ cell migration. *Dev. Biol.* **240,** 488–498.

Molyneaux, K. A., Zinszner, H., Kunwar, P. S., Schaible, K., Stebler, J., Sunshine, M. J., O'Brien, W., Raz, E., Littman, D., Wylie, C., and Lehmann, R. (2003). The chemokine SDF1/CXCL12 and its receptor CXCR4 regulate mouse germ cell migration and survival. *Development* **130,** 4279–4286.

Namekawa, S. H., Park, P. J., Zhang, L. F., Shima, J. E., McCarrey, J. R., Griswold, M. D., and Lee, J. T. (2006). Postmeiotic sex chromatin in the male germline of mice. *Curr. Biol.* **16,** 660–667.

Nayernia, K., Nolte, J., Michelmann, H. W., Lee, J. H., Rathsack, K., Drusenheimer, N., Dev, A., Wulf, G., Ehrmann, I. E., Elliott, D. J., Okpanyi, V., Zechner, U., et al. (2006). In vitro-differentiated embryonic stem cells give rise to male gametes that can generate offspring mice. *Dev. Cell* **11,** 125–302.

Nesterova, T. B., Mermoud, J. E., Hilton, K., Pehrson, J., Surani, M. A., McLaren, A., and Brockdorff, N. (2002). *Xist* expression and macroh2a1.2 localisation in mouse primordial and pluripotent embryonic germ cells. *Differentiation* **69,** 216–225.

Nieuwkoop, P. D. (1947). Experimental observations on the origin and determination of the germ cells, and on the development of the lateral plates and germ ridges in the urodeles. *Arch. Neerl. Zool.* **8,** 1–205.

Ohinata, Y., Payer, B., O'Carroll, D., Ancelin, K., Ono, Y., Sano, M., Barton, S. C., Obukhanych, T., Nussenzweig, M., Tarakhovsky, A., Saitou, M., and Surani, M. A. (2005). Blimp1 is a critical determinant of the germ cell lineage in mice. *Nature* **439,** 207–213.

Ozdzenski, W. (1967). Observations on the origin of primordial germ cells in the mouse. *Zoologica Pol.* **17,** 367–379.

Papaioannou, V. E., McBurney, M. W., Gardner, R. L., and Evans, M. J. (1975). Fate of teratocarcinoma cells injected into early mouse embryos. *Nature* **258,** 70–73.

Papaioannou, V. E., Gardner, R. L., McBurney, M. W., Babinet, C., and Evans, M. J. (1978). Participation of cultured teratocarcinoma cells in mouse embryogenesis. *J. Embryol. Exp. Morphol.* **44,** 93–104.

Payer, B., Saitou, M., Barton, S. C., Thresher, R., Dixon, J. P., Zahn, D., Colledge, W. H., Carlton, M. B., Nakano, T., and Surani, M. A. (2003). Stella is a maternal effect gene required for normal early development in mice. *Curr. Biol.* **13,** 2110–2117.

Pellas, T. C., Ramachandran, B., Duncan, M., Pan, S. S., Marone, M., and Chada, K. (1991). Germ-cell deficient (*gcd*), an insertional mutation manifested as infertility in transgenic mice. *Proc. Natl. Acad. Sci. USA* **88,** 8787–8791.

Pesce, M., Wang, X., Wolgemuth, D. J., and Schöler, H. R. (1998). Differential expression of the *Oct-4* transcription factor during mouse germ cell differentiation. *Mech. Dev.* **71,** 89–98.

Resnick, J. L., Bixler, L. S., Cheng, L., and Donovan, P. J. (1992). Long-term proliferation of mouse primordial germ cells in culture. *Nature* **359,** 550–551.

Ross, A., Munger, S., and Capel, B. (2007). *Bmp7* regulates germ cell proliferation in mouse fetal gonads. *Sex Dev.* **1,** 127–137.

Saitou, M., Barton, S. C., and Surani, M. A. (2002). A molecular programme for the specification of germ cell fate in mice. *Nature* **418,** 293–300.

Sato, M., Kimura, T., Kurokawa, K., Fujita, Y., Abe, K., Masuhara, M., Yasunaga, T., Ryo, A., Yamamoto, M., and Nakano, T. (2002). Identification of *Pgc7*, a new gene expressed specifically in preimplantation embryos and germ cells. *Mech. Dev.* **113,** 91–94.

Schöler, H. R., Dressler, G. R., Balling, R., Rohdewohld, H., and Gruss, P. (1990). *Oct-4*: A germline specific transcription factor mapping to the mouse t-complex. *EMBO J.* **9,** 2185–2195.

Seki, Y., Hayashi, K., Itoh, K., Mizugaki, M., Saitou, M., and Matsui, M. (2005). Extensive and orderly reprogramming of genome-wide chromatin modifications associated with specification and early development of germ cells in mice. *Dev. Biol.* **278,** 440–458.

Seki, Y., Yamaji, M., Yabuta, Y., Sano, M., Shigeta, M., Matsui, Y., Saga, Y., Tachibana, M., Shinkai, Y., and Saitou, M. (2007). Cellular dynamics associated with the genome-wide epigenetic reprogramming in migrating primordial germ cells in mice. *Development* **134,** 2627–2638.

Shovlin, T. C., Durcova-Hills, G., Surani, A., and McLaren, A. (2008). Heterogeneity in imprinted methylation patterns of pluripotent embryonic germ cells derived from pre-migratory mouse germ cells. *Dev. Biol.* **313,** 674–681.

Stevens, L. C. (1973). A new inbred subline of mice (129-tersv) with a high incidence of spontaneous congenital testicular teratomas. *J. Natl. Cancer Inst.* **50,** 235–242.

Stewart, C. L., Gadi, I., and Bhatt, H. (1994). Stem cells from primordial germ cells can reenter the germ line. *Dev. Biol.* **161,** 626–628.

Subramaniam, K., and Seydoux, G. (1999). *NOS-1* and *NOS-2*, two genes related to *Drosophila nanos*, regulate primordial germ cell development and survival in *Caenorhabditis elegans*. *Development* **126,** 4861–4871.

Sun, X., Meyers, E. N., Lewandoski, M., and Martin, G. R. (1999). Targeted disruption of *Fgf8* causes failure of cell migration in the gastrulating mouse embryo. *Genes Dev.* **13,** 1834–1846.

Suzuki, A., and Saga, Y. (2008). *Nanos2* suppresses meiosis and promotes male germ cell differentiation. *Genes Dev.* **22,** 430–435.

Suzuki, A., Tsuda, M., and Saga, Y. (2007). Functional redundancy among NANOS proteins and a distinct role of *Nanos2* during male germ cell development. *Development* **134,** 77–83.

Tada, M., Tada, T., Lefebvre, L., Barton, S. C., and Surani, M. A. (1997). Embryonic germ cells induce epigenetic reprogramming of somatic nucleus in hybrid cells. *Embo J.* **16**, 6510–6520.

Tada, T., Tada, M., Hilton, K., Barton, S. C., Sado, T., Takagi, N., and Surani, M. A. (1998). Epigenotype switching of imprintable loci in embryonic germ cells. *Dev. Genes Evol.* **207**, 551–561.

Tada, M., Takahama, Y., Abe, K., Nakatsuji, N., and Tada, T. (2001). Nuclear reprogramming of somatic cells by *in vitro* hybridization with ES cells. *Curr. Biol.* **11**, 1553–1558.

Takada, Y., Isono, K., Shinga, J., Turner, J. M., Kitamura, H., Ohara, O., Watanabe, G., Singh, P. B., Kamijo, T., Jenuwein, T., Burgoyne, P. S., and Koseki, H. (2007). Mammalian Polycomb *Scmh1* mediates exclusion of Polycomb complexes from the XY body in the pachytene spermatocytes. *Development* **134**, 579–590.

Takeuchi, Y., Molyneaux, K., Runyan, C., Schaible, K., and Wylie, C. (2005). The roles of FGF signaling in germ cell migration in the mouse. *Development* **132**, 5399–5409.

Tam, P. P., and Zhou, S. X. (1996). The allocation of epiblast cells to ectodermal and germ-line lineages is influenced by the position of the cells in the gastrulating mouse embryo. *Dev. Biol.* **178**, 124–132.

Tanaka, S. S., and Matsui, Y. (2002). Developmentally regulated expression of *Mil-1* and *Mil-2*, mouse interferon-induced transmembrane protein like genes, during formation and differentiation of primordial germ cells. *Gene Expr. Patterns* **2**, 297–303.

Tanaka, S. S., Toyooka, Y., Akasu, R., Katoh-Fukui, Y., Nakahara, Y., Suzuki, R., Yokoyama, M., and Noce, T. (2000). The mouse homolog of Drosophila *Vasa* is required for the development of male germ cells. *Genes Dev.* **14**, 841–853.

Tanaka, S. S., Nagamatsu, G., Tokitake, Y., Kasa, M., Tam, P. P., and Matsui, Y. (2004). Regulation of expression of mouse interferon-induced transmembrane protein like gene-3, Ifitm3 (mil-1, fragilis), in germ cells. *Dev. Dyn.* **230**, 651–659.

Tanaka, S. S., Yamaguchi, Y. L., Tsoi, B., Lickert, H., and Tam, P. P. (2005). IFITM/Mil/fragilis family proteins, IFITM1 and IFITM3, play distinct roles in mouse primordial germ cell homing and repulsion. *Dev. Cell* **6**, 745–756.

Tang, H., Ross, A., and Capel, B. (2008). Expression and functional analysis of *Gm114*, a putative mammalian ortholog of *Drosophila bam*. *Dev. Biol.* **318**, 73–81.

Tsuda, M., Sasaoka, Y., Kiso, M., Abe, K., Haraguchi, S., Kobayashi, S., and Saga, Y. (2003). Conserved role of nanos proteins in germ cell development. *Science* **301**, 1239–1241.

Turner, J. M., Aprelikova, O., Xu, X., Wang, R., Kim, S., Chandramouli, G. V., Barrett, J. C., Burgoyne, P. S., and Deng, C. X. (2004). BRCA1, histone H2AX phosphorylation, and male meiotic sex chromosome inactivation. *Curr. Biol.* **14**, 2135–2142.

Turner, J. M., Mahadevaiah, S. K., Fernandez-Capetillo, O., Nussenzweig, A., Xu, X., Deng, C. X., and Burgoyne, P. S. (2005). Silencing of unsynapsed meiotic chromosomes in the mouse. *Nat. Genet.* **37**, 41–47.

Wang, Y., Zayas, R. M., Guo, T., and Newmark, P. A. (2007). *Nanos* function is essential for development and regeneration of planarian germ cells. *Proc. Natl. Acad. Sci. USA* **104**, 5901–5906.

Western, P., Maldonado-Saldivia, J., van den Bergen, J., Hajkova, P., Saitou, M., Barton, S., and Surani, M. A. (2005). Analysis of *Esg1* expression in pluripotent cells and the germline reveals similarities with *Oct4* and *Sox2* and differences between human pluripotent cell lines. *Stem Cells* **23**, 1436–1442.

Western, P. S., Miles, D. C., van den Bergen, J. A., Burton, M., and Sinclair, A. H. (2008). Dynamic regulation of mitotic arrest in fetal male germ cells. *Stem Cells* **26**, 339–347.

Yabuta, Y., Kurimoto, K., Ohinata, Y., Seki, Y., and Saitou, M. (2006). Gene expression dynamics during germline specification in mice identified by quantitative single-cell gene expression profiling. *Biol. Reprod.* **75,** 705–716.

Yamaguchi, S., Kimura, H., Tada, M., Nakatsuji, N., and Tada, T. (2005). *Nanog* expression in mouse germ cell development. *Gene Expr. Patterns.* **5,** 639–646.

Ying, Y., Qi, X., and Zhao, G. Q. (2001). Induction of primordial germ cells from murine epiblasts by synergistic action of *Bmp4* and *Bmp8b* signaling pathways. *Proc. Natl. Acad. Sci. USA* **98,** 7858–7862.

Ying, Y., and Zhao, G. Q. (2001). Cooperation of endoderm-derived BMP2 and extraembryonic ectoderm- derived BMP4 in primordial germ cell generation in the mouse. *Dev. Biol.* **232,** 484–492.

Yoshimizu, T., Obinata, M., and Matsui, Y. (2001). Stage-specific tissue and cell interactions play key roles in mouse germ cells specification. *Development* **128,** 481–490.

Youngren, K. K., Coveney, D., Peng, X., Bhattacharya, C., Schmidt, L. S., Nickerson, M. L., Lamb, B. T., Deng, J. M., Behringer, R. R., Capel, B., Rubin, E. M., Nadeau, J. H., and Matin, A. (2005). The *Ter* mutation in the dead end gene causes germ cell loss and testicular germ cell tumours. *Nature* **435,** 360–364.

Yao, H. H., DiNapoli, L., and Capel, B. (2003). Meiotic germ cells antagonize mesonephric cell migration and testis cord formation in mouse gonads. *Development* **130,** 5895–5902.

Zamboni, L., and Upadhyay, S. (1983). Germ cell differentiation in mouse adrenal glands. *J. Exp. Zool.* **228,** 173–193.

CHAPTER SEVEN

The Neuroendocrine Control of Sex-Specific Behavior in Vertebrates: Lessons from Mammals and Birds

Margaret M. McCarthy[*] and Gregory F. Ball[†]

Contents

1. Introduction: General Principles and the Scope of the Chapter	214
2. Historical Overview	216
2.1. Hormones and the organizational/activational hypothesis	217
2.2. The importance of aromatization	219
2.3. Discovery of morphological sex differences in the vertebrate brain	221
3. Reproductive Behavior in Mammals	222
3.1. Sexual behavior—masculinization, feminization, and defeminization	222
3.2. Parental behavior	226
3.3. Aggression	228
4. Reproductive Behaviors in Birds	231
4.1. Organization of sex-typical reproductive behaviors in birds	232
4.2. Organization of parental and aggressive behaviors in birds	233
4.3. Aggression	234
4.4. Sexual differentiation of sex-typical sexual behaviors in quail	234
4.5. Brain sex differences in quail	236
4.6. Sex differences in the brain and behavior of zebra finches	237
4.7. Sexual differentiation of brain and behavior in zebra finches: Hormone and cell autonomous mechanisms	238
5. Unanswered Questions and Future Directions	241
References	242

[*] Departments of Physiology and Psychiatry, University of Maryland School of Medicine, Baltimore, Maryland 21201
[†] Department of Psychological and Brain Sciences, Johns Hopkins University, Baltimore, Maryland 21218

Current Topics in Developmental Biology, Volume 83   © 2008 Elsevier Inc.
ISSN 0070-2153, DOI: 10.1016/S0070-2153(08)00407-9   All rights reserved.

1. Introduction: General Principles and the Scope of the Chapter

The complex question of how the brain controls behavior is somewhat simplified by restricting the discussion to sex-specific behaviors, but the general task at hand remains daunting. Almost all behavior is influenced, in some way, by the sex of the individual producing it, but only a relatively small cohort of the entire behavioral repertoire is actually specific to one sex or the other. Not surprisingly, these sex-specific behaviors are most often related to reproduction (Kelley, 1988). Reproduction is initiated by the onset of courtship behaviors that are exhibited in a sex-biased fashion as one sex tries to attract members of the opposite sex for gamete transfer and therefore, can often be markedly dimorphic (Kelley and Brenowitz, 2002). Prominent examples of these sex-biased or sexually dimorphic courtship behaviors include singing by male birds, male-to-male combat in ungulates, visual displays often involving colorful appendages by lizards and fish among others. These mating dances can be variable in length and occur in fits and starts, but eventually if reproduction is successful, are followed by the act of copulation or in the case of species that rely on external fertilization, gamete release. In the case of copulation, the transfer of sperm from the male to the female usually imposes a physical constraint of some sort that requires the execution of distinct behavioral responses that are stereotyped, sex-typical as well as species-typical in nature. This general view of the organization of sexual behavior can be thought of as distinguishing courtship behaviors from copulatory behaviors *per se*. In traditional ethological terms (e.g., Tinbergen, 1951), this distinction has been described in more general terms as distinguishing between appetitive and consummatory aspects of a motivated behavior. This appetitive/consummatory distinction was adopted to explain systematic variation in the occurrence and frequency of certain behaviors. Some species-typical behaviors result in a functional outcome that is associated with a reduction in motivation (consummatory responses) while other more variable behaviors allow an individual to converge on this functional outcome (appetitive responses; Timberlake and Silva, 1995). This sort of dichotomy is potentially problematic when one tries to apply it to complex sequences of behavior (Sachs, 2007). The concept should be used cautiously but the distinction continues to be useful to both ethologists and experimental psychologists in the elucidation of neuroendocrine mechanisms mediating a range of behaviors (Ball and Balthazart, 2008; Sachs, 2007). The study of sexual behavior is an area where this distinction has been particularly useful in guiding mechanistic studies (Beach, 1956; Pfaus *et al.*, 1999). In the case of sexual behavior, appetitive sexual behavior consists in searching for and approaching a potential mate whereas the consummatory component includes the actual contact between the sexes culminating in copulation (Pfaus et al., 1999). Thus, most courtship behaviors that function to attract a mate and stimulate them to bring them into a sexually receptive condition are appetitive behaviors and this is followed by the consummatory act of copulation or gamete release.

The underlying motivation to mate can also be markedly different as a function of sex in that males in many species exhibit little restraint, in that mating with multiple partners is often the best way for them to maximize their reproductive fitness, whereas in the case of females their probability of pursuing or even being willing to engage in sexual activity is influenced by a variety of variables including the time of ovulation, fitness of the partner, her current and predicted nutritional status, safety of the environment, and the availability of resources for ensuring that safety (i.e., nests, burrows, lack of predators, etc.). The ultimate reasons as to why there are such prominent sex differences in sexual motivation and strategies to maximize reproductive fitness are generally explained by theories of sexual selection (Andersson, 1994; Darwin, 1871). A critical idea is that differences in gamete size that define what is a male or a female set up a fundamental difference as to how one might maximize fitness. If one mating type produces many small gametes (called males) and another mating type produces a more limited number of large gametes (called females), then all things being equal, the sex producing the small gametes will prosper by fertilizing as many large gametes as possible while the sex producing a few large gametes will maximize fitness by choosing to mate with mates producing high quality gametes and insure that the zygotes are formed at times auspicious for the development and survival of the progeny (Bateman, 1948). However, all things are not always equal and there are many exceptions to Bateman's principle in animals and plants (Synder and Gowaty, 2007). In sex role reversal species, females compete to be chosen by males. Even in species with traditional sex differences in morphology consistent with male competition and female choosiness there are more and more cases coming to light documenting female promiscuity that is advantageous to the female. Hence, the dynamics of male–female interactions related to sexual reproduction remains an important area of research among evolutionary biologists (Roughgarden, 2004; Synder and Gowaty, 2007). Such discussions are beyond the scope of this chapter where we will focus exclusively on questions related to the control of sex differences in behavior.

Once successful mating is achieved, the subsequent parental behaviors, if they occur at all, may or may not continue to be constrained by sex. In egg-laying species, with parental care such as birds and many teleost species, the absence of a sex-specific lactation process allows for either sex to build and guard a nest or incubate the eggs and even feed and defend the hatchlings. In mammals, due to the fact that only females can gestate and lactate, the burden is almost entirely one-sided and this is generally reflected in only females engaging in nest building and maternal care, although there are exceptions. Scientists who define themselves as behavioral neuroendocrinologists endeavor to elucidate the rules governing the control of these divergent sex-specific behaviors and our goal in this chapter is to review the current state-of-the-art and unifying principles as well as areas for future exploration. A comprehensive review of this vast literature is beyond the scope of a single chapter; thus, we will focus on sex differences in the reproductive behavior of select species of mammals and birds and the ontogenetic processes that regulate the development of neural substrates controlling sex-specific sexual behaviors. A good discussion of sex differences in a wide range of species can be found in the volume edited by Short and Balaban (1994).

2. Historical Overview

The observation that the sexes behave differently is so self-evident, it does not warrant description as a discovery. However, the proximate mechanisms explaining as to why the sexes behave differently were not nearly as obvious. The importance of the gonads has been recognized since antiquity. For example, in his writings Aristotle made remarkably accurate descriptions of the detailed effects of castration on male-typical behaviors in a variety of vertebrate species including humans (see for review Wilson and Roehrborn, 1999). Discussions of the similarities in morphology and behavior between females and castrated males can also be found in the writings of Aristotle and his contemporaries (Mayhew, 2004). The realization that the physiological control of these morphological and behavioral traits differ between the sexes and are the result of variation in internal secretions was not clarified until the classic experiments of Berthold (1849) on roosters in which he transplanted testes to castrated roosters and thereby rescued the birds from the effects of castration. This study established that the testis release a factor in the blood that regulates male-typical morphology (the cock's comb) and behavior (crowing, aggression, and copulatory behavior). This pioneering work was followed by studies in the early twentieth century that utilized approaches such as injecting the extract from testis of one animal into another and observing the effect of such manipulations on sex typical behavior and morphology (Jost, 1947; Lillie, 1916). Subsequent work that was focused on behavior continued to emphasize gonadal secretions, and in particular started to elucidate the role of gonadal steroid hormones *per se* (Beach, 1948). However, the emphasis at that time was on how adult steroid levels can be regulated by environmental variables such as day length, temperature, food and mate availability to influence sexual receptivity (Beach, 1948). The way in which adult behavioral patterns were constrained or modulated by early developmental effects or genetic sex, though hypothesized to play a role (Steinach, 1940) was not well understood. Much emphasis was placed on the gonad driving the appropriate behavioral response. While it is true that the gonad and the associated steroidogenesis are critical components of adult sex differences in behavior, and that the environment is an important modulating influence, we now know that development events are critical and there is an emerging view that genetic differences independent of the gonad might matter as well.

2.1. Hormones and the organizational/ activational hypothesis

Every field has its iconic papers and in the field of behavioral neuroendocrinology, it is without question the seminal 1959 report by Phoenix, Goy, Gerall, and Young in which they first postulated what came to be known as the Organizational/Activational Hypothesis of hormone action (Phoenix *et al.*, 1959). Building on their own and previous observations they proposed and then empirically determined that sex-specific gonadal steroids act on the developing brain to organize the neural substrate, which is in turn acted on by sex-specific gonadal steroids in the adult to induce the appropriate reproductive behaviors (Fig. 7.1). This general principle has stood the test of time and generalization across species, but has been refined and adjusted for different endpoints and circumstances. Parameters that vary across species are the timing of organizational events (i.e., entirely prenatal vs perinatal vs entirely postnatal), the magnitude of the differentiation between males and females, endpoints that are differentiated, and the relative importance of estrogens versus androgens. Parameters that vary within species but across endpoints also include the precise timing of events, which steroid is mediating the organizational effects, and the cellular mechanisms evoked to induce steroid-mediated differentiation.

The Organizational/Activational Hypothesis of hormone action is a useful heuristic model that provides a strong empirical framework from which questions of sex differences in the brain are addressed. Accelerating interest in sex differences research has prompted the development of how-to guides to assist in the proper design of experiments so that clear distinctions can be made between organizational versus activational effects of steroid hormones (Becker *et al.*, 2005; Greenspan *et al.*, 2007). However, like any dogma, the Organizational/Activational Hypothesis has been subject to abuse as well as challenge. It has been considered a foregone conclusion that subtle sex differences in complex traits such as cognition and emotionality are organized early in life, but there is in reality, little data to support this view (see McCarthy and Konkle, 2005). The concept of additional variables beyond early life exposure to gonadal steroids, impacting on adult sex differences in behavior is beginning to gain acceptance but

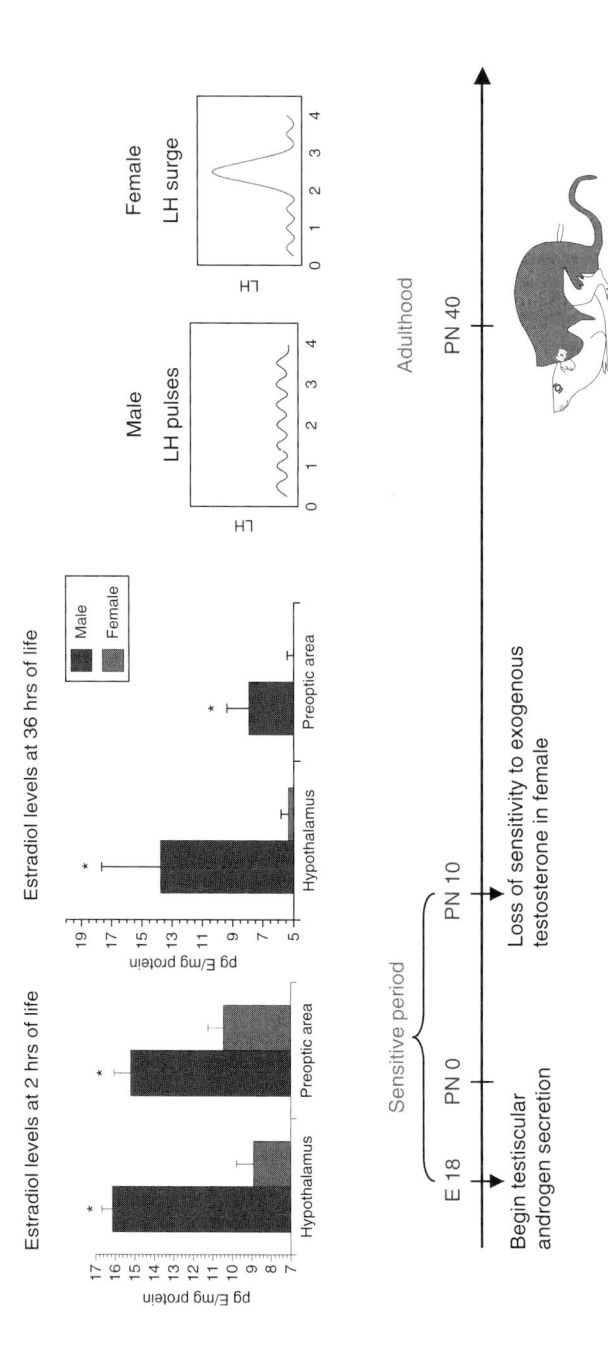

Figure 7.1 Organizational/activational hypothesis of hormone action on the developing brain. A well established dogma in the field of brain sexual differentiation is the concept that hormonal effects early in development organize the neural architecture and that this differentiated substrate is then activated in adulthood by the sex-specific hormonal milieu. Organizational effects occur during a restricted sensitive period, which in rodents is operationally defined as the onset of androgen synthesis by the fetal male testis and the loss of sensitivity in females to exogenous hormone administration after the first week to ten days of life. In rodents, the aromatization of testosterone to estradiol within neurons results in newborn males having 2- to 3-fold higher levels of estradiol in the preoptic area and hypothalamus than females. It is estradiol, as opposed to testosterone that initiates the cellular events leading to masculinization. Two major endpoints are sexually differentiated, or organized, in the rodent: (1) sexual behavior, such that females show lordosis and males mount and thrust, and (2) control of gonadotropin secretion resulting in pulsatile release of LH in males and a surge of LH to induce ovulation in females.

requires further development. Two important emerging principles include the potential contribution of genetic sex differences (Arnold, 2004) and a role for local steroidogenesis within subregions of the brain (Hojo et al., 2004) that may allow steroids to act more in the nature of neurotransmitters than hormones (Balthazart and Ball, 2006).

2.2. The importance of aromatization

The synthesis of steroid hormones is distinct from that of many other signaling molecules on many levels including that they are made on demand, never stored. The initial step in steroid synthesis requires transport of cholesterol out of the blood stream and into the cell, where it is converted into pregnenolone, a large 4-ringed molecule with an extensive carbon chain off the D ring. This precursor, which has biological activity itself, provides the substrate for all other steroids. Androgens and estrogens are refined products that require 6–9 enzymatic conversions. Estradiol, the most potent and prevalent of the estrogens, is synthesized directly from testosterone via a single enzymatic reaction involving the aromatization of the A ring by the p450 enzyme, aromatase. Testosterone also serves as the immediate precursor to the potent androgen, dihydrotestosterone via a reduction reaction of the 5-carbon (Fig. 7.2). The biological activity of dihydrotestosterone versus estradiol is profoundly different as they act via distinct nuclear receptors, the androgen receptor, and at least two isoforms of estrogen receptor. There is little to no cross reactivity between the receptors and they exert their actions via distinct palindromic response elements on the DNA or membrane anchored signaling molecules. Even the coactivators associated with androgen versus estrogen receptors are distinct (see for review Auger, 2004).

Early studies of the organizational effects of gonadal steroids on sexual differentiation of brain and behavior focused on the role of the so-called "male" steroids, androgens. Thus, pregnant dams or newborn female rat pups were treated with testosterone or dihydrotestosterone to observe if the brain was masculinized. Estradiol was included as a control steroid, predicted to have no effect since it was considered a "female" hormone. Surprisingly, estradiol was even more potent at masculinizing the brains of neonatal females than was testosterone or dihydrotesterone, the latter being notably weak or without effect. This suggested that estrogens, not androgens were the masculinizing hormone. But there was a problem, the level of estradiol in the maternal circulation is so high that there would be no means for varying the amount of exposure between male and female fetuses. Two pieces of information were key to solving the mystery. First was that the fetal circulation contains high levels of a steroid binding globulin, α-fetoprotein, which has a high affinity for estradiol. When estradiol is bound to this globulin, it is sequestered in the blood stream and therefore largely incapable of gaining access to the interior of neurons. Second was the

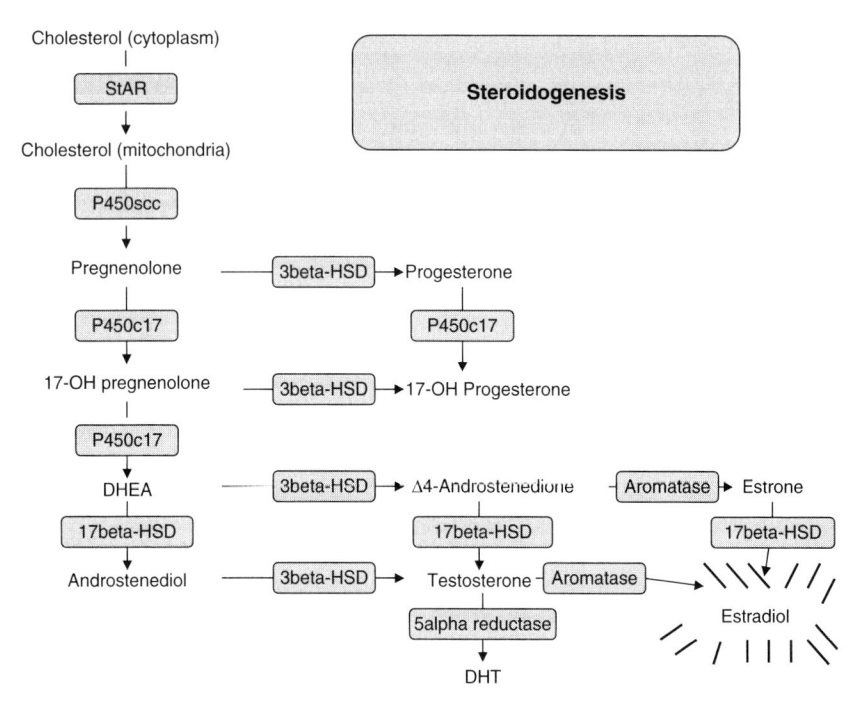

Figure 7.2 Steroidogenic pathway and the importance of aromatization. The synthesis of steroids begins with the transport of cholesterol into mitochondria via StAR (steroid acute regulatory protein). Many of the steroid metabolizing enzymes is located on the inner membrane of the mitochondria, whereas other are found in the cytoplasm, resulting in a complex compartmentalization of synthesis of particular steroids. Androgens and estrogens are refined products of a series of up to a dozen or more enzymatic conversions, with estradiol being a direct by-product of testosterone. The discovery of high concentrations of the aromatase enzyme in the brains of neonatal animals contributed to the elaboration of the Aromatization Hypothesis, the notion that estrogens, not androgens, mediate many of the cellular events critical in sexual differentiation of the brain. Recently, there has been increasing evidence indicating that androgens and estrogens can be synthesized entirely within neurons and therefore not dependent upon gonadal secretions for the necessary precursors. The importance of this process to regionally specific brain development and the establishment of sex differences outside of the diencephalon is an active area of current research.

observation that neurons in the neonatal brain express high levels of aromatase and are thus capable of synthesizing estradiol from testicularly derived androgens. Importantly, α-fetoprotein has little affinity for testosterone and so it can move freely from the fetal circulation into the brain to be locally aromatized to estradiol (reviewed in McCarthy, 2008).

Aromatization is now known to be a critical process in the sexual differentiation of the rodent brain, as well as in multiple bird species. This insight provides us the important step of knowing the first signaling molecule in the sequence of events that will lead to permanent organization. Recent work has been able to push beyond the estrogen receptor to begin to elucidate the complex sequelae of cellular events that will determine cell fate and morphology. The local synthesis of estradiol from peripheral androgens continues in the adult at a lesser rate but nonetheless importantly contributes to the expression of sexual behavior in males. However, the relative importance of estrogens to differentiation varies by brain region within as well as across species. To-date, no clear role for estrogens have been identified in the sexual differentiation of the primate brain. Instead, this appears to be an entirely androgen driven process.

2.3. Discovery of morphological sex differences in the vertebrate brain

After it was discovered in the early twentieth century that sex steroid hormones regulate many male-typical and female-typical reproductive behaviors, it was hypothesized that sex differences in plasma concentrations of steroids such as the androgens and the estrogens were responsible for these sex differences in behavior (e.g., Steinach, 1940; see review of these early ideas in Beach, 1948). It subsequently became clear though that gonadectomizing males and females and administering male-typical or female-typical concentrations of steroids was not sufficient to reverse sex differences in behavior in many cases (Beach, 1948). These types of results indicated that the male and female brain responds to steroid hormones differently and also that the brain areas involved in mediating the action of sex steroid hormones in the implementation of sexually dimorphic behaviors might even be morphologically different. As reviewed previously, the reality of the differential responsiveness to steroids of the male and female brain was demonstrated experimentally by Phoenix et al. (1959) in their study of steroid regulation of sex differences in reproductive behaviors in guinea pigs. Adult female guinea pigs can not be induced to produce male-typical sexual behaviors with androgen treatment. However, they found that by administering female guinea pigs high concentrations of androgens during the perinatal period, they could masculinize their brains so that the females would respond to exogenous testosterone with the activation of male-typical sexual behaviors. Phoenix et al. (1959) argued that the perinatal action of testosterone served to organize the brain in a sexually dimorphic manner and the adult action of sex steroid hormones served to activate sex-typical behaviors in the appropriate manner. The question raised by this research was what sort of differences might be present in the male and female brain that would be responsible for sex differences in the response to steroids and in the activation of various sex-typical behaviors related to reproduction.

Initial studies focused on rats and identified rather subtle sex differences. For example, Pfaff (1966) and Dörner and Staudt (1969) reported sex differences in cell nuclear size. In a very influential study, Raisman and Field (1973) investigated at the electron microscope level the types of synaptic inputs to the preoptic area that were present after lesion of the striatal inputs. They found more dendritic spine synapses in females as compared to males (Raisman and Field, 1973). These types of sex differences required very careful histological analyses to be revealed. It was therefore quite surprising to neuroscientists in the 1970s when Nottebohm and Arnold (1976b) described dramatic differences in brain nuclear volume among the song control nuclei of zebra finches and canaries. This was quickly followed by the description of the sexually-dimorphic nucleus in the rat preoptic region that also exhibits a remarkable male-biased difference in volume (Gorski et al., 1978). The emphasis on sex differences in the volume of particular brain regions continued and was extended to humans. A sexually dimorphic nucleus has also been described in the hypothalamus of humans, being larger in males (Swaab and Fliers, 1985), and reported by others to be smaller in homosexual men than heterosexual men (LeVay, 1991). However, in neither the rat nor the human has a functional significance been attributed to the sexually dimorphic nucleus, leaving unanswered the question of why it is larger in males. Some insight may be gained from another animal model, the domestic sheep. Approximately 10% of domestic rams show a sexual interest exclusively in other males, and interestingly these male-oriented rams have a female sized SDN (Perkins and Roselli, 2007). Thus interest in volumetric sex differences continues. However, an equally robust and as important of a sex difference can also be found in the pattern of synaptic connections within particular brain regions. Heroic studies conducted by Arai and his colleagues (Matsumoto, 2000) using quantitative EM on laboratory rats revealed that males have 2- to 3-fold more dendrtic spine synapses as females in some brain regions, while the opposite might be true in other regions. These findings have been confirmed by additional methods, such as Golgi-Cox impregnation, and advances on understanding the mechanistic basis of this sex difference are being made (see the following section). However, there is no information currently available regarding sex differences in synaptic profiles in primates, including humans, nor does there appear to be any information available on what is rapidly becoming a preferred animal model, the mouse. Given the obvious importance of sex differences in the density of synapses in a particular region, this is a gap in our knowledge that needs to be filled.

3. Reproductive Behavior in Mammals

3.1. Sexual behavior—masculinization, feminization, and defeminization

Analogous to the undifferentiated gonad that is destined to become an ovary in the absence of the Sry gene, the immature brain also is undifferentiated and destined to take on a female phenotype in the absence of hormonal induction of masculinization. The feminized brain is defined as one that supports the expression of female sexual receptivity, maternal behavior towards young and positive feedback to estradiol to regulate the release of gonadotropins from the pituitary to initiate ovulation. The masculinized brain is defined as one that supports the expression of male sexual behavior, is commonly ambivalent towards young and insensitive to the positive feedback effects of estradiol. However, the full compliment of male brain development also involves the process of defeminization, which is the active removal of the capacity to exhibit female sexual behavior (Fig. 7.3). If defeminization fails or is incomplete, it is possible to have an adult male that shows species typical mounting and thrusting towards receptive females but that if given the proper hormonal milieu will also adopt a sexually receptive posture in response to soliciting males. In rodents, defeminization is a distinct process from masculinization but is also actively induced by gonadal steroids, most prominently estradiol (Kudwa *et al.*, 2005).

The brain regions and cellular mechanisms that mediate defeminization are distinct from those controlling masculinization. The control of male sexual behavior in all species examined to-date involves the preoptic area (POA) of the diencephalon. A region so-named for its location posterior to the optic chiasm, not due to any role in the control of vision. The dendritic morphology of POA neurons is profoundly different in male versus female rats in which males have a 2- to 3-fold greater density of excitatory synapses found on dendritic spines. This sex difference is entirely determined during the perinatal sensitive period by the actions of locally aromatized estradiol from the testicularly secreted androgens. Early attempts to elucidate the cellular mechanism(s) by which estradiol permanently organizes the brain to control adult male sex behavior focused on the obvious candidates, neurotransmitters. Multiple studies investigated the possible role of the classic transmitters, acetycholine, dopamine, serotonin, and so on, but results were largely inconclusive in that disrupting any one of a number of neurotransmitter systems exerted deleterious effects on adult male sexual behavior but there was no clear role for any in the normal process of brain masculinization. It was generally concluded that estradiol must be acting through a multitude of systems simultaneously and that the combined effects were so complex as to have eluded our ability to clearly differentiate them. However, recent studies have required a rethinking of this view in light of the surprising discovery that the key mediator of brain masculinization is not a neurotransmitter at all, but is instead an entirely different class of signaling molecules, the prostaglandins. Prostaglandins are membrane lipid derived molecules synthesized from arachadonic acid by the cyclooxegenase enzymes, COX-1 and COX-2. Short-lived metabolites of arachadonic acid are further refined into a series of prostanoids and thromboxanes by additional enzymatic conversions. Prostaglandin E2 (PGE2) is one of the

Figure 7.3 Feminization, masculinization, and defeminization. The brain begins as bipotential. In rodents, and most mammals, the default pathway is to develop as female via a process denoted as feminization but which remains poorly understood. The discovery that the masculinizing hormone, estradiol (E2) decreases the protein FAK (focal adhesion kinase) and its associated signaling molecule, paxillin, suggests that this may be a part of the feminization process which includes reduction in the length and branching of dendrites on neurons in the hypothalamus. Masculinization is an active process in which E2 increases the transcription of the enzyme COX-2 which directly leads to increased production of the prostaglandin, PGE2, which initiates a series of cellular events that increases the density of dendritic spines synapses on the neurons of the male preoptic area which control male sexual behavior. An additional process, defeminization, is required for normal male development. The function of defeminization is to remove the capacity for the expression of female sexual behavior and in this case appears to involve a nongenomic activation of PI3 Kinase, which stimulates the release of glutamate from a presynaptic terminal which then acts on the postsynaptic neuron to induce the formation of dendritic spine synapses. This process occurs in the mediobasal hypothalamus and is independent of events occurring in the preoptic area. One of the great challenges ahead is to form a coherent picture in which the various mechanisms of estradiol action on the developing brain are understood in relation to each other.

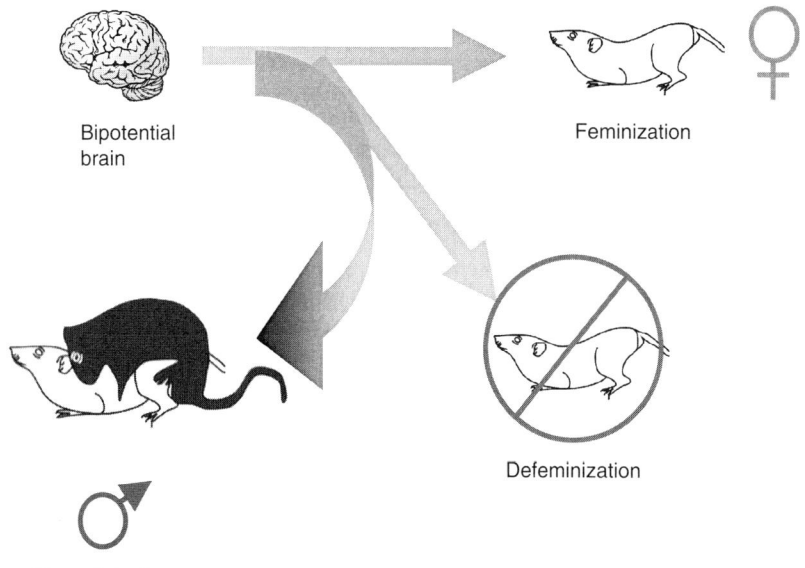

most biologically active prostanoids and exerts wide ranging effects throughout the body including the initiation of fever, closing of the ductus arteriosus, bone formation, and secretion of LHRH among others (see for review Narumiya et al., 1999; Sugimoto and Narumiya, 2007; Ushikubi et al., 2000). Receptors for PGE2 are found on neurons and glia, and the COX enzymes are also localized to these cell types throughout the brain but are at particularly high levels in the preoptic area. PGE2 induces glutamate release from astrocytes, a subtype of glia, and thereby can modulate neuronal activity. Newborn male rats have higher levels of COX-2, the inducible form of the enzyme, and estradiol potently up regulates its expression when given to females. Increased COX-2 is directly yoked to increased PGE2 and following estradiol treatment there is a 7-fold increase in levels of this prostanoid in the developing POA. Via a mechanism that is not entirely understood but at least partially involves activation of glutamate receptors, PGE2 induces the formation of dendritic spine synapses on developing neurons of the POA. The increase in spine synapses neonatally is directly and positively correlated with masculinization of behavior as an adult. The density of dendritic spines induced by PGE2 during the perinatal sensitive period endures into adulthood and is independent of adult circulating hormone levels (Amateau and McCarthy, 2002, 2004). Thus PGE2 organizes the neural substrate for masculine sexual behavior and this is then activated by androgens in adulthood. The effects of PGE2 are specific to masculinization; however, since newborn females treated with PGE2 to induce masculinization are not defeminized. This is most effectively demonstrated by treating a PGE2-masculinized female with testosterone before testing her for male sexual behavior, and then treating her with estradiol and progesterone and testing her for female sexual behavior. These animals will show perfectly normal male or female sexual behavior depending on the adult hormonal milieu, revealing that the neural circuits for both male and female sex behavior are intact. Conversely, males that are deprived of endogenous PGE2 neonatally by inhibition of COX-2, are not masculinized, meaning they show no male sex behavior as adults, but they are completely defeminized by their own gonadal steroids developmentally (Todd et al., 2005). These animals are entirely asexual, exhibiting neither male nor female sexual behavior. This raises the interesting question of the cellular mechanism of defeminization.

The POA is the major brain region controlling male sexual behavior and the ventromedial nucleus (VMN) of the mediobasal hypothalamus is the major brain region controlling female sexual behavior. There is also a profound sex difference in the dendritic morphology of neurons of the VMN, with males again having more spine synapses than females but this is in large part because the dendrites are longer and branch more frequently in males (Mong et al., 1999; Todd et al., 2007). In this brain region, the sex difference in neuronal morphology is not determined by prostaglandins but instead involves a nongenomic activation of PI3 Kinase by estradiol and enhanced release of glutamate to promote synaptogenesis (Schwarz et al., 2008). Distinguishing whether estradiol-induced changes in neuronal morphology are relevant to defeminization versus masculinization is complicated by our lack of knowledge regarding the normal course of events for feminization. Given that masculinization and defeminization are both induced processes initiated by estradiol, it is reasonable to assume that this steroid would induce gene expression and new proteins. If, on the other hand, estradiol suppresses the expression of a particular gene, it can be speculated that this gene might be a component of the feminization process. One such candidate gene is focal adhesion kinase (FAK) and its associated protein, paxillin, both of which are found at higher levels in the newborn female VMN than the males and are reduced by estradiol treatment (Speert et al., 2007). Both FAK and paxillin are negative regulators of dendritic growth and branching and suppressing the levels of each might be essential for the longer dendrites observed in males. The ability to directly tie the morphological changes induced by glutamate release and FAK/paxillin expression in the developing VMN to adult behavior is complicated by the intertwined nature of feminization, defeminization, and masculinization. Based on behaviors observed in transgenic mice, some have argued that there is no process of defeminization and that the neural circuitries for male and female sexual behavior coexist in the brains of all animals but are differentially suppressed by olfactory input (Kimchi et al., 2007). In the absence of clearly discernible circuits for either behavior, this remains a topic for debate.

3.2. Parental behavior

As discussed briefly above, parental behavior in mammals is constrained by the occurrence of lactation in females, thereby precluding the direct feeding of offspring by fathers. The social organization and mating strategies of all mammalian species are influenced by this division of labor, but not all species solve the problem in the same way. A relatively common strategy is group living females that raise their young in a cooperative fashion with a dominant male siring the majority of offspring. In rodents, this is called a deme, and in larger mammals such as lions or ungulates it might be called a harem. Males are generally tolerated in the group until reaching sexual maturity at which point they leave or are forced out to seek their fortunes elsewhere. An important distinction is whether the groups form and disband with the mating season or are stable throughout the year. Regardless, in both structures, the males do little, if any, parental care.

The hormonal and neural control of maternal behavior has been well characterized in the laboratory rat (Numan, 1994). Females that have not given birth will generally ignore newborns, or may even kill them as they are regarded with fear. Sequential increases in estradiol and progesterone, followed by a precipitous withdrawal at birth are essential triggers for the onset of a series of behaviors that include retrieving pups to a nest, licking and grooming them, and finally crouching to allow for nursing. In the rabbit, hormonal changes during late pregnancy sequentially regulate the gathering of nest material, its formation into a nest, the loosening of abdominal fur, and finally the plucking of this fur by the expectant female to line the now prepared nest (Gonzalez-Mariscal, 2001). The gonadal steroids of pregnancy act both directly on neurons of the maternal behavior neural circuit, beginning in the preoptic area, but also to induce the synthesis and release of critical neuropeptides, the most prominent being oxytocin and prolactin. Both of these neurohormones not only play essential roles in the process of parturition, breast development, and milk letdown, but also in coordinating the behavioral repertoire that is so critical for reproductive success.

Laboratory mice were considered a poor model for investigating the variables regulating maternal behavior due to spontaneous expression of pup retrieval by all females, regardless of reproductive experience. This is in notable contrast to mice trapped in the wild and then allowed to breed in a laboratory setting. Not surprisingly, these mice will breed quite readily under the highly favorable conditions of the lab, but the behavior towards pups of virgin mice, or even multiparous mice that have not recently given birth, is decidedly nonmaternal. The majority of females will commit infanticide toward any pups they encounter, even if the female is in the very late stages of pregnancy. But upon delivering their own young, the previously infanticidal female now exhibits complete and appropriate maternal care (McCarthy, 1990). This remarkable behavioral plasticity was lost in the laboratory mouse, most likely due to extreme genetic selection. The advent of transgenic mice bearing null mutations for select genes has rekindled interest in maternal behavior in rodents following the unexpected generation of extreme phenotypes in which newly parturient females completely ignore their own young. Initial excitement and claims of having discovered *the* maternal gene have been tempered by the frequency with which the phenotype occurs. The Jackson Laboratories catalogue of mouse lines currently lists almost 30 knockout strains which suffer from impaired maternal behavior, requiring the investigator to conduct costly cross fostering or heterozygous matings if the line is to be maintained. Why the behavioral sequelae required for appropriate maternal care has proven so fragile is unknown.

The study of sex differences in maternal behavior has not gained much attention given that males do not go through pregnancy or its associated hormonal changes. It has been argued by some that the neuroendocrine basis of sex differences in maternal care in rats is not the result of an ontogenetic sexual differentiation process but rather the result of sex differences in the experiences of endogenous hormone secretions and the appropriate stimuli in adulthood (Kelley, 1988). However, there is some evidence that the ability to express maternal behavior is organized developmentally in a similar fashion to that of sexual behavior (Lonstein and De Vries, 2000), but if this is true, it is still unclear whether the female phenotype is the default and the male induced, or vice versa. There are some species in which males do show active parental care and this has provided valuable insight to understand the function of specific neuroanatomical sex differences. Prairie voles are a monogamous species in which animals form a strong pair bond after mating, and males spend as much time on the nest and protecting the young as the female (although she is still the source of nutrients). Voles also have a prominent sex difference in the vasopressin neural network, with males showing much denser and extensive vasopressinergic projections in the septal region. Converging evidence suggests these vasopressin neurons regulate the parental behavior exhibited by males, leading to the speculation that this system has evolved in order for the sexes to be more similar as opposed to different (De Vries, 2004). This novel viewpoint is changing the way many sex differences in brain and behavior are interpreted and reminds us of the ease at which we sometimes ignore the obvious when guided by our preconceptions.

3.3. Aggression

Sex differences in aggression would seem something so self-evident as to be hardly worth mentioning. Just look at the any schoolyard, athletic field, or prison; the preponderance of males engaged in aggressive behaviors is overwhelming. The importance of testosterone as the driving force behind this sex difference seems equally self-evident. Both of these are true, males are more aggressive than females and androgens increase the frequency and intensity of aggressive encounters. This is best illustrated in seasonally breeding animals in which testosterone levels fluctuate widely, and male-to-male aggressive interactions increase dramatically as males compete for mates, territory, or other limiting resources relevant to reproductive success, and then drop precipitously with falling testosterone levels. But it is a bit disingenuous to call this a sex difference in aggression since the entire constellation of variables dictating a species specific mating strategy confines the behavioral repertoire of males and females such that it is advantageous for males to compete and females not to. Similarly, maternal aggression, a particularly vicious form of aggression, occurs only in females, only in response to certain threats and only during a restricted postpartum period when offspring are vulnerable. This can hardly be considered a sex difference in aggression since only females give birth, and it clearly illustrates that the capacity for exhibiting high levels of aggression exists in females. Displays of aggression are further constrained by variables extrinsic to the individual such as the social structure, availability of resources, artificial conditions associated with laboratory settings, etc.

Despite the many variables influencing expression of aggressive behavior, questions can still be asked regarding organizational versus activational hormonal control. Unlike sexual behavior that involves distinct motor patterns in males versus females, aggression is expressed in the same way by both sexes, but it is the frequency and intensity of the expression that varies. A naturally occurring source of variability in prenatal exposure to gonadal steroids provides some of the best evidence that at least some component of adult expressions of aggression are organized during early life. Species with large litter sizes, such as rats and mice, have a bifurcate uterus that allows fetuses to be packed in as peas in a pod. Because of the copious quantities of androgens produced by late gestation male fetuses, a female fetus that finds herself sandwiched between two males in the uterine horn experiences a distinctly different endocrine environment from a female that develops between two sisters. By delivering the rat or mice pups via ceasarean section, the relative position of a fetus in relation to other fetuses can be noted, the individual marked and then development monitored. Very modest changes in the external genitalia of females developing between males instead of females confirm that hormonal exposure has been different. When animals are monitored through to adulthood, it is apparent that females that developed between male littermates are more likely to exhibit aggression, and their responsiveness to the effects of androgen treatment on inducing aggression is also greater (vom Saal, 1984). This phenomenon is referred to as the intrauterine position effect or IUP, and has also been used to explain subtle individual variability in sexual and social behaviors (Ryan and Vandenbergh, 2002; Fig. 7.4). But in many ways what the IUP effect is best at illustrating is the tremendous sensitivity of the developing brain to even subtle perturbations in the hormonal milieu.

While there is good evidence for organizational effects of hormones on adult expression of aggression, there is equally good evidence that it is a relatively weak effect. This stems in part from observations of the importance of adult circulating hormone levels, which are tightly linked to fluctuating aggression in seasonal breeders, but can also be associated with changes in the perception of and response to a social threat in nonseasonal breeders, including rodents and humans. Anabolic steroids lower the threshold to respond to provocation in a laboratory setting, and cortical regions associated with aggression and impulsivity disorders show greater activation in females treated with testosterone compared to placebo (Bernhardt et al., 1998; Hermans et al., 2008; McGinnis, 2004). But the hegemony of testosterone begins to break down when one tries to correlate individual serum levels with behavior. This is in part because as with all steroids, testosterone levels vary throughout the day, and relative levels of free to bound hormone can impact access to the brain, something we still cannot assess accurately. Perhaps more importantly, behavior itself has an enormous influence on circulating androgen levels. Individuals who

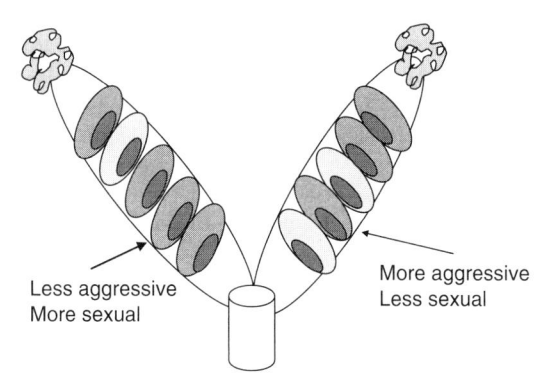

Figure 7.4 Impact of intrauterine position on the adult behavioral phenotypes. Rodents possess a bifurcate uterus in which fetuses are packed as peas-in-a-pod. Because the process of sexual differentiation begins *in utero*, female fetuses packed between two male siblings are exposed to the copious quantities of testosterone produced by their brothers as it leachs from the amniotic sac and into the contiguous bloodstream. Females that develop between other females experience a different hormonal milieu as a result of not being contiguous to males. The relative differences in hormone levels experienced by female littermates developing under these varying circumstances are very slight, but the impacts can be considerable and help to demonstrate the exquisite sensitivity of the developing brain to perturbations in the hormonal milieu. For instance, females developing between males are noticeably more aggressive and show reduced female sexual receptivity as adults compared to females that develop between females.

experience victory in a competitive situation will show a significant and sustained increase in serum testosterone, while those experiencing defeat show a decrease. Remarkably, even just observing ones favorite sports team win or lose can alter hormonal profiles as demonstrated by tracking salivary steroid content in fans watching a World Cup soccer match (Bernhardt et al., 1998). Even winning a contest that is purely a matter of chance and involves only a small reward ($5) will increase serum testosterone in young men (McCaul et al., 1992). However, woman do not respond to chance victories with an increase in serum testosterone, suggesting an interesting sex difference in parameters associated with reward value, dominance and obtainment, and defense of resources. Children as young as 5 years old exhibit sex differences in aggression, and while these do not neatly correlate with androgens, there is an effect in each sex, with androstenedione being a better predictor of provocation in males and testosterone of affectivity in females (Azurmendi et al., 2006).

A second major variable dissociating early organization actions of androgens and adult aggressive behavior is the importance of genetics. The notion that genetic differences originating with the X or Y chromosome might contribute to sex differences in behavior is a relatively new one (Arnold, 2004), but the empirical evidence that genetics is important to variability in aggression is long standing and unquestioned (Carlier *et al*., 1990). Interestingly, much of the attention on genetic variation is on how it influences an individual's sensitivity to steroids, in particular androgens and estrogens. Both the relative impact of organizational effects and the activational effects of androgens are substantially influenced by genetics when assessed across mouse strains (Sandnaabba *et al*., 1994), and no doubt contributes to individual variability within strains.

In summary, aggression is a sexually dimorphic trait not because it takes a different form in males versus females, but rather because it is expressed at different levels of intensity and frequency and can be evoked in response to distinct stimuli. Normally docile, or at least civil, females can become viciously aggressive when under the hormonal grip of early postpartum and lactation, but only when the perceived threat is toward their young, not themselves. Males are on average more likely to be aggressive than females, and this is a product of both early exposure to androgens during a perinatal sensitive period and the impact of testosterone on reactive thresholds as an adult. But prior experience, both immediate and long-term, combined with genetic predispositions, can introduce far more variability in responsiveness than hormone levels. Thus, sex differences in aggression are not so easily attacked from a mechanistic standpoint as those seen for reproductive behavior. Carrying this point a bit further, the common assumption that male aggressive and sexual behaviors share many of the same neurobiological underpinnings was challenged by examining sensitivity to pharmacological manipulation of each response and the pattern of neuronal activation following either an aggressive or sexual encounter. While there were several common brain areas activated in response to both fighting and, well, fornicating, there were also many distinct regions associated with aggression, such as the posterodorsal medial amygdala and premammillary hypothalamus. Moreover, the two behavioral programs had seperate and opposite responses to pharmacological manipulation of serotoninergic neurotransmission (Veening *et al*., 2005), thus its not surprising that the proximate factors regulating the behaviors differ as well.

4. Reproductive Behaviors in Birds

Studies of birds have raised intriguing complications that challenge broad theories of the development of sex differences based solely on research in mammals. Avian studies of sex differentiation have focused on two different model systems: gallinaceous birds as exemplified by Japanese quail (*Coturnix japonica*) and song birds as exemplified by zebra finches (*Taeniopygia guttata*). Studies of Japanese quail have in many respects confirmed and also broadened principles established in rodents. Work on zebra finches suggests that quite different principles are needed to explain the ontogeny of sex differences in this species.

4.1. Organization of sex-typical reproductive behaviors in birds

In males, sexual behavior consists in searching for and approaching a potential mate (i.e., appetitive components) whereas the consummatory component includes the actual contact between the sexes culminating in copulation (Ball and Balthazart, 2002, 2008). Appetitive male sexual behaviors include most courtship behaviors that function to attract females and stimulate them to bring them into a sexually receptive condition. In birds, males are known to exhibit a wide diversity of visual and vocal displays that function in this manner (Armstrong, 1947). As expected there is a substantial amount of species variability. In some cases, stereotyped movements involving elaborate plumage displays are used to attract and stimulate females. In other cases, vocal behavior is paramount and in many cases there is a combination of these two sorts of displays. What is clear is that communication involving other sensory modalities such as olfaction or touch is not particularly important. In female birds, ovarian cycles are not organized into periods of behavioral estrus as they are in mammals. However, females will solicit male copulation with specialized displays that will occur only during the breeding season when gonadal steroid hormone levels are high. Receptivity involves the behaviors exhibited just prior to and during copulation that are stimulated in response to the appropriate behavior by the male.

Field studies of avian female reproductive behavior have stressed the active role the female plays in soliciting male sexual behaviors and managing mating behavior in a way that is advantageous to the female's reproductive success (Gowaty, 1996). It is clear for example, in some species such as the dunnock (*Prunella modularis*) with a very complex mating system (some females will mate monogamously but others will mate polygynously or in a polyandrous fashion), that females will actively solicit extra-pair copulations in an apparently deceptive fashion (Davies, 1992). The complexity of female behaviors associated with copulation is illustrated by the fact that females in many species produce calls just prior to and during copulation (Birkhead and Moller, 1992). These calls are louder than would be expected if they were just communicating with their partner. It has been hypothesized that these calls signal to other males that the female is fertile and may also function as a solicitation of sorts for later extra-pair copulations that might be beneficial to the female's reproductive success.

The copulatory act itself requires that males mount females so that gamete transfer can be facilitated. In most avian species males lack an intromittent organ so gamete transfer involves what is sometimes referred to as the "cloacal kiss." These are cloacal contact movements that facilitate the deposition of sperm in the female. There are often other stereotyped motor patterns that precede cloacal contact movements *per se*. For example, in Japanese quail copulation consists of a sequence of stereotyped movements progressing from neck grab to mounts and cloacal contact movements (Adkins and Adler, 1972; Hutchison, 1978). In the select number of species that do have a penis-like intromittent organ, the copulatory act involves intromission as well. Many components of both appetitive and consummatory aspects of sexual behavior just summarized are sex-typical in organization in birds. Thus most studies on the sexual differentiation of behavior in birds have focused on sexual behaviors, either courtship activities such as song (studied in most detail in zebra finches) or sex-typical sexual behavior studied in the most detail in Japanese quail).

4.2. Organization of parental and aggressive behaviors in birds

The combination of universal oviparity and endothermy make it essential that all eggs receive some sort of attention after oviposition, although it is not necessarily the case that the newly hatched young will require any attention from its parents (Oring, 1982). Parental care, therefore, is a fundamental component of the life history of nearly all avian species. It appears to be a "primitive" life history characteristic that is a part of the group of morphological and behavioral adaptations that almost defines birds and differentiates them from their reptilian ancestors. Diversification of the class Aves has included patterns of parental behavior. Biparental care of some sort may be the primitive pattern and remains the most common (Lack, 1968; Silver *et al.*, 1985). But, the roles adopted by the sexes and the types of care provided to the offspring vary greatly among taxa. Although all birds lay eggs that require attendance of some sort for successful hatching, the type of posthatching care provided has coevolved with the mode of development of the young. Altricial development is most common in the class Aves though it appears to be the derived condition, precocial development being the primitive condition (Ricklefs, 1983). Altricial young are generally helpless at hatching, and therefore must be brooded and fed by their parents. Precocial young are somewhat independent at hatching and may only require supervision by the parents and no additional feeding. Given that biparental care is so widespread, the neuroendocrine basis of sex-specific parental care patterns has not been a focus of intensive research. It seems that males and females have the potential to perform all aspects of parental care including nest building, incubation, and progeny feeding and/or protection. Although the pattern of male and female sharing in these activities varies widely among taxa, the reasons for these sex differences may be related to more subtle effects of processes such as sex-specific responses to stimuli associated with responses to the eggs and young. The neuroendocrine basis of such differences is elusive. Even in so-called sex-reversed species such as the polyandrous sandpipers, the hormonal cycles of males and females are similar to what has been observed in species following more traditional parental care roles (Fivizzani and Oring, 1986; Schlinger et al., 1989).

4.3. Aggression

Male territorial aggression is a prominent feature of avian reproductive cycles (Wingfield et al., 1987) that at first glance might appear to be highly sex-typical. In temperate zone species, which have been studied in the most detail, male territorial aggression is very prominent and females appear to play little or no role in this aspect of the reproductive cycle. Male aggression in this context has been linked to testosterone (Wingfield et al., 1990). For example, the "challenge hypothesis," an influential idea developed based on studies of male birds has asserted that relationships between plasma concentrations of testosterone and aggression only become apparent when males are establishing territorial boundaries or dominance hierarchies. The proposed link between testosterone and aggression is most pronounced in temperate zone species (Wingfield et al., 1990). Comparative studies across a wide range of latitudes have found that in tropical species tesetosterone is almost undetectable in the blood despite the fact that there are prominent changes in the occurrence of aggressive behaviors (Hau et al., 2000). In species breeding in the far north, territorial aggression is temporally limited in scope due to the short breeding season and again the link is not always prominent (Hunt et al., 1995). Most studies of testosterone and territorial aggression have focused on males. This bias suggests that there might be a prominent sex difference in the occurrence of these behaviors and its hormonal control. However, as more work has concerned females, there is increasing evidence that this difference is more one of degree. For example, the administration of exogenous testosterone to male-darked juncos greatly increases male aggression (Ketterson et al., 1992), but it was also found that administering testosterone to female juncos increases female-typical aggression (Zysling et al., 2006). Thus sex differences in aggression in birds do not seem to necessarily involve prominent developmental differences relating to sexual differentiation of the brain early in ontogeny.

4.4. Sexual differentiation of sex-typical sexual behaviors in quail

Contrary to what is observed in rodents, in Japanese quail, female receptive behavior can be elicited in both males and females by an appropriate treatment with estrogens (Adkins, 1975); however, T cannot activate male-typical copulatory behavior (mount and cloacal contact movement) in females (Adkins, 1975; Balthazart et al., 1983). Thus, male-typical behavior is strongly differentiated in quail while female-typical behavior is dimorphic in rodents.

An extensive set of experimental data demonstrates that the sex difference in copulatory behavior of quail results mostly, if not exclusively, from the exposure of embryonic females to endogenous estradiol secreted by the ovary. As shown initially by Adkins-Regan (Adkins, 1975) injections of estrogens into male embryos produce adults that fail to mount and copulate in a male-typical fashion even if injected with high doses of T: these birds have acquired a female behavioral phenotype and are said to be demasculinized (Adkins-Regan, 1985a). This robust effect has been replicated many times (Schumacher et al., 1989). The effectiveness of estrogens is limited to a critical period during the ontogeny in that injections carried out after day 12 of incubation are ineffective (See Fig. 7.5). It was important to establish that this model of differentiation established based on exogenous hormone treatment was valid under physiological conditions. Radioimmunoassays of steroid hormones in the plasma of developing quail embryos of both sexes revealed that estradiol levels are indeed much higher in female than in male quail embryos during the entire period when estrogens are effective in demasculinizing females (Schumacher et al., 1988). The administration of aromatase inhibitors or antiestrogens that block estrogen production or action during the proposed critical period for the demasculinzing effects of estradiol were found to block the process of demasculinization in female embryos (Adkins, 1976; Balthazart et al., 1992a; Fig. 7.5). Males were unaffected.

Taken together, these studies are quite consistent with the model originally articulated by Adkins-Regan that proposes that masculine sexual behavior is lost in females quail under the influence of estrogens during the first two thirds of the incubation period. They also show that estrogens represent a necessary and sufficient stimulus responsible for the behavioral demasculinization of females because the blockade of their synthesis produces adult females that have a behavioral phenotype indistinguishable from that of normal males (Fig. 7.5). This differentiation process appears to be the opposite of the pattern observed in mammals (steroids secreted by the male induce the differentiation). This pattern of results correlates with the fact that females are homogametic in mammals (XX) while males are homogametic in birds (ZZ). Therefore in both groups, the phenotype of the homogametic sex develops in the (relative) absence of steroids (this is the so-called "neutral" sex) and the phenotype of the heterogametic sex is imposed by the early action of steroids. The homogametic sex is also "neutral" for morphological differentiation and these correlations between the patterns of morphological or behavioral differentiation on one hand and the sex chromosome type on the other hand appears to be common to most vertebrate classes, suggesting possible causality (Adkins-Regan, 1985b). Proximate mechanisms that might explain why this generalization occurs

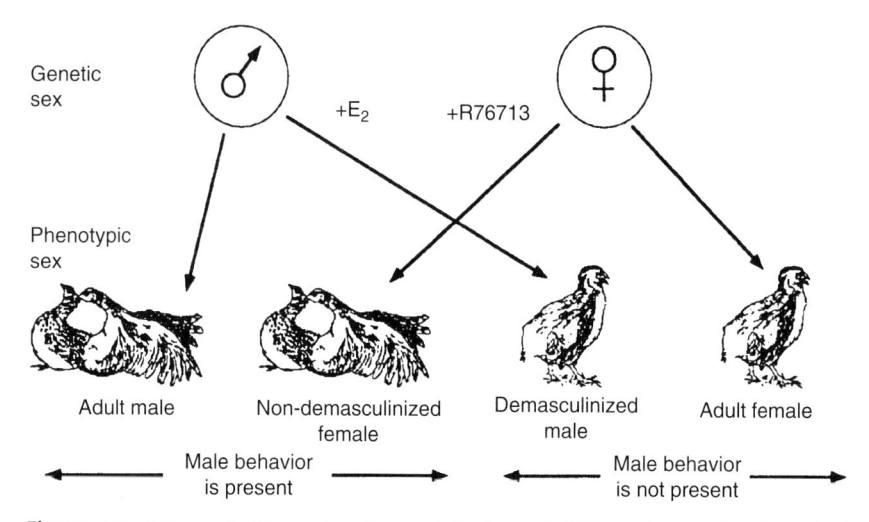

Figure 7.5 Schematic illustrating the model of sexual differentiation of male-typical sexual behavior in Japanese quail. Administering exogenous 17β estradiol of genotypic males leads to a demasculinization of behavior so that in adulthood such animals will not exhibit male-typical behavior even in the presence of high doses of testosterone. Administering the aromatase inhibitor R76713 (or Vorozole) in the egg will block the production of estradiol in the female embryo so that the demasculinization will not be induced. In adulthood these genotypic females will exhibit male-typical sexual behaviors in response to testosterone treatment. See text for more details. Adapted from Balthazart et al. (1996).

have not been identified to date. Nonetheless, it remains the case that taking into account the type of sex chromosomes present in males and females of birds and mammals allows one to formulate a single rule describing the sexual differentiation in both classes of vertebrates and therefore to generalize the rule that had been formulated with data derived from mammalian studies only. This rule does not generalize to all vertebrates because some species lack sex chromosomes and possess environmentally based sex determination mechanisms (Crews, 1993).

4.5. Brain sex differences in quail

These experiments on sex differentiation in quail clearly identified the ontogenetic process that leads to a form of androgen-insensitivity in female quail as reflected by a total absence of male-typical sexual behavior even when appropriate endocrine stimulation is provided. The proximate causes of this insensitivity to androgen that is present in adult females are, however, less clear; the brain mechanisms that control the sex differences in reproductive behavior of adult quail have not been formally identified so far. However, it is expected that sex differences in neurochemistry and/or neuroanatomy will be identified that mediate sex differences in the responsiveness to androgen. A number of morphological and neurochemical sex differences have been detected in the adult quail brain (for review: Balthazart and Adkins-Regan, 2002). However, most of these sex differences disappear when adult birds are placed in similar endocrine conditions. They presumably reflect a differential activation by steroids in adulthood, not an organizational effect of embryonic estrogens and they cannot be a sufficient explanation for the behavioral sex dimorphism which is still observed when gonadectomized males and females are submitted to the same replacement therapy with T.

A few brain differences have nevertheless been discovered that remain when the adult levels of steroids are controlled (e.g., neuronal size in the dorso-lateral part of the medial preoptic nucleus, POM: (Panzica *et al.*, 1991), aromatase activity in the preoptic area: (Schumacher and Balthazart, 1986) and dopamine turnover in the POM: (Balthazart *et al.*, 1992b), but there is no direct evidence that these differences directly control the behavioral sex dimorphism. We also do not know whether they result, like the dimorphism in behavior, from a differential exposure to estrogens of males and females during the embryonic period.

4.6. Sex differences in the brain and behavior of zebra finches

The oscine songbirds are well known for the production of elaborate vocalizations referred to as "songs" (Ball and Hulse, 1998). Song, in many species, is produced more often or even exclusively by males. One of the best-studied songbird species, that illustrates this sex difference, is the zebra finch (*Taeniopygia guttata*). Male zebra finches produce a stereotypic song while females never sing. The neuroendocrine traits mediating these sex differences in an adult behavior are relatively well understood. There are marked sex differences in the neural circuit controlling song learning and production (Arnold, 1992; Balthazart and Adkins-Regan, 2002; Nottebohm and Arnold, 1976a). The "song system" of zebra finches consists of a telencephalic component that includes four interconnected nuclei: HVC, the robust nucleus of the arcopallium (RA), area X of the medial striatum and the magnocellular nucleus of the anterior nidopallium (MAN). The circuit also includes midbrain and brainstem components that contain the nucleus intercollicularis complex and the tracheo-syringeal division of the nucleus of the XII motor nerve (nXIIts). Several telencephalic nuclei are substantially larger (3–5 times) in volume in males than in females (Arnold, 1992; Ball *et al.*, 1995; Nottebohm and Arnold, 1976a). Area X is not even observable in females. Another notable feature of these telencephalic nuclei is that many contain receptors for androgens (Arnold *et al.*, 1976; Balthazart *et al.*, 1992c). This appears to be an exception to the general rule that receptors for sex steroids are evolutionarily conserved in that they are present in the same limbic and diencephalic areas in most vertebrate groups (Arnold *et al.*, 1976). Furthermore, the very occurrence of these telencephalic nuclei appears to represent a specialization of the songbird brain. Other groups of birds that have very similar overall brain organization do not have the well-defined cellular groups that make up the song system (Ball, 1994; Kroodsma and Konishi, 1991) though other avian taxa that exhibit vocal learning such as parrots and hummingbirds do have brain areas similar in many respects (Jarvis, 2004).

Copulatory behavior is also sexually dimorphic in zebra finches (Adkins-Regan and Ascenzi, 1987). Testosterone does not activate mounting behavior and cloacal contact movements in gonadectomized adult females as it does in males. The neural basis of this sex difference in the action of T has not been investigated in finches, but presumably it involves similar neuroendocrine mechanisms as have been described in other birds.

4.7. Sexual differentiation of brain and behavior in zebra finches: Hormone and cell autonomous mechanisms

The mechanisms controlling the ontogeny of these sex differences have been the subject of intensive research over the last 20 years. Copulatory behavior in zebra finches differentiates through an endocrine process that appears to be similar to what has been described in quail. Perinatal treatment of males with estrogen demasculinizes the birds so that male-typical copulatory behavior is no longer observed in adulthood, even after the appropriate treatment with T (Adkins-Regan and Ascenzi, 1987). However, it was discovered in the early eighties that early treatment of females with estrogens masculinizes singing behavior and the associated song control nuclei: females that receive neonatal treatment with estrogen sing as adults and have larger song control nuclei than untreated females (Gurney and Konishi, 1980; Simpson and Vicario, 1991). Early estrogen action therefore appears to have two potentially incompatible effects in zebra finches: demasculinization of mounting behavior in males and masculinization of song in females. Because a male zebra finch normally sings and copulates while the female does not do either of these behaviors, it is difficult to understand how the same endocrine environment could lead to these opposite effects on two different male-typical behaviors (Adkins-Regan and Ascenzi, 1987, 1990). One obvious way to resolve this dilemma is to propose that there are two separate periods during ontogeny when estrogen produces these incompatible effects. This hypothesis was tested by administering estrogen to zebra finches at various ages during ontogeny. This approach failed to identify different sensitive periods for the masculinization of song or the masculinization of mounting: both behaviors were maximally affected by treatment with an estrogen during the first week after hatching (Adkins-Regan *et al.*, 1994).

Measurement of plasma concentrations of 17β-estradiol (E2) and T during ontogeny have not reliably detected a sex difference in the concentration of these hormones in zebra finches (see Balthazart and Ball, 1995 for a review). It has now been established that the brain of zebra finches contains high levels of aromatase activity and is therefore a potentially important source of endogenous estrogens. However, no sex difference in brain aromatase activity can be detected during ontogeny (Schlinger and Arnold, 1992). Similarly studies of the expression of the mRNA for aromatase and estrogen receptor α and β were not found to exhibit a sex specific pattern when investigated during embryonic development and just after hatching (Perlman and Arnold, 2003). A male-biased sex difference has been identified in the receptor coactivator L7/SPA in the developing zebra finch telencephalon (Duncan and Carruth, 2007). This is one of the few findings that could lead an endocrine signal that is different between the sexes and could plausibly act as the differentiating factor. It is always possible that the failure to identify other sex differences in steroid action could stem from technical inadequacies such as the lack of sensitivity of methods or unsatisfactory sampling regimens. However, the same methods have provided evidence for clear differences in other species, therefore this is somewhat unlikely.

A variety of experimental manipulations have been performed to test whether endogenous estrogen from whatever source is responsible for the masculinization of the vocal control system and/or the demasculinization of copulatory behavior in male zebra finches. None of these studies have clarified the paradoxical situation described above. Castration of young zebra finches, the administration of antiestrogens immediately after hatching (Mathews *et al.*, 1988) the administration of aromatase inhibitors to zebra finch nestlings (Balthazart *et al.*, 1994; Wade and Arnold, 1994; Wade *et al.*, 1994) have all been found to be ineffective. Thus all attempts to manipulate estrogen action during ontogeny in zebra finches using the traditional neuroendocrine pharmacological methods have been unsuccessful. Manipulation of estrogen synthesis early in embryonic development can sex reverse the development of the gonad. Wade, Arnold and colleagues (Wade and Arnold, 1996; Wade *et al.*, 1996) employed an aromatase inhibitor to block estrogen production in embryonic finches and induced the development of testicular tissue (along with ovarian tissue) in genetically female zebra finches. Despite the fact that this testicular tissue could produce androgens and partially masculinize peripheral structures such as the syrinx there was no effect on the development of neural song system (Wade and Arnold, 1996; Wade *et al.*, 1996).

These studies convincingly demonstrated that hormones of gonadal origin do not seem to be involved in the differentiation of the neural circuit controlling song. However, the brain itself can produce steroids *de novo* from cholesterol so the hormonal regulation of the brain could be regulated autonomously within the brain (London et al., 2006). It is also possible that sex specific gene expression in the brain could sculpt a neural circuit regulating dimorphic behavior without working via gonadal hormone actions (Arnold, 2004). There is evidence for both processes. Studies of cultured brain slices derived from juvenile male and female finches involving the administration and blockade of estrogen revealed that brain synthesized estrogen was necessary and sufficient for the development of certain aspects of the sexually dimorphic neural substrate controlling song (Holloway and Clayton, 2001). Sex differences in the brain in the expression of candidate genes related to neural development and not the result of sex differences in steroid hormone have been identified (Chen et al., 2005) thought not in all cases (Kim and Arnold, 2005). Perhaps the most convincing case of sex differences in gene expression in the brain was described in the gynandromorphic zebra finch (Agate et al., 2003). The gynandromorphic phenotype is an animal that is female on one side and male on the other side. It is a rare trait in birds but one such case was analyzed by Agate and colleagues (Agate et al., 2003; Fig. 7.6). This bird was striking because it had male plumage on one side and female plumage on the other. Its behavior was male-typical, including song. It has a testis on one side and an ovary on the other. Several genes that are Z or W chromosome specific were found

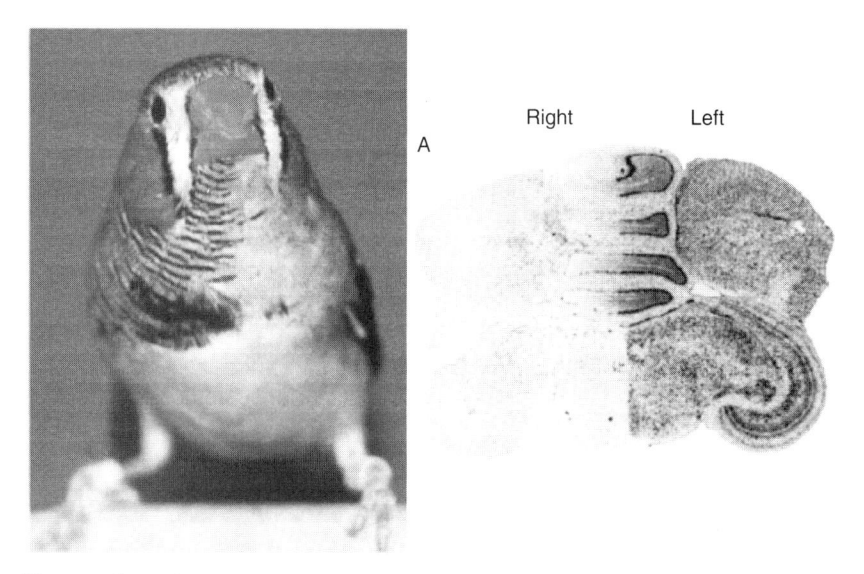

Figure 7.6 Left pane, a gynandromorphic zebra finch, male-typical plumage on the right side and female-typical plumage on the left side. Right pane, photomicrographs illustrating *in situ* hybridization histochemistry for Z and W specific probes. The mRNA encoding the W chromosome gene, ASW, exhibits a widespread distribution on the left side but is virtually absent on the right side. Note the strikingly sharp dividing line that follows the brain midline. Taken from Agate et al. (2003).

to be limited in expression to the male or female side of the brain (Fig. 7.6). The song nuclei tended to be male-like in morphology though there was a strong left-right asymmetry with song nuclei such as HVC being much larger on the male-side of the brain. These findings further support the conclusion that there is no role for hormone effects of gonadal origin in the development of these sex differences since both sides of the brain would have equal exposure to hormones of gonadal origin (Agate *et al.*, 2003). The fact that the male side of the brain is much larger than the female side of the brain is also consistent with the notion that cell autonomous differentiation occurring via genes specific to the Z or W chromosome (Agate *et al.*, 2003). However, the female side of the brain is also larger than one would expect from a female-typical brain. This finding suggests that a diffusible factor such as a hormone is produced within the brain that can also exert effects on sexual differentiation (Agate *et al.*, 2003). Thus in zebra finches the neural basis of sex-typical copulatory behaviors appear to be regulated by steroid hormone action of gonadal origin while in the brain cell autonomous and diffusible factors work together to regulate the sexually dimorphic circuit controlling male-typical song behavior.

5. Unanswered Questions and Future Directions

Substantial advances are being made on multiple fronts in our quest to understand the etiology and the functional significance of sex differences in the brain. These include identification of specific cellular mechanisms mediated by steroids and an emerging appreciation of the importance of genetics in contributing to adult phenotypic differences. The continued use of comparative approaches has allowed for the elucidation of common principles across vertebrates and could ultimately provide insight into the evolutionary basis for organizational sex differences as opposed to purely adult behavioral differences. Despite these advances, there remains much to be learned and there is an essential need for integration of findings from different approaches and different research groups. The current understanding is analogous to the six-blind-men-and-an-elephant approach, with each investigator independently focusing on their own specific part and describing it in extensive detail but failing to place it in the context of the entire elephant. We still do not understand how the complete neural circuitry controlling any sex difference in behavior is sexually differentiated in a vertebrate species. We know a great deal about specific links in the chain, but how changes in one link impact on the range of potential changes in another, is not known. An important emerging principle is that each link is differentiated by a distinct mechanism, in other words there is no uniform mechanism by which estrogens or androgens act across brain regions to induce masculinization, defeminization, or demasculinization. One ultimate consequence of the multitude of mechanisms is the enhanced potential for variable phenotypes. In other words, a male brain might be particularly strongly masculinized in response to one component of male behavior, such as response to olfactory or visual cues, but less so in response to another, say the motivation to mount or to sing. Similarly, females might vary independently in the sensitivity threshold for sexual behavior versus maternal behavior. Alternatively, these two endpoints might be connected in some way that we do not currently understand. Does sexual differentiation of one brain region preclude or direct the magnitude or type of differentiation that will occur in another? To answer this question we need a combined whole animal and cellular approach that will allow for integration of analyses across multiple brain regions and multiple behavioral and physiologic endpoints. We have not yet achieved this level of sophistication but can hope that the next generation of behavioral neuroendocrinologists will be prepared to meet the challenge and advance our understanding to achieve the goal of a full appreciation of the importance of sex differences in the brain.

REFERENCES

Adkins, E. K. (1975). Hormonal basis of sexual differentiation in the Japanese quail. *J. Comp. Physiol. Psychol.* **89,** 61–71.

Adkins, E. K. (1976). Embryonic exposure to an antiestrogen masculinizes behavior of female quail. *Physiol. Behav.* **17,** 357–359.

Adkins, E. K., and Adler, N. T. (1972). Hormonal control of behavior in the Japanese quail. *J. Comp. Physiol. Psychol.* **81,** 27–36.

Adkins-Regan, E. (1985a). Exposure of embryos to an aromatization inhibitor increases copulatory behaviour of male quail. *Behav. Proc.* **11,** 153–158.

Adkins-Regan, E. (1985b). Nonmammalian psychosexual differentiation. In "Handbook of Neurobiology" (N. T. Adler, R. Goy, and D. Pfaff, eds.), Vol. 7, pp. 43–76. Plenum Press, New York.

Adkins-Regan, E., and Ascenzi, M. (1987). Social and sexual behaviour of male and female zebra finches treated with oestradiol during the nestling period. *Anim. Behav.* **35,** 1100–1112.

Adkins-Regan, E., and Ascenzi, M. (1990). Sexual differentiation of behavior in the zebra finch: Effect of early gonadectomy or androgen treatment. *Horm. Behav.* **24,** 114–127.

Adkins-Regan, E., Mansukhani, V., Seiwert, C., and Thompson, R. (1994). Sexual differentiation of brain and behavior in the zebra finch: Critical periods for effects of early estrogen treatment. *J. Neurobiol.* **25,** 865–877.

Agate, R. J., Grisham, W., Wade, J., Mann, S., Wingfield, J. C., Schanen, C., Palotie, A., and Arnold, A. P. (2003). Neural, not gonadal, origin of brain sex differences in a gynadromorphic finch. *Proc. Natl. Acad. Sci. USA* **100,** 4873–4878.

Amateau, S. K., and McCarthy, M. M. (2002). A novel mechanism of dendritic spine plasticity involving estradiol induction of prostglandin-E2. *J. Neurosci.* **22,** 8586–8596.

Amateau, S. K., and McCarthy, M. M. (2004). Induction of PGE(2) by estradiol mediates developmental masculinization of sex behavior. *Nat. Neurosci.* **7,** 643–650.

Andersson, M. (1994). "Sexual Selection." Princeton University Press, Princeton, NJ.
Armstrong, E. A. (1947). "Courtship and Display Amongst Birds." Linsay Drummond, London.
Arnold, A. P. (1992). Developmental plasticity in neural circuits controlling birdsong: Sexual differentiation and the neural basis of learning. *J. Neurobiol.* **23**, 1506–1528.
Arnold, A. P. (2004). Sex chromosomes and brain gender. *Nat. Rev. Neurosci.* **5**, 701–708.
Arnold, A. P., Nottebohm, F., and Pfaff, D. W. (1976). Hormone concentrating cells in vocal control areas of the brain of the zebra finch (*Poephila guttata*). *J. Comp. Neurol.* **165**, 487–512.
Auger, A. P. (2004). Steroid receptor control of reproductive behavior. *Horm. Behav.* **45**, 168–172.
Azurmendi, A., Braza, F., Garcia, A., Braza, P., Munoz, J. M., and Sanchez-Martin, J. R. (2006). Aggression, dominance, and affiliation: Their relationships with androgen levels and intelligence in 5-year-old children. *Horm. Behav.* **50**, 132–140.
Ball, G. F. (1994). Neurochemical specializations associated with vocal learning in songbirds and budgerigars. *Brain Behav. Evol.* **44**, 234–246.
Ball, G. F., and Balthazart, J. (2002). Neuroendocrine mechanisms regulating reproductive cycles and reproductive behavior in birds. *In* "Hormones, Brain, and Behavior" (D. W. Pfaff, A. P. Arnold, A. M. Etgen, S. E. Fahrbach, and R. T. Rubin, eds.), Vol. 2, pp. 649–798. Academic Press, San Diego, CA.
Ball, G. F., and Balthazart, J. (2008). How useful is the appetitive and consummatory disctinction for our understanding of the neuroendocrine control of sexual behavior? *Horm. Behav.* **53**, 307–311.
Ball, G. F., and Hulse, S. H. (1998). Bird song. *Am. Psych.* **53**, 37–58.
Ball, G. F., Absil, P., and Balthazart, J. (1995). Peptidergic delineations of nucleus interface reveal a sex difference in volume. *Neuroreport* **6**, 957–960.
Balthazart, J., and Adkins-Regan, E. (2002). Sexual differentiation of brain and behavior in birds. *In* "Hormones, Brain, and Behavior" (D. W. Pfaff, A. P. Arnold, A. M. Etgen, S. E. Fahrbach, and R. T. Rubin, eds.), Vol. 4, pp. 223–301. Academic Press, San Diego, CA.
Balthazart, J., and Ball, G. (1995). Sexual differentiation of brain and behavior in birds. *Trends Endocrinol. Metab.* **6**, 21–29.
Balthazart, J., and Ball, G. F. (2006). Is brain estradiol a hormone or a neurotransmitter? *Trends Neurosci.* **29**, 241–249.
Balthazart, J., Schumacher, M., and Ottinger, M. A. (1983). Sexual differences in the Japanese quail: Behavior, morphology and intracellular metabolism of testosterone. *Gen. Comp. Endocrinol.* **51**, 191–207.
Balthazart, J., De Clerck, A., and Foidart, A. (1992a). Behavioral demasculinization of female quail is induced by estrogens: Studies with the new aromatase inhibitor, R76713. *Horm. Behav.* **26**, 179–203.
Balthazart, J., Foidart, A., Sante, P., and Hendrick, J. C. (1992b). Effects of alpha-methyl-para-tyrosine on monoamine levels in the Japanese quail: Sex differences and testosterone effects. *Brain Res. Bull.* **28**, 275–288.
Balthazart, J., Foidart, A., Wilson, E. M., and Ball, G. F. (1992c). Immunocytochemical localization of androgen receptors in the male songbird and quail brain. *J. Comp. Neurol.* **317**, 407–420.
Balthazart, J., Absil, P., Fiasse, V., and Ball, G. F. (1994). Effects of the aromatase inhibitor R76713 on sexual differentiation of brain and behavior in zebra finches. *Behaviour* **131**, 225–260.
Balthazart, J., Tlemçani, O., and Ball, G. F. (1996). Do sex differences in the brain explain sex differences in the hormonal induction of reproductive behavior? What 25 years of research on the Japanese quail tells us. *Horm. Behav.* **30**, 627–661.

Bateman, A. J. (1948). Intra-sexual selection in *Drosophila*. *Heredity* **2**, 349–368.
Beach, F. A. (1948). "Hormones and Behavior." Paul B. Hoeber, New York.
Beach, F. A. (1956). Characteristics of masculine "sex drive" *Nebr. Symp. Motiv.* **4**, 1–32.
Becker, J. B., Arnold, A. P., Berkley, K. J., Blaustein, J. D., Eckel, L. A., Hampson, E., Herman, J. P., Marts, S., Sadee, W., Steiner, M., Taylor, J., and Young, E. (2005). Strategies and methods for research on sex differences in brain and behavior. *Endocrinology* **146**, 1650–1673.
Bernhardt, P. C., Dabbs, J. M., Jr., Fielden, J. A., and Lutter, C. D. (1998). Testosterone changes during vicarious experiences of winning and losing among fans at sporting events. *Physiol. Behav.* **65**, 59–62.
Berthold, A. A. (1849). Transplantation der Hoden. *Arch. F. Anat. U. Physiol.* **16**, 42–46.
Birkhead, T. R., and Moller, A. P. (1992). "Sperm Competition in Birds." Academic Press, San Diego, CA.
Carlier, M., Roubertoux, P. L., Kottler, M. L., and Degrelle, H. (1990). Y chromosome and aggression in strains of laboratory mice. *Behav. Genet.* **20**, 137–156.
Chen, X., Agate, R. J., Itoh, Y., and Arnold, A. P. (2005). Sexually dimorphic expression of trkb, a Z-linked gene, in early posthatch zebra finch brain. *Proc. Natl. Acad. Sci. USA* **102**, 7730–7735.
Crews, D. (1993). The organizational concept and vertebrates without sex chromosomes. *Brain Behav. Evol.* **42**, 202–214.
Darwin, C. (1871). "The Descent of Man, and Selection in Relation to Sex." Murry, London.
Davies, N. B. (1992). "Dunnock Behaviour and Social Evolution." Oxford University Press, Oxford.
De Vries, G. J. (2004). Minireview: Sex differences in adult and developing brains: Compensation, compensation, compensation. *Endocrinology* **145**, 1063–1068.
Dörner, G., and Staudt, J. (1969). Structural changes in the hypothalamic ventromedial nucleus of the male rat, following neonatal castration and androgen treatment. *Neuroendocrinology* **4**, 278–281.
Duncan, K. A., and Carruth, L. L. (2007). The sexually dimorphic expression of L7/SPA, an estrogen receptor coactivator, in zebra finch telencephalon. *Dev. Neurobiol.* **67**, 1852–1866.
Fivizzani, A. J., and Oring, L. W. (1986). Plasma steroid hormones in relation to behavioral sex role reversal in the spotted sandpiper, Actitis macularia. *Biol. Reprod.* **35**, 1195–1201.
Gonzalez-Mariscal, G. (2001). Neuroendocrinology of maternal behavior in the rabbit. *Horm. Behav.* **40**, 125–132.
Gorski, R. A., Gordon, J. H., Shryne, J. E., and Southam, A. M. (1978). Evidence for a morphological sex difference within the medial preoptic area of the rat brain. *Brain Res.* **148**, 333–346.
Gowaty, P. A. (1996). Battles of the sexes and the origins of monogamy. *In* "Partnership in Birds" (J. L. Black, ed.), pp. 21–52. Oxford University Press, Oxford.
Greenspan, J. D., Craft, R. M., LeResche, L., Arendt-Nielsen, L., Berkley, K. J., Fillingim, R. B., Gold, M. S., Holdcroft, A., Lautenbacher, S., Mayer, E. A., Mogil, J. S., Murphy, A. Z., *et al.* (2007). Studying sex and gender differences in pain and analgesia: A consensus report. *Pain* **132**(Suppl. 1), S26–S45.
Gurney, M. E., and Konishi, M. (1980). Hormone-induced sexual differentiation of brain and behavior in zebra finches. *Science* **208**, 1380–1383.
Hau, M., Wikelski, M., Soma, K. K., and Wingfield, J. C. (2000). Testosterone and year-round territorial agression in a tropical birds. *Gen. Comp. Endocrinol.* **117**, 20–33.
Hermans, E. J., Ramsey, N. F., and van Honk, J. (2008). Exogenous testosterone enhances responsiveness to social threat in the neural circuitry of social aggression in humans. *Biol. Psychiatry* **63**, 263–270.

Hojo, Y., Hattori, T. A., Enami, T., Furukawa, A., Suzuki, K., Ishii, H. T., Mukai, H., Morrison, J. H., Janssen, W. G., Kominami, S., Harada, N., Kimoto, T., *et al.* (2004). Adult male rat hippocampus synthesizes estradiol from pregnenolone by cytochromes p45017α and P450 aromatase localized in neurons. *Proc. Natl. Acad. Sci. USA* **101**, 865–870.

Holloway, C. C., and Clayton, D. E. (2001). Estrogen synthesis in the male brain triggers development of the avian song control pathway *in vitro*. *Nat. Neurosci.* **4**, 170–175.

Hunt, K. E., Wingfield, J. C., Astheimer, L. B., Buttemer, W. A., and Hahn, T. P. (1995). Temporal patterns of territorial behavior and circulating testosterone in the Lapland longspur and other arctic passerines. *Am. Zool.* **35**, 274–284.

Hutchison, R. E. (1978). Hormonal differentiation of sexual behavior in Japanese quail. *Horm. Behav.* **11**, 363–387.

Jarvis, E. D. (2004). Brains and birdsong. *In* "Nature's Music: The Science of Birdsong" (P. Marler and H. Slabbekoorn, eds.), pp. 226–271. Elsevier, Amsterdam.

Jost, A. (1947). Reserches sur la différenciation sexuelle de l'embryon de lapin. *Arch. Anat. Microsc. Morphol. Exp.* **36**, 271–315.

Kelley, D. (1988). Sexually dimorphic behaviors. *Ann. Rev. Neurosci.* **11**, 225–251.

Kelley, D. B., and Brenowitz, E. A. (2002). Hormonal influences on courtship behaviors. *In* "Behavioral Endocrinology" (J. B. Becker, S. M. Breedlove, D. Crews, and M. M. McCarthy, eds.), pp. 289–329. MIT Press, Cambridge, MA.

Ketterson, E. D., Nolan, V., Jr., Wolf, L., and Zeigenfus, C. (1992). Testosterone and avian life histories: Effects of experimentally elevated testosterone on behavior and correlates of fitness in dark-eyed junco (Junco hyemalis). *Am. Nat.* **140**, 980–999.

Kim, Y. H., and Arnold, A. P. (2005). Distribution and onset of retinaldehyde dehydrogenase (zraldh) expression in zebra finch brain: Lack of sex difference in HVC and RA at early posthatch ages. *J. Neurobiol.* **65**, 260–268.

Kimchi, T., Xu, J., and Dulac, C. (2007). A functional circuit underlying male sexual behaviour in the female mouse brain. *Nature* **448**, 1009–1014.

Kroodsma, D., and Konishi, M. (1991). A suboscine bird (Eastern phoebe, Sayornis phoebe) develops normal song without auditory feedback. *Anim. Behav.* **42**, 477–487.

Kudwa, A. E., Bodo, C., Gustafsson, J. A., and Rissman, E. F. (2005). A previously uncharacterized role for estrogen receptor β: Defeminization of male brain and behavior. *Proc. Natl. Acad. Sci. USA* **102**, 4608–4612.

Lack, D. (1968). "Ecological Adaptations for Breeding in Birds." Methuen, London.

LeVay, S. (1991). A difference in hypothalamic structure between heterosexual and homosexual men. *Science* **253**, 1034–1037.

Lillie, F. R. (1916). The theory of the freemartin. *Science* **43**, 611–613.

London, S. E., Monks, D. A., Wade, J., and Schlinger, B. A. (2006). Widespread capacity for steroid synthesis in the avian brain and song system. *Endocrinology* **147**, 5975–5987.

Lonstein, J. S., and De Vries, G. J. (2000). Sex differences in the parental behavior of rodents. *Neurosci. Biobehav. Rev.* **24**, 669–686.

Mathews, G. A., Brenowitz, E. A., and Arnold, A. P. (1988). Paradoxical hypremasculinization of the zebra finch song system by an antiestrogen. *Horm. Behav.* **22**, 540–551.

Matsumoto, A. (2000). "Sexual Differentiation of the Brain." CRC Press, Boca Raton.

Mayhew, R. (2004). "The Female in Aristotle's Biology: Reason and Rationalization." University of Chicago Press, Chicago.

McCarthy, M. M. (1990). Oxytocin inhibits infanticide in female house mice. *Horm. Behav.* **24**, 365–375.

McCarthy, M. M. (2008). Estradiol and the developing brain. *Physiol. Rev.* **88**, 91–124.

McCarthy, M. M., and Konkle, A. T. (2005). When is a sex difference not a sex difference? *Front Neuroendocrinol.* **26**, 85–102.

McCaul, K. D., Gladue, B. A., and Joppa, M. (1992). Winning, losing, mood, and testosterone. *Horm. Behav.* **26,** 486–504.

McGinnis, M. Y. (2004). Anabolic androgenic steroids and aggression: Studies using animal models. *Ann. N. Y. Acad. Sci.* **1036,** 399–415.

Mong, J. A., Glaser, E., and McCarthy, M. M. (1999). Gonadal steroids promote glial differentiation and alter neuronal morphology in the developing hypothalamus in a regionally specific manner. *J. Neurosci.* **19,** 1464–1472.

Narumiya, S., Sugimoto, Y., and Ushikubi, F. (1999). Prostanoid receptors: Structures, properties, and functions. *Physiol. Rev.* **79,** 1193–1226.

Nottebohm, F., and Arnold, A. P. (1976). Sexual dimorphism in vocal control areas of the songbird brain. *Science* **194,** 211–213.

Numan, M. (1994). Maternal behavior. *In* "Physiology of Reproduction" (E. Knobil and J. D. Neill, eds.), Vol. 2, pp. 108–302. Raven Press, New York.

Oring, L. W. (1982). Avian mating systems. *In* "Avian Biology" (D. S. Farner, J. R. King, and K. C. Parkes, eds.), Vol. VI, pp. 1–92. Academic Press, New York.

Panzica, G. C., Viglietti-Panzica, C., Sanchez, F., Sante, P., and Balthazart, J. (1991). Effects of testosterone on a selected neuronal population within the preoptic sexually dimorphic nucleus of the Japanese quail. *J. Comp. Neurol.* **303,** 443–456.

Perkins, A., and Roselli, C. E. (2007). The ram as a model for behavioral neuroendocrinology. *Horm. Behav.* **52,** 70–77.

Perlman, W. R., and Arnold, A. P. (2003). Expression of estrogen receptor and aromatase mrnas in embryonic and posthatch zebra finch brain. *J. Neurobiol.* **55,** 204–219.

Pfaff, D. W. (1966). Morphological changes in the brains of adult male rats after neonatal castration. *J. Endocrinol.* **36,** 415–416.

Pfaus, J. G., Smith, W. J., and Coopersmith, C. B. (1999). Appetitive and consummatory sexual behaviors of female rats in bilevel chambers. I. A correlational and factor analysis and the effects of ovarian hormones. *Horm. Behav.* **35,** 224–240.

Phoenix, C. H., Goy, R. W., Gerall, A. A., and Young, W. C. (1959). Organizing action of prenatally administered testosterone proprionate on the tissues mediating mating behavior in the female guinea pig. *Endocrinology* **65,** 369–382.

Raisman, G., and Field, P. M. (1973). Sexual dimorphism in the neuropil of the preoptic area of the rat and its dependence on neonatal androgens. *Brain Res.* **54,** 1–29.

Ricklefs, R. E. (1983). Avian postnatal development avian biology, (D. S. Farner, J. R. King, and K. C. Parkes, eds.), Vol. VII, pp. 1–83. Academic Press, New York.

Roughgarden, J. (2004). "Evolution's Rainbow." University of California Press, Los Angeles, CA.

Ryan, B. C., and Vandenbergh, J. G. (2002). Intrauterine position effects. *Neurosci. Biobehav. Rev.* **26,** 665–678.

Sachs, B. D. (2007). A contextual definition of male sexual arousal. *Horm. Behav.* 569–578.

Sandnabba, N. K., Lagerspetz, K. M., and Jensen, E. (1994). Effects of testosterone exposure and fighting experience on the aggressive behavior of female and male mice selectively bred for intermale aggression. *Horm. Behav.* **28,** 219–231.

Schlinger, B. A., and Arnold, A. P. (1992). Plasma sex steroids and tissue aromatization in hatchling zebra finches: Implications for the sexual differentiation of singing behavior. *Endocrinology* **130,** 289–299.

Schlinger, B. A., Fivizzani, A. J., and Callard, G. V. (1989). Aromatase, 5α- and 5β-reductase in brain, pituitary and skin of the sex-role reversed Wilson's phalarope. *J. Endocrinol.* **122,** 573–581.

Schumacher, M., and Balthazart, J. (1986). Testosterone-induced brain aromatase is sexually dimorphic. *Brain Res.* **370,** 285–293.

Schumacher, M., Sulon, J., and Balthazart, J. (1988). Changes in serum concentrations of steroids during embryonic and post-hatching development of male and female Japanese quail (Coturnix coturnix japonica). *J. Endocrinol.* **118,** 127–134.

Schumacher, M., Hendrick, J. C., and Balthazart, J. (1989). Sexual differentiation in quail: Critical period and hormonal specificity. *Horm. Behav.* **23,** 130–149.

Schwarz, J. M., Liang, S. L., Thompson, S. M., and McCarthy, M. M. (2008). Estradiol induces dendritic spines on developing hypothalamic neurons by enhancing glutamate release independent of transcription: A mechanism for organizational sex differences. *Neuron* **58,** 584–598.

Short, R. V., and Balaban, E. (1994). "The Differences Between the Sexes." p. 479. Cambridge University Press, Cambridge.

Silver, R., Andrews, H., and Ball, G. F. (1985). Parental care in an ecological perspective: A quantitative analysis of avian subfamilies. *Am. Zool.* **25,** 823–840.

Simpson, H. B., and Vicario, D. S. (1991). Early estrogen treatment of female zebra finches masculinizes the brain pathway for learned vocalizations. *J. Neurobiol.* **22,** 777–793.

Speert, D. B., Konkle, A. T. M., Zup, S. L., Schwarz, J. A., Shiroor, C., and Taylor, M. (2007). Focal adhesion kinase and paxillin: Novel regulators of brain sexual differentiation? *Endocrinology* **148,** 3391–3401.

Steinach, E. (1940). "Sex and Life." Viking Press, New York.

Sugimoto, Y., and Narumiya, S. (2007). Prostaglandin E receptors. *J. Biol. Chem.* **282,** 11613–11617.

Swaab, D. F., and Fliers, E. (1985). A sexually dimorphic nucleus in the human brain. *Science* **228,** 1112–1115.

Synder, B. F., and Gowaty, P. A. (2007). A reappraisal of Bateman's classic study of intrasexual selection. *Evolution* **61,** 2457–2468.

Timberlake, W., and Silva, K. M. (1995). Appetitive behavior in ethology, psychology, and behavior systems. *In* "Perspectives in Ethology: Behavioral Design" (N. S. Thompson, ed.), Vol. 11, pp. 211–253. Plenum Press, New York.

Tinbergen, N. (1951). "The Study of Instinct." Clarendon Press, Oxford.

Todd, B. J., Schwarz, J. M., and McCarthy, M. M. (2005). Prostaglandin-E2: A point of divergence in estradiol-mediated sexual differentiation. *Horm. Behav.* **48,** 512–521.

Todd, B. J., Schwarz, J. M., Mong, J. A., and McCarthy, M. M. (2007). Glutamate AMPA/kainate receptors, not GABAA receptors, mediate estradiol-induced sex differences in the hypothalamus. *Dev. Neurobiol.* **67,** 304–315.

Ushikubi, F., Sugimoto, Y., Ichikawa, A., and Narumiya, S. (2000). Roles of prostanoids revealed from studies using mice lacking specific prostanoid receptors. *Jpn. J. Pharmacol.* **83,** 279–285.

Veening, J. G., Coolen, L. M., de Jong, T. R., Joosten, H. W., deBoer, S. F., Koolhaas, J. M., and Olivier, B. (2005). Do similar neural systems subserve aggressive and sexual behaviour in male rats? Insights from c-Fos and pharmacological studies. *Eur. J. Pharmacol.* **526,** 226–239.

vom Saal, F. S. (1984). The intrauterine position phenomenon: Effects on physiology, aggressive behavior and population dynamics in house mice. *Prog. Clin. Biol. Res.* **169,** 135–179.

Wade, J., and Arnold, A. P. (1994). Post-hatching inhibition of aromatase activity does not alter sexual differentiation of the zebra finch song system. *Brain Res.* **639,** 347–350.

Wade, J., and Arnold, A. P. (1996). Functional testicular tissue does not masculinize development of the zebra finch song system. *Proc. Natl. Acad. Sci. USA* **93,** 5264–5268.

Wade, J., Schlinger, B. A., Hodges, L., and Arnold, A. P. (1994). Fadrozole: A potent and specific inhibitor of aromatase in the zebra finch brain. *Gen. Comp. Endocrinol.* **94,** 53–61.

Wade, J., Springer, M. L., Wingfield, J. C., and Arnold, A. P. (1996). Neither testicular androgens nor embryonic aromatase activity alters morphology of the neural song system in zebra finches. *Biol. Reprod.* **55,** 1126–1132.

Wilson, J. D., and Roehrborn, C. (1999). Long-term consequences of castration in men: Lessons from the Skoptzy and eunuchs of the Chinese and Ottomona courts. *J. Clin. Endocrinol. Metab.* **84,** 4324–4331.

Wingfield, J. C., Ball, G. F., Dufty, A. M., Hegner, R. E., and Ramenofsky, M. (1987). Testosterone and aggression in birds: Tests of the "challenge" hypothesis. *Am. Sci.* **75,** 602–608.

Wingfield, J. C., Hegner, R. E., Dufty, A. M., and Ball, G. F. (1990). The "challenge hypothesis": Theoretical implications for patterns of testosterone secretion, mating systems and breedings strategies. *Am. Nat.* **136,** 829–846.

Zysling, D. A., Greives, T. J., Breuner, C. W., Casto, J. M., Demas, G. E., and Ketterson, E. D. (2006). Behavioral and physiological resopnses to experimentally elevated testosterone in female dark-eyed juncos (Junco hyemalis carolinensis). *Horm. Behav.* **50,** 200–207.

Index

A

Adkins-Regan injections, 235–236
Aggression. *See also* Territorial aggression
   anabolic steroids, 229–230
   genetics, 230–231
   testosterone behavior, 228–229
Androgens, 218
Aromatase activity, 239
Aromatization. *See also* Steroidogenic pathway
   definition, 220
   steroid hormone synthesis, 218
Attention deficit disorder (ADHD), 175

B

β–catenin signaling pathway. *See* Ovarian pathway
Birds, reproductive behavior
   parental and aggressive behaviors, 233–234
   sex-typical organization, 232–233
   sexual behaviors in quail
      Adkins-Regan injections, 235–236
      brain sex differences, 236–237
      territorial aggression, 234
Blimp1 gene, 190
Bone morphogenetic protein (BMP), 191
Brain
   feminized and masculinized, 222
   intrauterine position effect, 229
   sex differences in quail, 236–237
   sex-specific differentiation
      hormones role, 175–176
      X-inactivation, 174–175
   sexually dimorphic behaviors, 220–222

C

Caenorhabditis elegans. *See also* Hermaphrodites
   epigenetic regulation
      germline reprogramming, 199
      histone modifications, 200
      somatic genes, 198–199
   germ cell stages, 42–43
   HER-1 protein
      FEM protein regulation, 46–47
      spermatogenesis, 45–46
      TRA-2 receptor inhibition, 46
   sex determination
      DM genes, 29–30
      male and hermaphrodite, 2–3
      pathways involved, 5–6
      SDC proteins, 6–7
      signal elements, 4–5
      TRA factors, 7–10
   sexual dimorphism
      cell death, 25
      hermaphrodites *vs.* male, 10–12
      male mating behavior, 23–25
      male tail, 18–23
      sex muscles, 14–16
      somatic gonad, 12–14
      ventrolateral hypodermis, 16–18
   TRA-1A protein
      fog-1 and fog-3 promotors, 49
      oogenesis, 47–48
      spermatogenesis, 49–50
   tra-2 function
      RPN-10 regulation, 53
      TRA-1 regulation, 52–53
      translational repression, 50–52
      transcriptional program, 187–190
Castration, 216–217
Centrosomin (Cnn), 114
Coelomocytes, 10
Copulatory spicules, 19–20
Covert sexual specialization, 26
CUL-2 proteins, 9–10

D

Defeminization, 222
Dimorphic cell cycle
   meiosis-inducing factor, 195–196
   retinoic acid role, 198
   Stra8 expression, 196–197
   testis-specific pathways, 196
Dimorphic courtship behaviors, 214
Distal tip cell (DTC), 14
Dosage compensation, sex determination, 6–7
Doublesex/mab-3 (DM) gene, 29–30
DSX Female (DSX^F), 70–71
dsx gene
   abdominal pigmentation
      Abd-B expression, 91–92
      bab1/2 expression, 91
      expression, 72–73
   fat body sex dimorphism, 92
   genital disc

dsx gene (*cont.*)
   mesodermal cells, 88
   sex-specific signaling, 87
  and msSGPs, 78–79
  pigment cells development, 79
  sex-specific control, 70–71
  sexually dimorphic tissues, 74
  SGP development
   expression pattern, 77–78
   male-specific expression, 77
  spontaneous mutation, 70
  stem cell niche formation, 84–85
  target genes, 71–72
DSX Male (DSX^M), 70–71

E

early gene, 72
EGFR signaling pathway
  female germ cell, 122
  male germ cell, 117–118
Epigenetic regulation
  germline reprogramming, 199
  global demethylation, 200–201
  histone modifications, 200
  meiotic sex chromosome inactivation, 202
  somatic genes, 198–199
  X-chromosome inactivation, 201–202
Escort cells, 118, 135
Estradiol, 218, 220

F

Fat body enhancer (FBE), 71
Female germ cells in male soma
  cell–cell communication, 139
  male signal, 141–142
  pseudonurse cells, 141
  signaling pathways, 138–139
  spermatocytes formation, 140–141
  spermatogonia, 140
  yb and *hh* genes, 139–140
Feminized brain, 222
FEM proteins
  inhibition, 8
  proteolysis, 47
  regulators role, 54–55
  TRA-1 degradation, 46–47
  TRA regulation, 9–10
  vs. TRA-2, 53–54
Focal adhesion kinase (FAK), 226
Fragilis, 188–189
fused (*fu*) gene, 127

G

Gametes, 215
Genital disc in *Drosophila*
  development

   homeotic genes, 86
   segments, 87
  dsx mutants and
   mesoderm cells, 88
   sex-specific signaling, 87
   primordia, 85–86
Genital tubercle (GT) differentiation, 173–174
Germarium
  location, 118
  signaling pathways, 121, 138
  testis apex, 119
Germ cell lineage
  epigenetic and regulative mechanism, 188
  origin, 183
  sex-specific development, 193–195
Germ cells
  developmental stages, 42–43
  fog-1 gene, 55
  fshr-1 gene, 58
  mitosis role and translational regulation, 57
  sexual fate, 57–58
  sperm/oocyte switch, 55–56
  TRA-1 transcriptional control
   fog-1 and fog-3 promotors, 49
   oogenesis, 47–48
   sprematogenesis, 49–50
Germline sex determination
  ovo gene, 125, 127
  pathway, 126
  Sxl and *fu* mutant, 127–128
Germline stem cells (GSCs), in *Drosophila*
  adult gonad, 110
  EGFR signaling pathway
   female niche, 122
   male niche, 117–118
  female germ cells and male soma
   cell–cell communication, 139
   male signal, 141–142
   pseudonurse cells, 141
   signaling pathways, 138–139
   spermatocytes formation, 140–141
   spermatogonia, 140
   yb and *hh* genes, 139–140
  female niche and proliferative zone
   asymmetric division, 119
   cyst cells, 119
   escort cells, 118
   gap-junctional proteins, 119–121
   ovary structure, 118
  germline sex determination
   ovo gene, 125, 127
   pathway, 126
   Sxl and *fu* mutant, 127–128
  Jak/STAT signaling pathway
   female niche, 121
   male niche, 116
  male germ cells and female soma
   escort cells, 135

Index

fusome, 134–135
mitotic cystocyte divisions, 133–134
mosaic intersexes, 128–132
ring canal, 134–135
source regions, 136–137
spectrosomes, 132–133
tumorous follicles, 135–136
male niche and proliferative zone
  asymmetric division, 114
  cyst cells, 114–115
  germinal proliferation center, 112
  mitotic divisions, 115
  signaling pathways, 137–138
  somatic stem cells, 111
  testis structure, 110
sexual dimorphism in gonad, 80–81
somatic sex determination
  Dsx^F and Dsx^M factors, 125
  Sxl genes, 122, 125
TGF-β signaling pathway
  female niche, 121–122
  male niche, 116–117
GLI protein, 9, 27
Gonadal bipotential primordium
  bipotential fate, 157–158
  cell proliferation, 156–157
Gonads, 156–157

H

her-1 gene, 7–8
HER-1 protein
  FEM protein regulation, 46–47
  spermatogenesis, 45–46
  TRA-2 receptor inhibition, 46
Hermaphrodites
  chromosomes, 2, 4
  gonad and germ cells, 42
  male's sensory rays contact, 24–25
  neurons, 10
  self-fertilization, 43
  sex muscles, 14–15
Hermaphrodite-specific neurons (HSNs), 25
Homeotic genes, 86
Hormone organizational/activational hypothesis, 217–219
Hox proteins
  motor neurons, 18
  sex-specific activity, 17
  V rays identity, 22–23

I

Intrauterine position effect (IUP), 229

J

Jak/STAT signaling pathway
  female germ cells, 121, 138–139
  male germ cells, 16, 138

L

late genes, 72
Leydig cells, 165

M

Maintenance promoter *(Pm)*, 68
Male germ cells and female soma
  escort cells, 135
  fusome, 134–135
  mitotic cystocyte divisions, 133–134
  mosaic intersexes, 128–132
  ring canal, 134–135
  source regions, germ cells, 136–137
  spectrosomes and germarium, 132–133
  tumorous follicles, 135–136
Male-specific somatic gonad precursors (msSGP)
  development, 78
  dsx mutants, 78–79
Masculinized brain, 222
Meiotic sex chromosome inactivation (MSCI), 202
Morphological sex differences, 220–222
Mosaic intersexes, 128–132
Muscle of Lawrence (MOL)
  sex determination in *Drosophila*
    cell autonomous process, 93
    signaling pathways, 94

N

Nematodes
  females and males, 44
  sex-determination process, 45
Neurotransmitters, 222–223
Nuage. *See* Germ cell lineage

O

Ovarian morphogenesis
  Bmp2 role, 168–169
  fetal somatic pathway, 169–170
Ovarian pathway
  β–catenin signaling pathway, 166–167
  Sry gene role, 155–156
  Wnt4 function, 166
  Z/Od gene role, 165–166
ovo gene, 125–128

P

Pigment cells
  development, 79
  and *dsx* gene, 79–80
Postdeirid, 21
Preoptic area (POA)
  neurotransmitters, 222–223
  prostaglandins, 224–225
  ventromedial nucleus, 225–226

Primordial germ cells (PGCs)
  Blimp1 role, 190
  Bmp signaling, 191–192
  cell culture, 203–204
  EG cell lines, 203–204
  epigenetic regulation
    germline reprogramming, 199
    global demethylation, 200–201
    histone modifications, 200
    meiotic sex chromosome inactivation, 202
    somatic genes, 198–199
    X-chromosome inactivation, 201–202
  ES cells to germ cells, 205
  gastrulation and migration, 190–191
  germline-associated genes, 192–193
  libraries, 185
  neonatal testis cells, 204
  pluripotency genes, 192
  pluripotent stem cells generation, 202–204
  tissue nonspecific alkaline phosphatase, 184
Proctodeum, 19–20
Prostaglandins, 224–225
Pseudo nurse cell
  female germ cells, 141
  male germ cells, 136

Q

Quail, sexual behavior
  Adkins-Regan injections, 235–236
  brain sex differences, 236–237

R

Reproductive behavior in birds
  parental and aggressive, 233–234
  in quail
    Adkins-Regan injections, 235–236
    brain sex differences, 236–237
  sex-typical organization, 232–233
  territorial aggression, 234

S

SDC proteins, 6–7
Secondary sex differentiation
  brain, role of, 174–176
  duct development, 170–172
  external genitalia, 173–174
SEL-10 proteins, 8
Sex determination in *C. elegans*
  DM genes, 29–30
  male and hermaphrodite, 2–3
  pathway, 5–6
  SDC proteins, 6–7
  and sex differentiation
    downstream regulation, 28
    TRA-1 and transcription, 27–28
  signal elements, 4–5
  TRA factors
    FEM protein inhibition, 8
    her-1 gene, 7–8
    regulation, 9–10
Sex determination in *Drosophila*
  dsx gene
    expression, 72–73
    sex-specific control, 70–71
    sexually dimorphic tissues, 74
    spontaneous mutation, 70
    target genes, 71–72
  female-promoting character, 66
  Sxl gene
    activation, 68
    RNA binding protein, 67–68
    TRA splicing factor, 69–70
Sex-determination pathway
  HER-1 protein
    FEM protein regulation, 46–47
    spermatogenesis, 45–46
    TRA-2 receptor inhibition, 46
  tra-2 function
    RPN-10 regulation, 53
    TRA-1 regulation, 52–53
    translational repression, 50–52
Sex lethal (*Sxl*) gene
  activation, 68
  RNA binding protein, 67–68
  targets of, 69
Sex-specific lactation process, 215–216
Sexual behavior in mammals
  aggression
    anabolic steroids, 229–230
    genetics, 230–231
    testosterone behavior, 228–229
  parental behavior
    hormonal and neural control, 226–227
    maternal care, 227–228
  preoptic area
    neurotransmitters, 222–223
    prostaglandins, 224–225
    ventromedial nucleus, 225–226
Sexual behaviors in quail
  Adkins-Regan injections, 235–236
  brain sex differences, 236–237
Sexual differentiation
  C. elegans
    downstream regulation, 28
    transcription regulation, 27–28
  zebra finches
    copulatory behavior, 238
    hormone and cell autonomous mechanisms, 238–2241
    vocalizations, 237–238
Sexual dimorphism in *C. elegans*

Index

covert sexual specialization, 26
hermephrodites *vs.* male, 10–12
male mating behavior
   hermaphrodite contact and sensory rays, 24–25
   mate searching, 23–24
male tail
   proctodeum and spicules, 19
   rectal epithelial cells, 19–20
   sensory rays, 20–21
   signaling, 20
   V rays, 21–23
sex muscles
   hermaphrodite, 14–15
   male, 15–16
sex-specific cell death, 25
SGP
   distal daughter and DTC, 14
   division and features, 12–14
   proximal-distal axis, 12
ventrolateral hypodermis
   Hox proteins, 17
   motor neurons, 18
   P blast cells, 16
   vulva development, 17–18
Sexual dimorphism in *Drosophila* gonad
   abdominal pigmentation
      cuticular structures, 90
      dsx gene, 91–92
   fat body, 92
   gonad formation
      embryonic stages, 74–75
      female-specific gene expression, 76
      testis development, 74
   male-specific SGP
      development, 78
      dsx mutants, 78–79
   muscle of Lawrence
      cell autonomous process, 93
      signaling pathways, 94
      patterning genes expression, 95
   pigment cells
      development, 79
      dsx gene, 79–80
   somatic gonad precursors
      development, 76–77
      dsx gene, 77–78
   stem cell niches
      dsx gene, 84–85
      female cell and ovary development, 81–82
      gametes, 80–81
      male cell development, 80
      male *vs.* female cell, 82–84
Somatic gonad precursors (SGP)
   in *C. elegans*
      distal daughter and DTC, 14
      division and features, 12–14

      proximal–distal axis, 12
   development
      JAK/STAT pathway, 76–77
      mesoderm, 76
      sex-specific signaling, 77
   dsx gene
      expression pattern, 77–78
      male-specific expression, 77
Somatic sex determination
   active ovarian pathway, 155–156
   bipotential primordium, 156–158
   critical experiments, 153–155
   genetic control, 152–153
   in germline stem cells
      Dsx^F and Dsx^M factors, 125
      Sxl genes, 122, 125
   molecular specification
      ovarian morphogenesis, 168–170
      ovarian pathway, 165–167
      testis morphogenesis, 162–165
      testis pathway, 160–162
Soma-to-germline signaling pathways, 137
Sox9 expression, 160–161. *See also* Testis pathway
Sry gene
   expression, 155
   ovarian pathway, 155–156
   teatis pathway, 161–162
Stella, 186–187
Stem cell niche
   dsx gene, 84–85
   female cell and ovary development, 81–82
   gametes, 80–81
   male cell development, 80
   male *vs.* female cell, 82–84
Steroid hormones synthesis, 218
Steroidogenic pathway, 223
Stra8 gene expression, 196–197

T

Territorial aggression, 234
Testis morphogenesis
   migration, 163–164
   Sox9 expression, 164–165
   testis cord formation, 162–163
Testis pathway
   sertoli cell differentiation, 160–161
   Sry activity, 161–162
Testosterone
   aggression, 228–229
   androgen synthesis, 218
   estradiol, 223
TGF-β signaling pathway
   female germ cells and male soma, 138–139
   female niche, 121–122

TGF-β signaling pathway (cont.)
　male germ cells and female
　　　soma, 137
　male niche, 116–117
Tissue nonspecific alkaline phosphatase
　(TNAP), 184
Touch receptor neurons (TRNs), 25
TRA-1A protein
　fog-1 and fog-3 promotors, 49
　oogenesis, 47–48
　sprematogenesis, 49–50
tra-1 gene, 9
TRA-1 transcription factor
　FEM protein
　　inhibition role, 8
　　regulation, 9–10
　and her-1 gene, 7–8
TRA splicing factor, 69–70
Turner's syndrome, 152

V

Vasa protein, 193
Ventromedial nucleus (VMN)
　　morphology, 225–226

V rays
　identity and Hox genes, 22–23
　neurogenesis of, 21–22

W

Wnt4 function, 166

X

XOL-1 regulator, 5–6

Z

Zebra finches sex differentiation
　copulatory behavior, 238
　hormone and cell autonomous mechanisms
　　aromatase activity, 239
　　estrogen action, 238–239
　　neural circuit differentiation, 239–241
　vocalizations, 237–238
Z hypothesis, 156

Contents of Previous Volumes

Volume 47

1. **Early Events of Somitogenesis in Higher Vertebrates: Allocation of Precursor Cells during Gastrulation and the Organization of a Moristic Pattern in the Paraxial Mesoderm**
   Patrick P. L. Tam, Devorah Goldman, Anne Camus, and Gary C. Shoenwolf

2. **Retrospective Tracing of the Developmental Lineage of the Mouse Myotome**
   Sophie Eloy-Trinquet, Luc Mathis, and Jean-François Nicolas

3. **Segmentation of the Paraxial Mesoderm and Vertebrate Somitogenesis**
   Olivier Pourqulé

4. **Segmentation: A View from the Border**
   Claudio D. Stern and Daniel Vasiliauskas

5. **Genetic Regulation of Somite Formation**
   Alan Rawls, Jeanne Wilson-Rawls, and Eric N. Olsen

6. **Hox Genes and the Global Patterning of the Somitic Mesoderm**
   Ann Campbell Burke

7. **The Origin and Morphogenesis of Amphibian Somites**
   Ray Keller

8. **Somitogenesis in Zebrafish**
   Scott A. Halley and Christiana Nüsslain-Volhard

9. **Rostrocaudal Differences within the Somites Confer Segmental Pattern to Trunk Neural Crest Migration**
   Marianne Bronner-Fraser

Volume 48

1. **Evolution and Development of Distinct Cell Lineages Derived from Somites**
   Beate Brand-Saberi and Bodo Christ

2. **Duality of Molecular Signaling Involved in Vertebral Chondrogenesis**
   Anne-Hélène Monsoro-Burq and Nicole Le Douarin

3. **Sclerotome Induction and Differentiation**
   Jennifer L. Docker

4. **Genetics of Muscle Determination and Development**
   Hans-Henning Arnold and Thomas Braun

5. **Multiple Tissue Interactions and Signal Transduction Pathways Control Somite Myogenesis**
   Anne-Gaëlle Borycki and Charles P. Emerson, Jr.

6. **The Birth of Muscle Progenitor Cells in the Mouse: Spatiotemporal Considerations**
   Shahragim Tajbakhsh and Margaret Buckingham

7. **Mouse–Chick Chimera: An Experimental System for Study of Somite Development**
   Josiane Fontaine-Pérus

8. **Transcriptional Regulation during Somitogenesis**
   Dennis Summerbell and Peter W. J. Rigby

9. **Determination and Morphogenesis in Myogenic Progenitor Cells: An Experimental Embryological Approach**
   Charles P. Ordahl, Brian A. Williams, and Wilfred Denetclaw

Volume 49

1. **The Centrosome and Parthenogenesis**
   Thomas Küntziger and Michel Bornens

2. **γ-Tubulin**
   Berl R. Oakley

3. **γ-Tubulin Complexes and Their Role in Microtubule Nucleation**
   Ruwanthi N. Gunawardane, Sofia B. Lizarraga, Christiane Wiese, Andrew Wilde, and Yixian Zheng

4. **γ-Tubulin of Budding Yeast**
   Jackie Vogel and Michael Snyder

5. **The Spindle Pole Body of *Saccharomyces cerevisiae*: Architecture and Assembly of the Core Components**
   Susan E. Francis and Trisha N. Davis

6. **The Microtubule Organizing Centers of** *Schizosaccharomyces pombe*
   Iain M. Hagan and Janni Petersen
7. **Comparative Structural, Molecular, and Functional Aspects of the** *Dictyostelium discoideum* **Centrosome**
   Ralph Gräf, Nicole Brusis, Christine Daunderer, Ursula Euteneuer, Andrea Hestermann, Manfred Schliwa, and Masahiro Ueda
8. **Are There Nucleic Acids in the Centrosome?**
   Wallace F. Marshall and Joel L. Rosenbaum
9. **Basal Bodies and Centrioles: Their Function and Structure**
   Andrea M. Preble, Thomas M. Giddings, Jr., and Susan K. Dutcher
10. **Centriole Duplication and Maturation in Animal Cells**
    B. M. H. Lange, A. J. Faragher, P. March, and K. Gull
11. **Centrosome Replication in Somatic Cells: The Significance of the G_1 Phase**
    Ron Balczon
12. **The Coordination of Centrosome Reproduction with Nuclear Events during the Cell Cycle**
    Greenfield Sluder and Edward H. Hinchcliffe
13. **Regulating Centrosomes by Protein Phosphorylation**
    Andrew M. Fry, Thibault Mayor, and Erich A. Nigg
14. **The Role of the Centrosome in the Development of Malignant Tumors**
    Wilma L. Lingle and Jeffrey L. Salisbury
15. **The Centrosome-Associated Aurora/Ipl-like Kinase Family**
    T. M. Goepfert and B. R. Brinkley
16. **Centrosome Reduction during Mammalian Spermiogenesis**
    G. Manandhar, C. Simerly, and G. Schatten
17. **The Centrosome of the Early** *C. elegans* **Embryo: Inheritance, Assembly, Replication, and Developmental Roles**
    Kevin F. O'Connell
18. **The Centrosome in** *Drosophila* **Oocyte Development**
    Timothy L. Megraw and Thomas C. Kaufman
19. **The Centrosome in Early** *Drosophila* **Embryogenesis**
    W. F. Rothwell and W. Sullivan

20. Centrosome Maturation

   Robert E. Palazzo, Jacalyn M. Vogel, Bradley J. Schnackenberg, Dawn R. Hull, and Xingyong Wu

Volume 50

1. Patterning the Early Sea Urchin Embryo

   Charles A. Ettensohn and Hyla C. Sweet

2. Turning Mesoderm into Blood: The Formation of Hematopoietic Stem Cells during Embryogenesis

   Alan J. Davidson and Leonard I. Zon

3. Mechanisms of Plant Embryo Development

   Shunong Bai, Lingjing Chen, Mary Alice Yund, and Zinmay Rence Sung

4. Sperm-Mediated Gene Transfer

   Anthony W. S. Chan, C. Marc Luetjens, and Gerald P. Schatten

5. Gonocyte–Sertoli Cell Interactions during Development of the Neonatal
   Rodent Testis

   Joanne M. Orth, William F. Jester, Ling-Hong Li, and Andrew L. Laslett

6. Attributes and Dynamics of the Endoplasmic Reticulum in Mammalian Eggs

   Douglas Kline

7. Germ Plasm and Molecular Determinants of Germ Cell Fate

   Douglas W. Houston and Mary Lou King

Volume 51

1. Patterning and Lineage Specification in the Amphibian Embryo

   Agnes P. Chan and Laurence D. Etkin

2. Transcriptional Programs Regulating Vascular Smooth Muscle Cell Development and Differentiation

   Michael S. Parmacek

3. Myofibroblasts: Molecular Crossdressers

   Gennyne A. Walker, Ivan A. Guerrero, and Leslie A. Leinwand

Contents of Previous Volumes

4. Checkpoint and DNA-Repair Proteins Are Associated with the Cores of Mammalian Meiotic Chromosomes
   Madalena Tarsounas and Peter B. Moens

5. Cytoskeletal and Ca^{2+} Regulation of Hyphal Tip Growth and Initiation
   Sara Torralba and I. Brent Heath

6. Pattern Formation during *C. elegans* Vulval Induction
   Minqin Wang and Paul W. Sternberg

7. A Molecular Clock Involved in Somite Segmentation
   Miguel Maroto and Olivier Pourquié

Volume 52

1. Mechanism and Control of Meiotic Recombination Initiation
   Scott Keeney

2. Osmoregulation and Cell Volume Regulation in the Preimplantation Embryo
   Jay M. Baltz

3. Cell–Cell Interactions in Vascular Development
   Diane C. Darland and Patricia A. D'Amore

4. Genetic Regulation of Preimplantation Embryo Survival
   Carol M. Warner and Carol A. Brenner

Volume 53

1. Developmental Roles and Clinical Significance of Hedgehog Signaling
   Andrew P. McMahon, Philip W. Ingham, and Clifford J. Tabin

2. Genomic Imprinting: Could the Chromatin Structure Be the Driving Force?
   Andras Paldi

3. Ontogeny of Hematopoiesis: Examining the Emergence of Hematopoietic Cells in the Vertebrate Embryo
   Jenna L. Galloway and Leonard I. Zon

4. Patterning the Sea Urchin Embryo: Gene Regulatory Networks, Signaling Pathways, and Cellular Interactions
   Lynne M. Angerer and Robert C. Angerer

Volume 54

1. **Membrane Type-Matrix Metalloproteinases (MT-MMP)**
   Stanley Zucker, Duanqing Pei, Jian Cao, and Carlos Lopez-Otin

2. **Surface Association of Secreted Matrix Metalloproteinases**
   Rafael Fridman

3. **Biochemical Properties and Functions of Membrane-Anchored Metalloprotease-Disintegrin Proteins (ADAMs)**
   J. David Becherer and Carl P. Blobel

4. **Shedding of Plasma Membrane Proteins**
   Joaquín Arribas and Anna Merlos-Suárez

5. **Expression of Meprins in Health and Disease**
   Lourdes P. Norman, Gail L. Matters, Jacqueline M. Crisman, and Judith S. Bond

6. **Type II Transmembrane Serine Proteases**
   Qingyu Wu

7. **DPPIV, Seprase, and Related Serine Peptidases in Multiple Cellular Functions**
   Wen-Tien Chen, Thomas Kelly, and Giulio Ghersi

8. **The Secretases of Alzheimer's Disease**
   Michael S. Wolfe

9. **Plasminogen Activation at the Cell Surface**
   Vincent Ellis

10. **Cell-Surface Cathepsin B: Understanding Its Functional Significance**
    Dora Cavallo-Medved and Bonnie F. Sloane

11. **Protease-Activated Receptors**
    Wadie F. Bahou

12. **Emmprin (CD147), a Cell Surface Regulator of Matrix Metalloproteinase Production and Function**
    Bryan P. Toole

13. **The Evolving Roles of Cell Surface Proteases in Health and Disease: Implications for Developmental, Adaptive, Inflammatory, and Neoplastic Processes**
    Joseph A. Madri

14. Shed Membrane Vesicles and Clustering of Membrane-Bound Proteolytic Enzymes
   M. Letizia Vittorelli

Volume 55

1. The Dynamics of Chromosome Replication in Yeast
   Isabelle A. Lucas and M. K. Raghuraman
2. Micromechanical Studies of Mitotic Chromosomes
   M. G. Poirier and John F. Marko
3. Patterning of the Zebrafish Embryo by Nodal Signals
   Jennifer O. Liang and Amy L. Rubinstein
4. Folding Chromosomes in Bacteria: Examining the Role of Csp Proteins and Other Small Nucleic Acid-Binding Proteins
   Nancy Trun and Danielle Johnston

Volume 56

1. Selfishness in Moderation: Evolutionary Success of the Yeast Plasmid
   Soundarapandian Velmurugan, Shwetal Mehta, and Makkuni Jayaram
2. Nongenomic Actions of Androgen in Sertoli Cells
   William H. Walker
3. Regulation of Chromatin Structure and Gene Activity by Poly(ADP-Ribose) Polymerases
   Alexei Tulin, Yurli Chinenov, and Allan Spradling
4. Centrosomes and Kinetochores, Who needs 'Em? The Role of Noncentromeric Chromatin in Spindle Assembly
   Priya Prakash Budde and Rebecca Heald
5. Modeling Cardiogenesis: The Challenges and Promises of 3D Reconstruction
   Jeffrey O. Penetcost, Claudio Silva, Maurice Pesticelli, Jr., and Kent L. Thornburg
6. Plasmid and Chromosome Traffic Control: How ParA and ParB Drive Partition
   Jennifer A. Surtees and Barbara E. Funnell

Volume 57

1. **Molecular Conservation and Novelties in Vertebrate Ear Development**
   B. Fritzsch and K. W. Beisel

2. **Use of Mouse Genetics for Studying Inner Ear Development**
   Elizabeth Quint and Karen P. Steel

3. **Formation of the Outer and Middle Ear, Molecular Mechanisms**
   Moisés Mallo

4. **Molecular Basis of Inner Ear Induction**
   Stephen T. Brown, Kareen Martin, and Andrew K. Groves

5. **Molecular Basis of Otic Commitment and Morphogenesis: A Role for Homeodomain-Containing Transcription Factors and Signaling Molecules**
   Eva Bober, Silke Rinkwitz, and Heike Herbrand

6. **Growth Factors and Early Development of Otic Neurons: Interactions between Intrinsic and Extrinsic Signals**
   Berta Alsina, Fernando Giraldez, and Isabel Varela-Nieto

7. **Neurotrophic Factors during Inner Ear Development**
   Ulla Pirvola and Jukka Ylikoski

8. **FGF Signaling in Ear Development and Innervation**
   Tracy J. Wright and Suzanne L. Mansour

9. **The Roles of Retinoic Acid during Inner Ear Development**
   Raymond Romand

10. **Hair Cell Development in Higher Vertebrates**
    Wei-Qiang Gao

11. **Cell Adhesion Molecules during Inner Ear and Hair Cell Development, Including Notch and Its Ligands**
    Matthew W. Kelley

12. **Genes Controlling the Development of the Zebrafish Inner Ear and Hair Cells**
    Bruce B. Riley

13. **Functional Development of Hair Cells**
    Ruth Anne Eatock and Karen M. Hurley

14. The Cell Cycle and the Development and Regeneration of Hair Cells
    Allen F. Ryan

Volume 58

1. A Role for Endogenous Electric Fields in Wound Healing
   Richard Nuccitelli
2. The Role of Mitotic Checkpoint in Maintaining Genomic Stability
   Song-Tao Liu, Jan M. van Deursen, and Tim J. Yen
3. The Regulation of Oocyte Maturation
   Ekaterina Voronina and Gary M. Wessel
4. Stem Cells: A Promising Source of Pancreatic Islets for Transplantation in Type 1 Diabetes
   Cale N. Street, Ray V. Rajotte, and Gregory S. Korbutt
5. Differentiation Potential of Adipose Derived Adult Stem (ADAS) Cells
   Jeffrey M. Gimble and Farshid Guilak

Volume 59

1. The Balbiani Body and Germ Cell Determinants: 150 Years Later
   Malgorzata Kloc, Szczepan Bilinski, and Laurence D. Etkin
2. Fetal–Maternal Interactions: Prenatal Psychobiological Precursors to Adaptive Infant Development
   Matthew F. S. X. Novak
3. Paradoxical Role of Methyl-CpG-Binding Protein 2 in Rett Syndrome
   Janine M. LaSalle
4. Genetic Approaches to Analyzing Mitochondrial Outer Membrane Permeability
   Brett H. Graham and William J. Craigen
5. Mitochondrial Dynamics in Mammals
   Hsiuchen Chen and David C. Chan
6. Histone Modification in Corepressor Functions
   Judith K. Davie and Sharon Y. R. Dent
7. Death by Abl: A Matter of Location
   Jiangyu Zhu and Jean Y. J. Wang

Volume 60

1. **Therapeutic Cloning and Tissue Engineering**
   Chester J. Koh and Anthony Atala

2. **α-Synuclein: Normal Function and Role in Neurodegenerative Diseases**
   Erin H. Norris, Benoit I. Giasson, and Virginia M.-Y. Lee

3. **Structure and Function of Eukaryotic DNA Methyltransferases**
   Taiping Chen and En Li

4. **Mechanical Signals as Regulators of Stem Cell Fate**
   Bradley T. Estes, Jeffrey M. Gimble, and Farshid Guilak

5. **Origins of Mammalian Hematopoiesis: *In Vivo* Paradigms and *In Vitro* Models**
   M. William Lensch and George Q. Daley

6. **Regulation of Gene Activity and Repression: A Consideration of Unifying Themes**
   Anne C. Ferguson-Smith, Shau-Ping Lin, and Neil Youngson

7. **Molecular Basis for the Chloride Channel Activity of Cystic Fibrosis Transmembrane Conductance Regulator and the Consequences of Disease-Causing Mutations**
   Jackie F. Kidd, Ilana Kogan, and Christine E. Bear

Volume 61

1. **Hepatic Oval Cells: Helping Redefine a Paradigm in Stem Cell Biology**
   P. N. Newsome, M. A. Hussain, and N. D. Theise

2. **Meiotic DNA Replication**
   Randy Strich

3. **Pollen Tube Guidance: The Role of Adhesion and Chemotropic Molecules**
   Sunran Kim, Juan Dong, and Elizabeth M. Lord

4. **The Biology and Diagnostic Applications of Fetal DNA and RNA in Maternal Plasma**
   Rossa W. K. Chiu and Y. M. Dennis Lo

5. **Advances in Tissue Engineering**
   Shulamit Levenberg and Robert Langer

6. Directions in Cell Migration Along the Rostral Migratory Stream: The Pathway for Migration in the Brain
   Shin-ichi Murase and Alan F. Horwitz

7. Retinoids in Lung Development and Regeneration
   Malcolm Maden

8. Structural Organization and Functions of the Nucleus in Development, Aging, and Disease
   Leslie Mounkes and Colin L. Stewart

Volume 62

1. Blood Vessel Signals During Development and Beyond
   Ondine Cleaver

2. HIFs, Hypoxia, and Vascular Development
   Kelly L. Covello and M. Celeste Simon

3. Blood Vessel Patterning at the Embryonic Midline
   Kelly A. Hogan and Victoria L. Bautch

4. Wiring the Vascular Circuitry: From Growth Factors to Guidance Cues
   Lisa D. Urness and Dean Y. Li

5. Vascular Endothelial Growth Factor and Its Receptors in Embryonic Zebrafish Blood Vessel Development
   Katsutoshi Goishi and Michael Klagsbrun

6. Vascular Extracellular Matrix and Aortic Development
   Cassandra M. Kelleher, Sean E. McLean, and Robert P. Mecham

7. Genetics in Zebrafish, Mice, and Humans to Dissect Congenital Heart Disease: Insights in the Role of VEGF
   Diether Lambrechts and Peter Carmeliet

8. Development of Coronary Vessels
   Mark W. Majesky

9. Identifying Early Vascular Genes Through Gene Trapping in Mouse Embryonic Stem Cells
   Frank Kuhnert and Heidi Stuhlmann

Volume 63

1. **Early Events in the DNA Damage Response**
   Irene Ward and Junjie Chen

2. **Afrotherian Origins and Interrelationships: New Views and Future Prospects**
   Terence J. Robinson and Erik R. Seiffert

3. **The Role of Antisense Transcription in the Regulation of X-Inactivation**
   Claire Rougeulle and Philip Avner

4. **The Genetics of Hiding the Corpse: Engulfment and Degradation of Apoptotic Cells in *C. elegans* and *D. melanogaster***
   Zheng Zhou, Paolo M. Mangahas, and Xiaomeng Yu

5. **Beginning and Ending an Actin Filament: Control at the Barbed End**
   Sally H. Zigmond

6. **Life Extension in the Dwarf Mouse**
   Andrzej Bartke and Holly Brown-Borg

Volume 64

1. **Stem/Progenitor Cells in Lung Morphogenesis, Repair, and Regeneration**
   David Warburton, Mary Anne Berberich, and Barbara Driscoll

2. **Lessons from a Canine Model of Compensatory Lung Growth**
   Connie C. W. Hsia

3. **Airway Glandular Development and Stem Cells**
   Xiaoming Liu, Ryan R. Driskell, and John F. Engelhardt

4. **Gene Expression Studies in Lung Development and Lung Stem Cell Biology**
   Thomas J. Mariani and Naftali Kaminski

5. **Mechanisms and Regulation of Lung Vascular Development**
   Michelle Haynes Pauling and Thiennu H. Vu

6. **The Engineering of Tissues Using Progenitor Cells**
   Nancy L. Parenteau, Lawrence Rosenberg, and Janet Hardin-Young

7. Adult Bone Marrow-Derived Hemangioblasts, Endothelial Cell Progenitors, and EPCs
   Gina C. Schatteman

8. Synthetic Extracellular Matrices for Tissue Engineering and Regeneration
   Eduardo A. Silva and David J. Mooney

9. Integrins and Angiogenesis
   D. G. Stupack and D. A. Cheresh

Volume 65

1. Tales of Cannibalism, Suicide, and Murder: Programmed Cell Death in *C. elegans*
   Jason M. Kinchen and Michael O. Hengartner

2. From Guts to Brains: Using Zebrafish Genetics to Understand the Innards of Organogenesis
   Carsten Stuckenholz, Paul E. Ulanch, and Nathan Bahary

3. Synaptic Vesicle Docking: A Putative Role for the Munc18/Sec1 Protein Family
   Robby M. Weimer and Janet E. Richmond

4. ATP-Dependent Chromatin Remodeling
   Corey L. Smith and Craig L. Peterson

5. Self-Destruct Programs in the Processes of Developing Neurons
   David Shepherd and V. Hugh Perry

6. Multiple Roles of Vascular Endothelial Growth Factor (VEGF) in Skeletal Development, Growth, and Repair
   Elazar Zelzer and Bjorn R. Olsen

7. G-Protein Coupled Receptors and Calcium Signaling in Development
   Geoffrey E. Woodard and Juan A. Rosado

8. Differential Functions of 14-3-3 Isoforms in Vertebrate Development
   Anthony J. Muslin and Jeffrey M. C. Lau

9. Zebrafish Notochordal Basement Membrane: Signaling and Structure
   Annabelle Scott and Derek L. Stemple

10. Sonic Hedgehog Signaling and the Developing Tooth
    Martyn T. Cobourne and Paul T. Sharpe

Volume 66

1. **Stepwise Commitment from Embryonic Stem to Hematopoietic and Endothelial Cells**
   Changwon Park, Jesse J. Lugus, and Kyunghee Choi

2. **Fibroblast Growth Factor Signaling and the Function and Assembly of Basement Membranes**
   Peter Lonai

3. **TGF-β Superfamily and Mouse Craniofacial Development: Interplay of Morphogenetic Proteins and Receptor Signaling Controls Normal Formation of the Face**
   Marek Dudas and Vesa Kaartinen

4. **The Colors of Autumn Leaves as Symptoms of Cellular Recycling and Defenses Against Environmental Stresses**
   Helen J. Ougham, Phillip Morris, and Howard Thomas

5. **Extracellular Proteases: Biological and Behavioral Roles in the Mammalian Central Nervous System**
   Yan Zhang, Kostas Pothakos, and Styliana-Anna (Stella) Tsirka

6. **The Genetic Architecture of House Fly Mating Behavior**
   Lisa M. Meffert and Kara L. Hagenbuch

7. **Phototropins, Other Photoreceptors, and Associated Signaling: The Lead and Supporting Cast in the Control of Plant Movement Responses**
   Bethany B. Stone, C. Alex Esmon, and Emmanuel Liscum

8. **Evolving Concepts in Bone Tissue Engineering**
   Catherine M. Cowan, Chia Soo, Kang Ting, and Benjamin Wu

9. **Cranial Suture Biology**
   Kelly A Lenton, Randall P. Nacamuli, Derrick C. Wan, Jill A. Helms, and Michael T. Longaker

Volume 67

1. **Deer Antlers as a Model of Mammalian Regeneration**
   Joanna Price, Corrine Faucheux, and Steve Allen

2. The Molecular and Genetic Control of Leaf Senescence and Longevity in *Arabidopsis*
   Pyung Ok Lim and Hong Gil Nam
3. Cripto-1: An Oncofetal Gene with Many Faces
   Caterina Bianco, Luigi Strizzi, Nicola Normanno, Nadia Khan, and David S. Salomon
4. Programmed Cell Death in Plant Embryogenesis
   Peter V. Bozhkov, Lada H. Filonova, and Maria F. Suarez
5. Physiological Roles of Aquaporins in the Choroid Plexus
   Daniela Boassa and Andrea J. Yool
6. Control of Food Intake Through Regulation of cAMP
   Allan Z. Zhao
7. Factors Affecting Male Song Evolution in *Drosophila montana*
   Anneli Hoikkala, Kirsten Klappert, and Dominique Mazzi
8. Prostanoids and Phosphodiesterase Inhibitors in Experimental Pulmonary Hypertension
   Ralph Theo Schermuly, Hossein Ardeschir Ghofrani, and Norbert Weissmann
9. 14-3-3 Protein Signaling in Development and Growth Factor Responses
   Daniel Thomas, Mark Guthridge, Jo Woodcock, and Angel Lopez
10. Skeletal Stem Cells in Regenerative Medicine
    Wataru Sonoyama, Carolyn Coppe, Stan Gronthos, and Songtao Shi

Volume 68

1. Prolactin and Growth Hormone Signaling
   Beverly Chilton and Aveline Hewetson
2. Alterations in cAMP-Mediated Signaling and Their Role in the Pathophysiology of Dilated Cardiomyopathy
   Matthew A. Movsesian and Michael R. Bristow
3. Corpus Luteum Development: Lessons from Genetic Models in Mice
   Anne Bachelot and Nadine Binart
4. Comparative Developmental Biology of the Mammalian Uterus
   Thomas E. Spencer, Kanako Hayashi, Jianbo Hu, and Karen D. Carpenter

5. Sarcopenia of Aging and Its Metabolic Impact
   Helen Karakelides and K. Sreekumaran Nair

6. Chemokine Receptor CXCR3: An Unexpected Enigma
   Liping Liu, Melissa K. Callahan, DeRen Huang, and Richard M. Ransohoff

7. Assembly and Signaling of Adhesion Complexes
   Jorge L. Sepulveda, Vasiliki Gkretsi, and Chuanyue Wu

8. Signaling Mechanisms of Higher Plant Photoreceptors: A Structure-Function Perspective
   Haiyang Wang

9. Initial Failure in Myoblast Transplantation Therapy Has Led the Way Toward the Isolation of Muscle Stem Cells: Potential for Tissue Regeneration
   Kenneth Urish, Yasunari Kanda, and Johnny Huard

10. Role of 14-3-3 Proteins in Eukaryotic Signaling and Development
    Dawn L. Darling, Jessica Yingling, and Anthony Wynshaw-Boris

Volume 69

1. Flipping Coins in the Fly Retina
   Tamara Mikeladze-Dvali, Claude Desplan, and Daniela Pistillo

2. Unraveling the Molecular Pathways That Regulate Early Telencephalon Development
   Jean M. Hébert

3. Glia–Neuron Interactions in Nervous System Function and Development
   Shai Shaham

4. The Novel Roles of Glial Cells Revisited: The Contribution of Radial Glia and Astrocytes to Neurogenesis
   Tetsuji Mori, Annalisa Buffo, and Magdalena Götz

5. Classical Embryological Studies and Modern Genetic Analysis of Midbrain and Cerebellum Development
   Mark Zervas, Sandra Blaess, and Alexandra L. Joyner

6. Brain Development and Susceptibility to Damage; Ion Levels and Movements
   Maria Erecinska, Shobha Cherian, and Ian A. Silver

7. Thinking about Visual Behavior; Learning about Photoreceptor Function
Kwang-Min Choe and Thomas R. Clandinin

8. Critical Period Mechanisms in Developing Visual Cortex
Takao K. Hensch

9. Brawn for Brains: The Role of MEF2 Proteins in the Developing Nervous System
Aryaman K. Shalizi and Azad Bonni

10. Mechanisms of Axon Guidance in the Developing Nervous System
Céline Plachez and Linda J. Richards

Volume 70

1. Magnetic Resonance Imaging: Utility as a Molecular Imaging Modality
James P. Basilion, Susan Yeon, and René Botnar

2. Magnetic Resonance Imaging Contrast Agents in the Study of Development
Angelique Louie

3. $^{1}H/^{19}F$ Magnetic Resonance Molecular Imaging with Perfluorocarbon Nanoparticles
Gregory M. Lanza, Patrick M. Winter, Anne M. Neubauer, Shelton D. Caruthers, Franklin D. Hockett, and Samuel A. Wickline

4. Loss of Cell Ion Homeostasis and Cell Viability in the Brain: What Sodium MRI Can Tell Us
Fernando E. Boada, George LaVerde, Charles Jungreis, Edwin Nemoto, Costin Tanase, and Ileana Hancu

5. Quantum Dot Surfaces for Use In Vivo and In Vitro
Byron Ballou

6. In Vivo Cell Biology of Cancer Cells Visualized with Fluorescent Proteins
Robert M. Hoffman

7. Modulation of Tracer Accumulation in Malignant Tumors: Gene Expression, Gene Transfer, and Phage Display
Uwe Haberkorn

8. Amyloid Imaging: From Benchtop to Bedside
   Chungying Wu, Victor W. Pike, and Yanming Wang

9. *In Vivo* Imaging of Autoimmune Disease in Model Systems
   Eric T. Ahrens and Penelope A. Morel

Volume 71

1. The Choroid Plexus-Cerebrospinal Fluid System: From Development to Aging
   Zoran B. Redzic, Jane E. Preston, John A. Duncan, Adam Chodobski, and Joanna Szmydynger-Chodobska

2. Zebrafish Genetics and Formation of Embryonic Vasculature
   Tao P. Zhong

3. Leaf Senescence: Signals, Execution, and Regulation
   Yongfeng Guo and Susheng Gan

4. Muscle Stem Cells and Regenerative Myogenesis
   Iain W. McKinnell, Gianni Parise, and Michael A. Rudnicki

5. Gene Regulation in Spermatogenesis
   James A. MacLean II and Miles F. Wilkinson

6. Modeling Age-Related Diseases in *Drosophila:* Can this Fly?
   Kinga Michno, Diana van de Hoef, Hong Wu, and Gabrielle L. Boulianne

7. Cell Death and Organ Development in Plants
   Hilary J. Rogers

8. The Blood-Testis Barrier: Its Biology, Regulation, and Physiological Role in Spermatogenesis
   Ching-Hang Wong and C. Yan Cheng

9. Angiogenic Factors in the Pathogenesis of Preeclampsia
   Hai-Tao Yuan, David Haig, and S. Ananth Karumanchi

Volume 72

1. Defending the Zygote: Search for the Ancestral Animal Block to Polyspermy
   Julian L. Wong and Gary M. Wessel

2. **Dishevelled: A Mobile Scaffold Catalyzing Development**
   Craig C. Malbon and Hsien-yu Wang

3. **Sensory Organs: Making and Breaking the Pre-Placodal Region**
   Andrew P. Bailey and Andrea Streit

4. **Regulation of Hepatocyte Cell Cycle Progression and Differentiation by Type I Collagen Structure**
   Linda K. Hansen, Joshua Wilhelm, and John T. Fassett

5. **Engineering Stem Cells into Organs: Topobiological Transformations Demonstrated by Beak, Feather, and Other Ectodermal Organ Morphogenesis**
   Cheng-Ming Chuong, Ping Wu, Maksim Plikus, Ting-Xin Jiang, and Randall Bruce Widelitz

6. **Fur Seal Adaptations to Lactation: Insights into Mammary Gland Function**
   Julie A. Sharp, Kylie N. Cane, Christophe Lefevre, John P. Y. Arnould, and Kevin R. Nicholas

Volume 73

1. **The Molecular Origins of Species-Specific Facial Pattern**
   Samantha A. Brugmann, Minal D. Tapadia, and Jill A. Helms

2. **Molecular Bases of the Regulation of Bone Remodeling by the Canonical Wnt Signaling Pathway**
   Donald A. Glass II and Gerard Karsenty

3. **Calcium Sensing Receptors and Calcium Oscillations: Calcium as a First Messenger**
   Gerda E. Breitwieser

4. **Signal Relay During the Life Cycle of *Dictyostelium***
   Dana C. Mahadeo and Carole A. Parent

5. **Biological Principles for *Ex Vivo* Adult Stem Cell Expansion**
   Jean-François Paré and James L. Sherley

6. **Histone Deacetylation as a Target for Radiosensitization**
   David Cerna, Kevin Camphausen, and Philip J. Tofilon

7. **Chaperone-Mediated Autophagy in Aging and Disease**
   Ashish C. Massey, Cong Zhang, and Ana Maria Cuervo

8. Extracellular Matrix Macroassembly Dynamics in Early Vertebrate Embryos
   Andras Czirok, Evan A. Zamir, Michael B. Filla, Charles D. Little, and Brenda J. Rongish

Volume 74

1. Membrane Origin for Autophagy
   Fulvio Reggiori

2. Chromatin Assembly with H3 Histones: Full Throttle Down Multiple Pathways
   Brian E. Schwartz and Kami Ahmad

3. Protein–Protein Interactions of the Developing Enamel Matrix
   John D. Bartlett, Bernhard Ganss, Michel Goldberg, Janet Moradian-Oldak, Michael L. Paine, Malcolm L. Snead, Xin Wen, Shane N. White, and Yan L. Zhou

4. Stem and Progenitor Cells in the Formation of the Pulmonary Vasculature
   Kimberly A. Fisher and Ross S. Summer

5. Mechanisms of Disordered Granulopoiesis in Congenital Neutropenia
   David S. Grenda and Daniel C. Link

6. Social Dominance and Serotonin Receptor Genes in Crayfish
   Donald H. Edwards and Nadja Spitzer

7. Transplantation of Undifferentiated, Bone Marrow-Derived Stem Cells
   Karen Ann Pauwelyn and Catherine M. Verfaillie

8. The Development and Evolution of Division of Labor and Foraging Specialization in a Social Insect (*Apis mellifera* L.)
   Robert E. Page Jr., Ricarda Scheiner, Joachim Erber, and Gro V. Amdam

Volume 75

1. Dynamics of Assembly and Reorganization of Extracellular Matrix Proteins
   Sarah L. Dallas, Qian Chen, and Pitchumani Sivakumar

2. Selective Neuronal Degeneration in Huntington's Disease
   Catherine M. Cowan and Lynn A. Raymond

3. RNAi Therapy for Neurodegenerative Diseases
   Ryan L. Boudreau and Beverly L. Davidson

4. Fibrillins: From Biogenesis of Microfibrils to Signaling Functions
   Dirk Hubmacher, Kerstin Tiedemann, and Dieter P. Reinhardt

5. Proteasomes from Structure to Function: Perspectives from Archaea
   Julie A. Maupin-Furlow, Matthew A. Humbard, P. Aaron Kirkland, Wei Li, Christopher J. Reuter, Amy J. Wright, and G. Zhou

6. The Cytomatrix as a Cooperative System of Macromolecular and Water Networks
   V. A. Shepherd

7. Intracellular Targeting of Phosphodiesterase-4 Underpins Compartmentalized cAMP Signaling
   Martin J. Lynch, Elaine V. Hill, and Miles D. Houslay

Volume 76

1. BMP Signaling in the Cartilage Growth Plate
   Robert Pogue and Karen Lyons

2. The CLIP-170 Orthologue Bik1p and Positioning the Mitotic Spindle in Yeast
   Rita K. Miller, Sonia D'Silva, Jeffrey K. Moore, and Holly V. Goodson

3. Aggregate-Prone Proteins Are Cleared from the Cytosol by Autophagy: Therapeutic Implications
   Andrea Williams, Luca Jahreiss, Sovan Sarkar, Shinji Saiki, Fiona M. Menzies, Brinda Ravikumar, and David C. Rubinsztein

4. Wnt Signaling: A Key Regulator of Bone Mass
   Roland Baron, Georges Rawadi, and Sergio Roman-Roman

5. Eukaryotic DNA Replication in a Chromatin Context
   Angel P. Tabancay, Jr. and Susan L. Forsburg

6. The Regulatory Network Controlling the Proliferation–Meiotic Entry Decision in the *Caenorhabditis elegans* Germ Line
   Dave Hansen and Tim Schedl

7. Regulation of Angiogenesis by Hypoxia and Hypoxia-Inducible Factors
   Michele M. Hickey and M. Celeste Simon

Volume 77

1. **The Role of the Mitochondrion in Sperm Function: Is There a Place for Oxidative Phosphorylation or Is this a Purely Glycolytic Process?**
   Eduardo Ruiz-Pesini, Carmen Díez-Sánchez, Manuel José López-Pérez, and José Antonio Enríquez

2. **The Role of Mitochondrial Function in the Oocyte and Embryo**
   Rémi Dumollard, Michael Duchen, and John Carroll

3. **Mitochondrial DNA in the Oocyte and the Developing Embryo**
   Pascale May-Panloup, Marie-Françoise Chretien, Yves Malthiery, and Pascal Reynier

4. **Mitochondrial DNA and the Mammalian Oocyte**
   Eric A. Shoubridge and Timothy Wai

5. **Mitochondrial Disease—Its Impact, Etiology, and Pathology**
   R. McFarland, R. W. Taylor, and D. M. Turnbull

6. **Cybrid Models of mtDNA Disease and Transmission, from Cells to Mice**
   Ian A. Trounce and Carl A. Pinkert

7. **The Use of Micromanipulation Methods as a Tool to Prevention of Transmission of Mutated Mitochondrial DNA**
   Helena Fulka and Josef Fulka, Jr.

8. **Difficulties and Possible Solutions in the Genetic Management of mtDNA Disease in the Preimplantation Embryo**
   J. Poulton, P. Oakeshott, and S. Kennedy

9. **Impact of Assisted Reproductive Techniques: A Mitochondrial Perspective from the Cytoplasmic Transplantation**
   A. J. Harvey, T. C. Gibson, T. M. Quebedeaux, and C. A. Brenner

10. **Nuclear Transfer: Preservation of a Nuclear Genome at the Expense of Its Associated mtDNA Genome(s)**
    Emma J. Bowles, Keith H. S. Campbell, and Justin C. St. John

Volume 78

1. **Contribution of Membrane Mucins to Tumor Progression Through Modulation of Cellular Growth Signaling Pathways**
   Kermit L. Carraway III, Melanie Funes, Heather C. Workman, and Colleen Sweeney

2. **Regulation of the Epithelial Na^+ Channel by Peptidases**
   Carole Planès and George H. Caughey

3. **Advances in Defining Regulators of Cementum Development and Periodontal Regeneration**
   Brian L. Foster, Tracy E. Popowics, Hanson K. Fong, and Martha J. Somerman

4. **Anabolic Agents and the Bone Morphogenetic Protein Pathway**
   I. R. Garrett

5. **The Role of Mammalian Circadian Proteins in Normal Physiology and Genotoxic Stress Responses**
   Roman V. Kondratov, Victoria Y. Gorbacheva, and Marina P. Antoch

6. **Autophagy and Cell Death**
   Devrim Gozuacik and Adi Kimchi

Volume 79

1. **The Development of Synovial Joints**
   I. M. Khan, S. N. Redman, R. Williams, G. P. Dowthwaite, S. F. Oldfield, and C. W. Archer

2. **Development of a Sexually Differentiated Behavior and Its Underlying CNS Arousal Functions**
   Lee-Ming Kow, Cristina Florea, Marlene Schwanzel-Fukuda, Nino Devidze, Hosein Kami Kia, Anna Lee, Jin Zhou, David MacLaughlin, Patricia Donahoe, and Donald Pfaff

3. **Phosphodiesterases Regulate Airway Smooth Muscle Function in Health and Disease**
   Vera P. Krymskaya and Reynold A. Panettieri, Jr.

4. Role of Astrocytes in Matching Blood Flow to Neuronal Activity
   Danica Jakovcevic and David R. Harder

5. Elastin-Elastases and Inflamm-Aging
   Frank Antonicelli, Georges Bellon, Laurent Debelle, and William Hornebeck

6. A Phylogenetic Approach to Mapping Cell Fate
   Stephen J. Salipante and Marshall S. Horwitz

Volume 80

1. Similarities Between Angiogenesis and Neural Development: What Small Animal Models Can Tell Us
   Serena Zacchigna, Carmen Ruiz de Almodovar, and Peter Carmeliet

2. Junction Restructuring and Spermatogenesis: The Biology, Regulation, and Implication in Male Contraceptive Development
   Helen H. N. Yan, Dolores D. Mruk, and C. Yan Cheng

3. Substrates of the Methionine Sulfoxide Reductase System and Their Physiological Relevance
   Derek B. Oien and Jackob Moskovitz

4. Organic Anion-Transporting Polypeptides at the Blood–Brain and Blood–Cerebrospinal Fluid Barriers
   Daniel E. Westholm, Jon N. Rumbley, David R. Salo, Timothy P. Rich, and Grant W. Anderson

5. Mechanisms and Evolution of Environmental Responses in *Caenorhabditis elegans*
   Christian Braendle, Josselin Milloz, and Marie-Anne Félix

6. Molluscan Shell Proteins: Primary Structure, Origin, and Evolution
   Frédéric Marin, Gilles Luquet, Benjamin Marie, and Davorin Medakovic

7. Pathophysiology of the Blood–Brain Barrier: Animal Models and Methods
   Brian T. Hawkins and Richard D. Egleton

8. Genetic Manipulation of Megakaryocytes to Study Platelet Function
   Jun Liu, Jan DeNofrio, Weiping Yuan, Zhengyan Wang, Andrew W. McFadden, and Leslie V. Parise

9. Genetics and Epigenetics of the Multifunctional Protein CTCF
   Galina N. Filippova

Volume 81

1. **Models of Biological Pattern Formation: From Elementary Steps to the Organization of Embryonic Axes**
   Hans Meinhardt

2. **Robustness of Embryonic Spatial Patterning in *Drosophila Melanogaster***
   David Umulis, Michael B. O'Connor, and Hans G. Othmer

3. **Integrating Morphogenesis with Underlying Mechanics and Cell Biology**
   Lance A. Davidson

4. **The Mechanisms Underlying Primitive Streak Formation in the Chick Embryo**
   Manli Chuai and Cornelis J. Weijer

5. **Grid-Free Models of Multicellular Systems, with an Application to Large-Scale Vortices Accompanying Primitive Streak Formation**
   T. J. Newman

6. **Mathematical Models for Somite Formation**
   Ruth E. Baker, Santiago Schnell, and Philip K. Maini

7. **Coordinated Action of N-CAM, N-cadherin, EphA4, and ephrinB2 Translates Genetic Prepatterns into Structure during Somitogenesis in Chick**
   James A. Glazier, Ying Zhang, Maciej Swat, Benjamin Zaitlen, and Santiago Schnell

8. **Branched Organs: Mechanics of Morphogenesis by Multiple Mechanisms**
   Sharon R. Lubkin

9. **Multicellular Sprouting during Vasculogenesis**
   Andras Czirok, Evan A. Zamir, Andras Szabo, and Charles D. Little

10. **Modelling Lung Branching Morphogenesis**
    Takashi Miura

11. **Multiscale Models for Vertebrate Limb Development**
    Stuart A. Newman, Scott Christley, Tilmann Glimm, H. G. E. Hentschel, Bogdan Kazmierczak, Yong-Tao Zhang, Jianfeng Zhu, and Mark Alber

12. **Tooth Morphogenesis *in vivo*, *in vitro* and *in silico***
    Isaac Salazar-Ciudad

13. Cell Mechanics with a 3D Kinetic and Dynamic Weighted Delaunay-Triangulation
   Michael Meyer-Hermann

14. Cellular Automata as Microscopic Models of Cell Migration in Heterogeneous Environments
   H. Hatzikirou and A. Deutsch

15. Multiscale Modeling of Biological Pattern Formation
   Ramon Grima

16. Relating Biophysical Properties Across Scales
   Elijah Flenner, Francoise Marga, Adrian Neagu, Ioan Kosztin, and Gabor Forgacs

17. Complex Multicellular Systems and Immune Competition: New Paradigms Looking for a Mathematical Theory
   N. Bellomo and G. Forni

Volume 82

1. Ontogeny of Erythropoiesis in the Mammalian Embryo
   Kathleen McGrath and James Palis

2. The Erythroblastic Island
   Deepa Manwani and James J. Bieker

3. Epigenetic Control of Complex Loci During Erythropoiesis
   Ryan J. Wozniak and Emery H. Bresnick

4. The Role of the Epigenetic Signal, DNA Methylation, in Gene Regulation During Erythroid Development
   Gordon D. Ginder, Merlin N. Gnanapragasam, and Omar Y. Mian

5. Three-Dimensional Organization of Gene Expression in Erythroid Cells
   Wouter de Laat, Petra Klous, Jurgen Kooren, Daan Noordermeer, Robert-Jan Palstra, Marieke Simonis, Erik Splinter, and Frank Grosveld

6. Iron Homeostasis and Erythropoiesis
   Diedra M. Wrighting and Nancy C. Andrews

7. **Effects of Nitric Oxide on Red Blood Cell Development and Phenotype**
   Vladan P. Čokić and Alan N. Schechter

8. **Diamond Blackfan Anemia: A Disorder of Red Blood Cell Development**
   Steven R. Ellis and Jeffrey M. Lipton

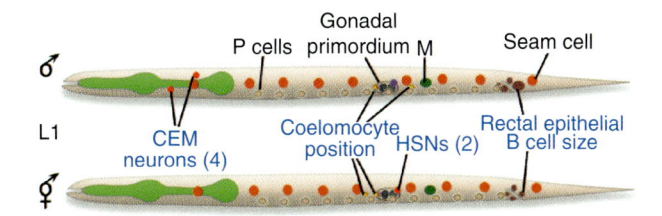

Jennifer Ross Wolff and David Zarkower, Figure 1.3 Comparison of the newly hatched L1 male and hermaphrodite. Blue labels indicate sex-specific differences established in the embryo. Black labels indicate cells that express different post-embryonic fates or lineages in the male and hermaphrodite. (Lints and Hall, 2005. Adapted with permission from WormAtlas, http://wormatlas.org/).

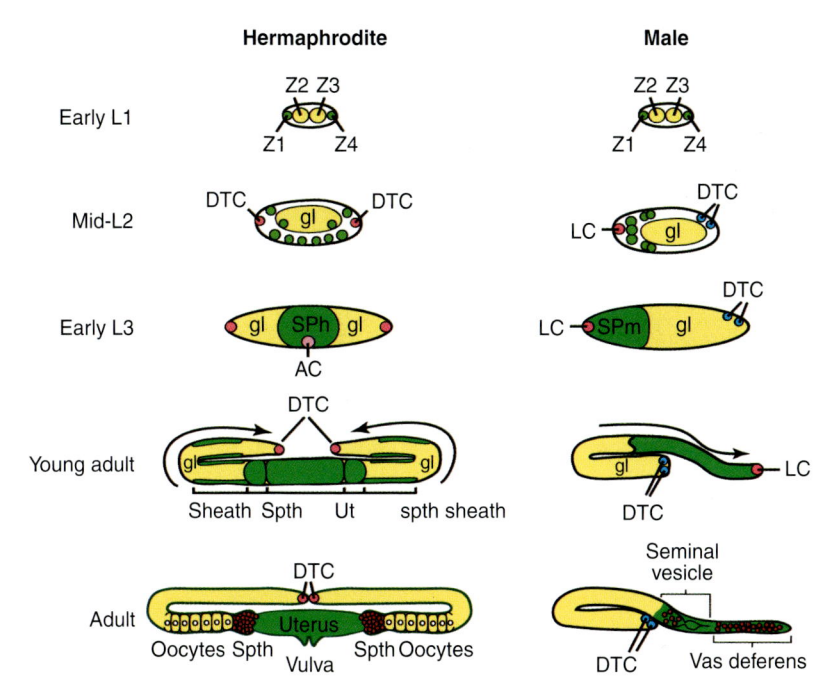

Jennifer Ross Wolff and David Zarkower, Figure 1.5 Sexually dimorphic gonadal development. At hatching, the four cell gonadal primordia of the male and hermaphrodite appear identical, but undergo sexually dimorphic division, migration, and fate specification to create dissimilar adult structures. The asymmetrical, J-shaped testis of the male includes an anterior mitotic germ line (gl) and posterior vas deferens and seminal vesicle (sv). The bilaterally symmetrical hermaphrodite ovotestes are surrounded by sheath cells, and include a distal mitotic germline (gl) and proximal spermatheca (spth) and uterus (ut). DTC, distal tip cell; LC, linker cell; AC, anchor cell. After Miskowski *et al.* (2001) with permission.

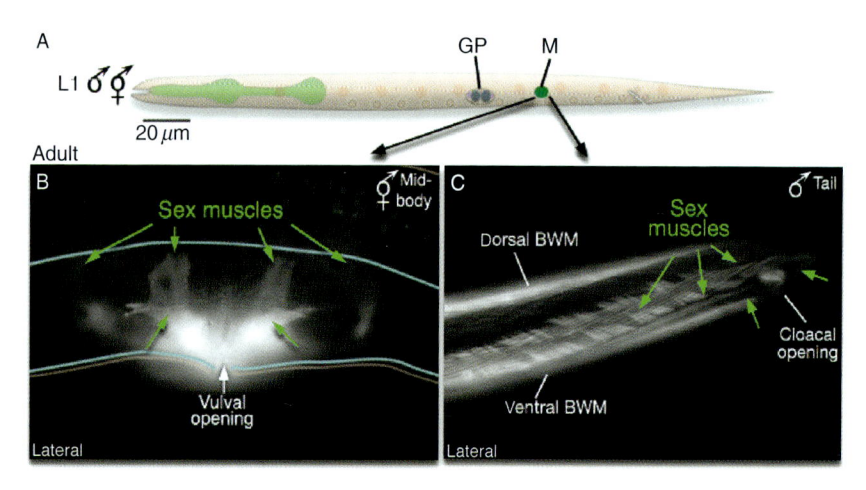

Jennifer Ross Wolff and David Zarkower, Figure 1.6 Sex muscles. (A) The sex muscles of both sexes are derived from the mesoblast M, which generates two sex myoblasts (SMs) in hermaphrodites and six SMs in males. (B) Vulval and uterine muscles of the adult hermaphrodite, expressing *hlh-1/hlh-2::gfp* (Strain source: B. Harfe, M. Krause, A. Fire). (C) Diagonal sex muscles of the adult male, expressing *unc-27::gfp* (Strain source: L. Jia and S. Emmons). BWM, body wall muscle; GP gonadal primordium. (Lints and Hall, 2005. Adapted with permission from WormAtlas, http://wormatlas.org/).

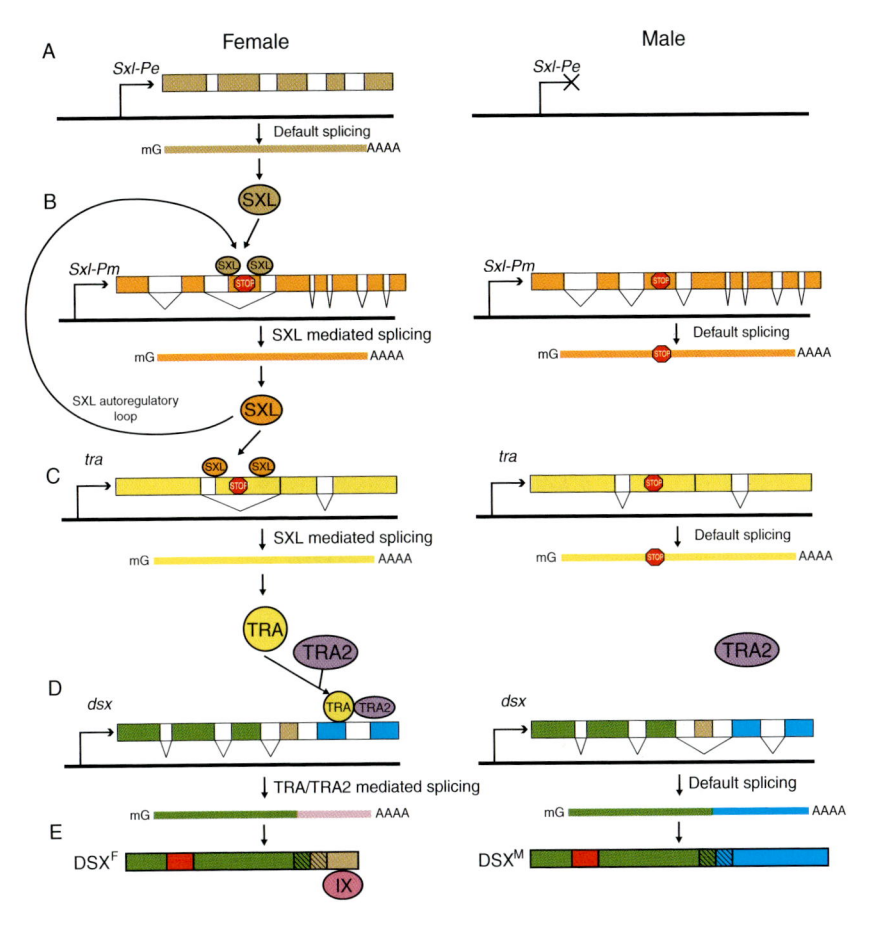

Nicole Camara et al., Figure 3.1 The *D. melanogaster* sex determination heirarchy. (A) Transcription of *Sxl* from the early promoter (Pe) is activated in females, but remains inactive in males. (B) SXL protein, found only in females, directs alternative splicing of the *Sxl Pm* transcript to an active form via removal of an early stop codon. This autoregulatory loop maintains SXL expression in females. In males, default splicing retains the early stop codon. (C) SXL regulates splicing of *tra* in females, resulting in removal of an early stop codon and production of TRA protein. In males, default splicing of *tra* retains the early stop condon. (D) In females, TRA and TRA2 direct splicing of the *doublesex* transcript into the female-specific isoform (dsx^F). The lack of functional TRA protein in males results in default splicing of the *dsx* transcript into the male-specific isoform (DSX^M). (E) Schematic of the two DSX isoforms. The DNA binding domain (DM domain) and the N-terminal dimerization domain are indicated in red. The C-terminal dimerization domain is indicated by stripes and consists of a non sex-specific region and a sex-specific region. Also shown is IX, which interacts specifically with the N-terminal tail of DSX^F.

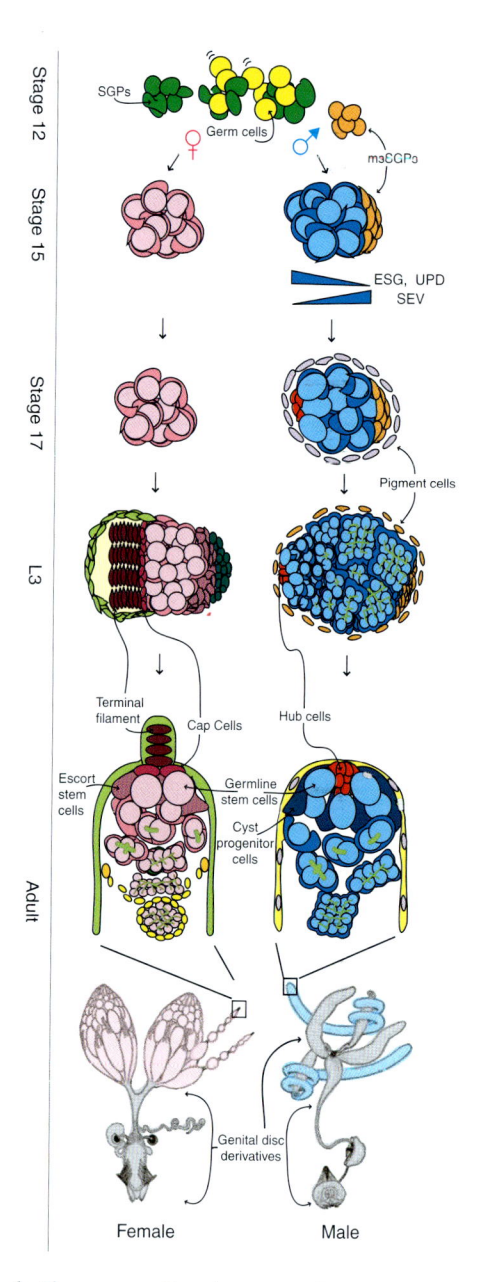

Nicole Camara et al., Figure 3.2 Development of sexual dimorphism in the gonad. Embryonic stages are as described (Campos-Ortega and Hartenstein, 1985). L3, 3rd instar larvae. The adult stage depicts the apical end of a single ovariole in the female and testis in the male, as indicated. The gonad begins as a bi-potential organ. Germ cells and somatic gonadal precursors (SGPs) interact and form an embryonic gonad by stage 15.

Both the germline and somatic gonad are sexually dimorphic at this time. msSGPs are initially specified in both sexes and join only the posterior of the male gonad. In the male, pigment cells and the embryonic hub form during stage 17. The female gonad undergoes ovary morphogenesis during late L3 to make individual ovarioles and establish the female germline stem cell niche at the L3-pupal transition. In adults, germline stem cells (GSCs) contact the somatic niche formed by cap cells in females and hub cells in males. Somatic stem cells (cyst progenitor cells in male and escort stem cells in females) also contact the niche. germ cells (yellow at st 12, then light pink or blue); SGPs (green, st 12, then pink or blue; msSGPs (orange); hub cells (red); pigment cells (gray), terminal filaments (dark purple); cap cells (magenta); escort stem cells (purple); hub (red), cyst progenitor cells (navy blue). Adult ovaries (pink), testes (blue), and genital disk derivatives (gray).

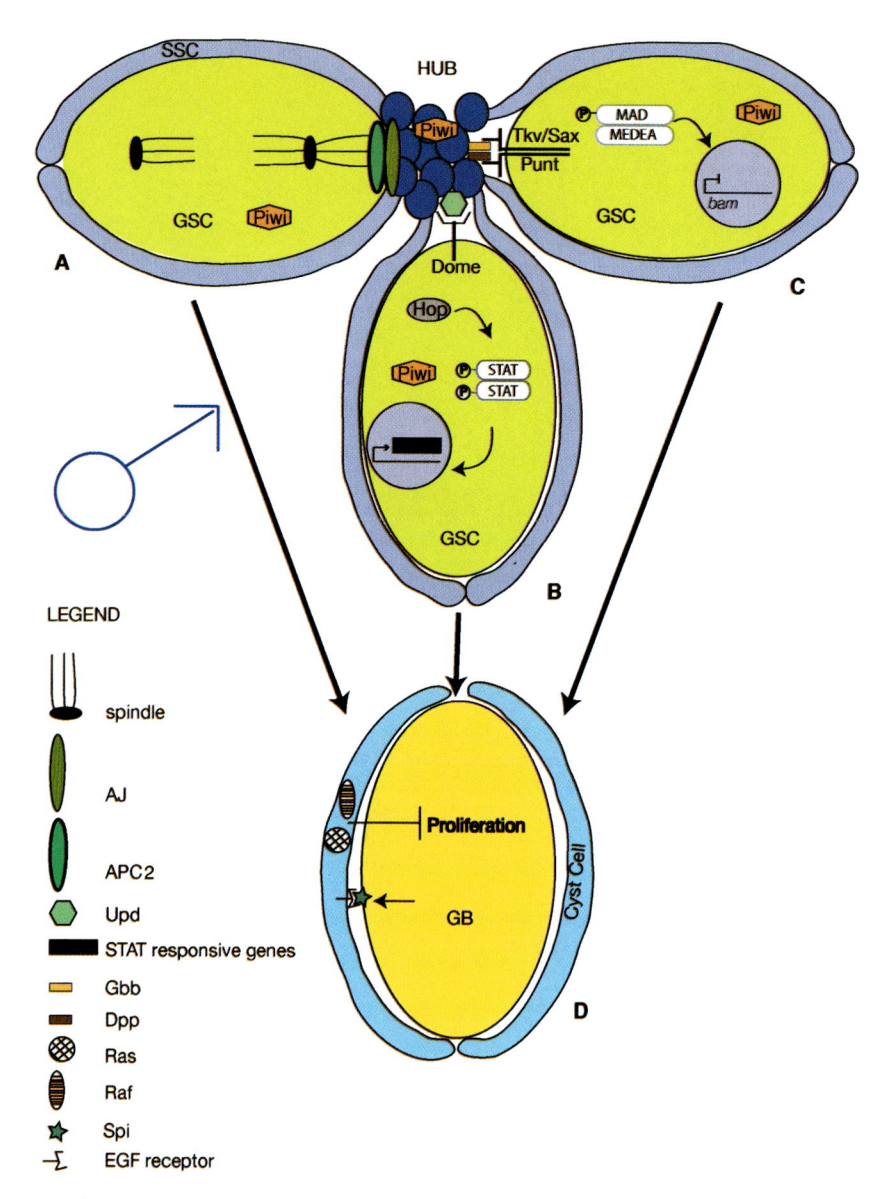

Leonie U. Hempel et al., Figure 4.2 Male Niche. The germinal proliferation center of the male consists of a central core of somatic cells (the HUB) around which germline stem cells (GSC) are arranged radially. Each GSC is surrounded by two somatic stem cells (SSCs). Signaling cascades are shown separately for clarity as are spaces between the various cell types. (A) GSCs attach to the hub via adherens junctions (AJ). Orthogonally oriented mitotic spindles anchor the GSC cortex to the hub via APC2 and help ensure asymmetric division of the GSCs. (B) Secretion of the ligand Upd from hub cells activates JAK-STAT signaling in GSCs, which is required for their self-renewal. STAT phosphorylation leads to transcriptional activation of STAT responsive genes (shown as a black box since they are unknown). (C) Gbb and Dpp expressed in hub cells control GSC self-renewal by activation of BMP signaling, which leads to repression of *bam* transcription through the Mad/Med complexes. (D) EGF signaling is activated in somatic cyst cells by the ligand Spi secreted by the gonialblast (GB). Activated EGF signaling (through Ras and Raf) is required for the differentiation of GBs.

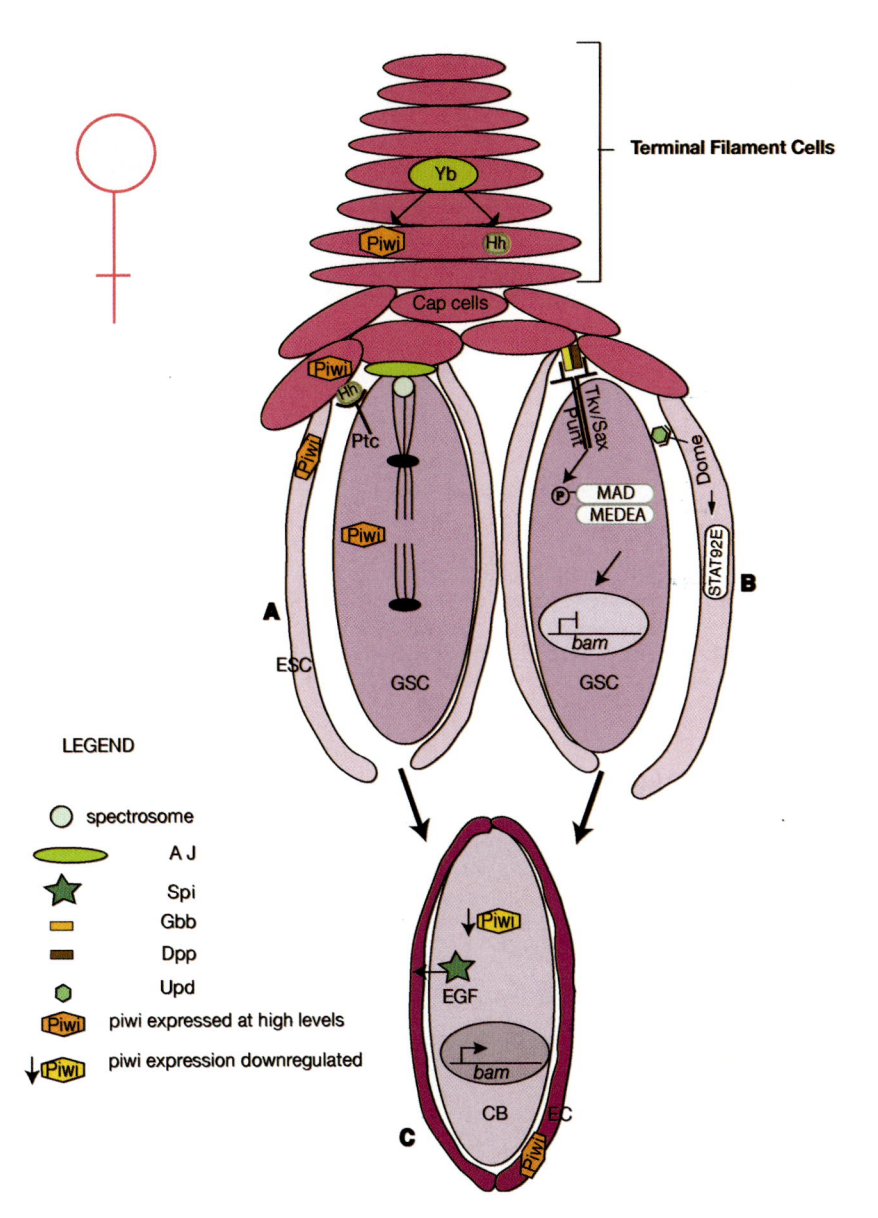

Leonie U. Hempel et al., Figure 4.4 Female niche. The apex of an ovariole consists of terminal filament cells and cap cells which constitute the niche. GSCs are surrounded by escort stem cells (ESCs). Signaling cascades are shown separately for clarity as are spaces between the various cell types. (A) GSCs attach to the niche cells via adherens junctions (AJ). Orthogonally oriented mitotic spindles anchor the GSC cortex to the cap cells by the spectrosome. Yb, Piwi, Hh, Gbb and Dpp are expressed in the niche and act to control GSC self-renewal extrinsically. Piwi is also expressed in the GSCs and controls GSC division intrinsically. (B) Dpp/Gbb signaling leads to repression of *bam* in GSCs through Mad/Med complexes. The JAK-STAT pathway is activated in ESCs and controls their maintenance. (C) Bam is expressed in cystoblasts (CBs) and is required for their differentiation.

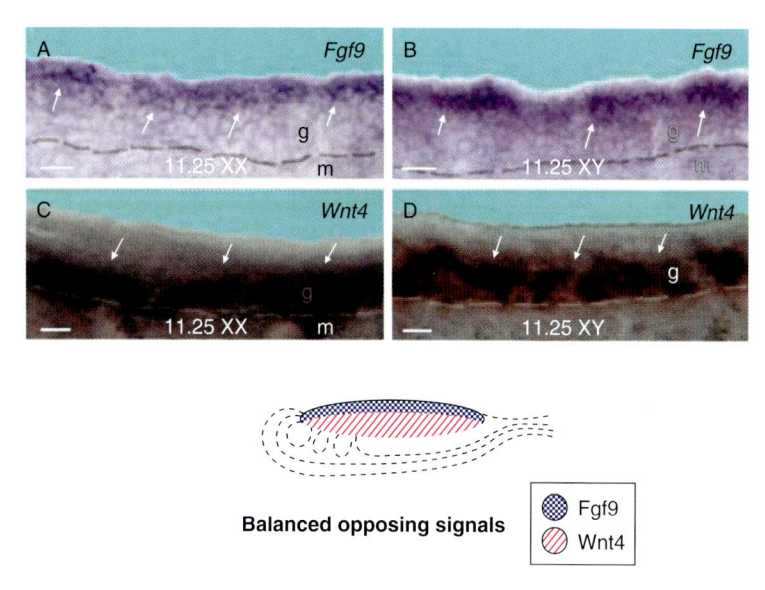

Danielle M. Maatouk and Blanche Capel, Figure 5.5 The bipotential gonad is balanced between two fates. Prior to E11.5, male and female gonads express genes which later become sex-specific. For example, the male-specific gene *Fgf9* is expressed along the surface of the gonad, while the female-specific gene *Wnt4*, is expressed along the border of the gonad and mesonephros. Figure modified from Kim *et al*., 2006.

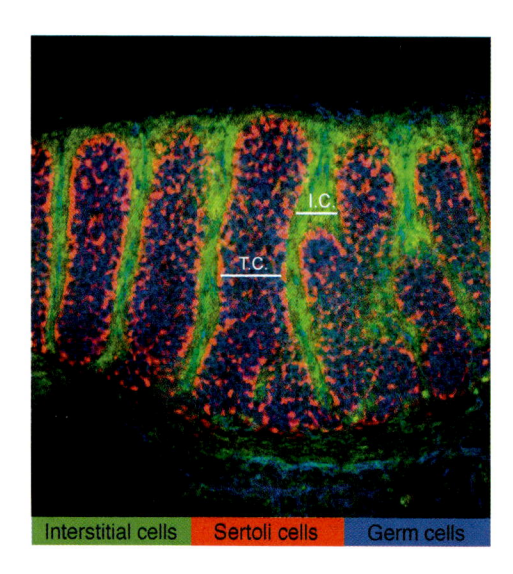

Danielle M. Maatouk and Blanche Capel, Figure 5.7 Distinct morphology of the XY gonad following sex determination. Shortly after *Sry* is expressed, the cells of the XY gonad reorganize into testis cords, (comprised of Sertoli cells and germ cells) or interstitial domains. The confocal microscopy image was taken of a mouse gonad at E13.5 after immunostaining for germ cells (blue), Sertoli cells (red), and interstitial cells (green). The gonad is oriented with the coelomic epithelium at the top and the mesonephric border at the bottom. T.C., testis cords; I.C., interstitial cells.

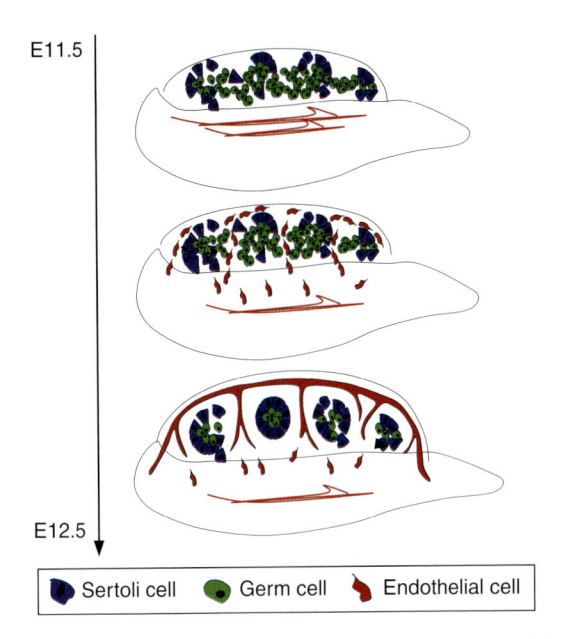

Danielle M. Maatouk and Blanche Capel, Figure 5.8 Formation of the male specific coelomic blood vessel in XY gonads. At E11.5, vasculature is apparent at the gonad/mesonephric border of both XX and XY gonads. Shortly after *Sry* is expressed in XY gonads, this vascular bed dissociates and endothelial cells migrate into the gonad. By E12.5 a newly formed blood vessel has formed along the coelomic surface of the gonad. Formation of the coelomic blood vessel occurs concurrent with the formation of testis cords.

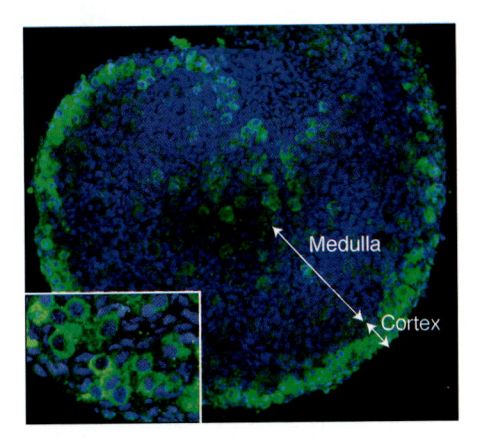

Danielle M. Maatouk and Blanche Capel, Figure 5.10 A distinct ovarian morphology is observed just prior to birth. While expression of *Sry* leads to a dramatic reorganization of the testis, changes in ovarian structure are subtle and are not apparent until near birth. A confocal microscopy image of an E18.5 mouse ovary shows germ cells (green) accumulating in the cortex, while those in the medulla die by apoptosis. The inset shows an enlargement of germ cells in the cortex. Blue staining indicates the nuclei of all cells.

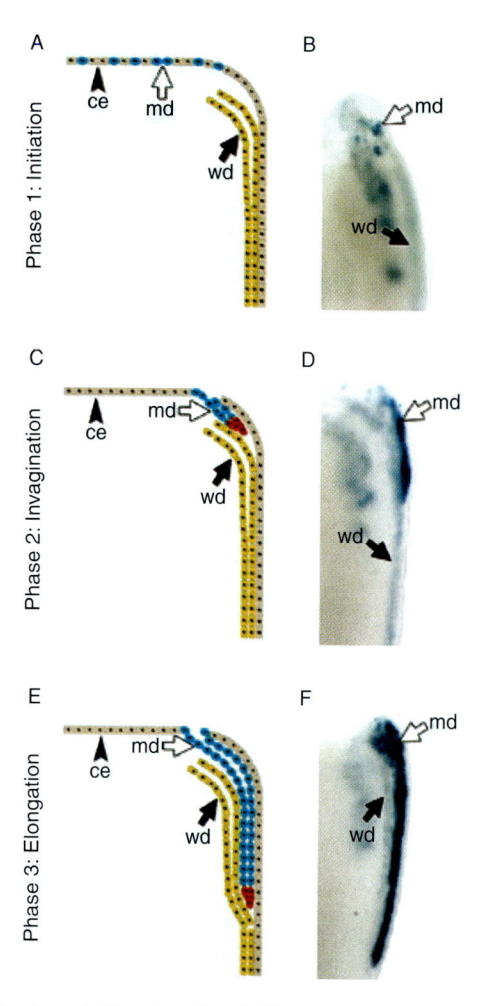

Danielle M. Maatouk and Blanche Capel, Figure 5.12 Morphogenesis of the sex ducts. Formation of the Müllerian duct can be divided into three phases. (A, B) During the initiation phase, Müllerian duct cells become specified. (C, D) In the second phase, these cells invaginate towards the Wolffian duct, with proliferation of the precursor cells at the leading edge of the duct. (E, F) In the third phase, the Müllerian duct elongates along the Wolffian duct. Müllerian duct cells (blue), proliferating Müllerian duct precursor cells (red), coelomic epithelial cells (brown), Wolffian epithelial cells (yellow). ce; coelomic epithelium, md; Müllerian duct, wd, Wolffian duct. Reproduced with permission (Orvis and Behringer, 2007).

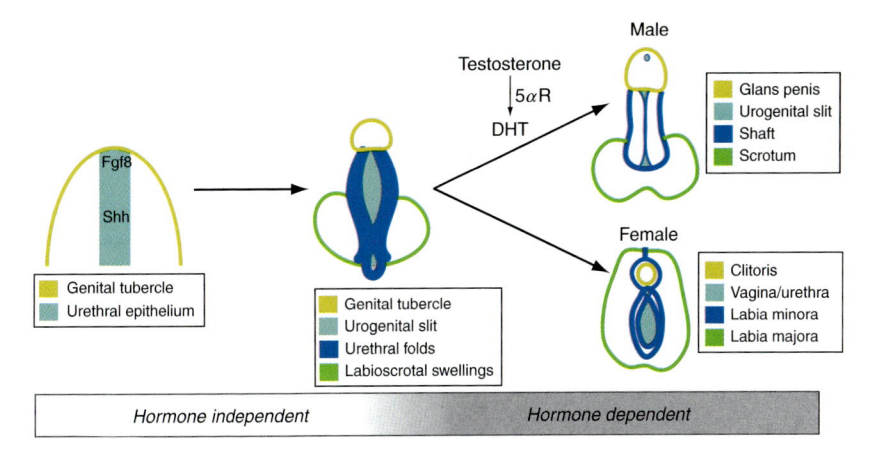

Danielle M. Maatouk and Blanche Capel, Figure 5.14 Morphogenesis of the external genitalia. Development of the external genitalia occurs in two phases; a hormone independent phase, followed by a hormone dependent phase. During the hormone independent phase, outgrowth of the genital tubercle is controlled by *Fgf8*, expressed in the urethral epithelium, and outgrowth is maintained by *Shh*. This process occurs in an identical fashion in both male and female embryos. After E16.5, sex-specific differentiation of the external genitalia begins and is dependent on the presence or absence of testosterone. Testosterone is converted to the active derivative, dihydrotestosterone (DHT) by the 5α-reductase enzyme (5αR).

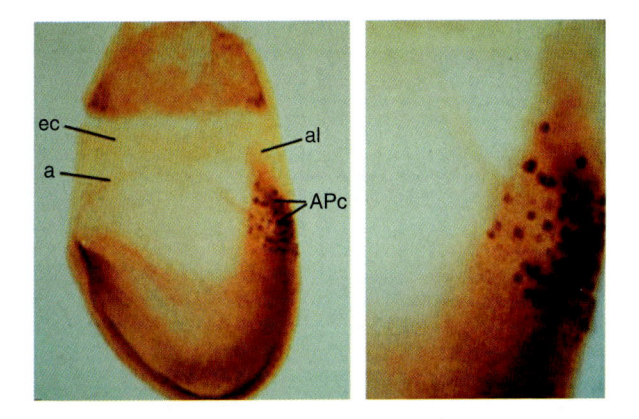

Durcova-Hills and Blanche Capel, Figure 6.1 Alkaline phosphatase is strongly expressed in a cluster of cells at the base of the allantois. Alkaline Positive cells (APc); allantois (al); extraembryonic ectoderm (ec); anterior (a). Used by permission (Ginsburg *et al.*, 1990).

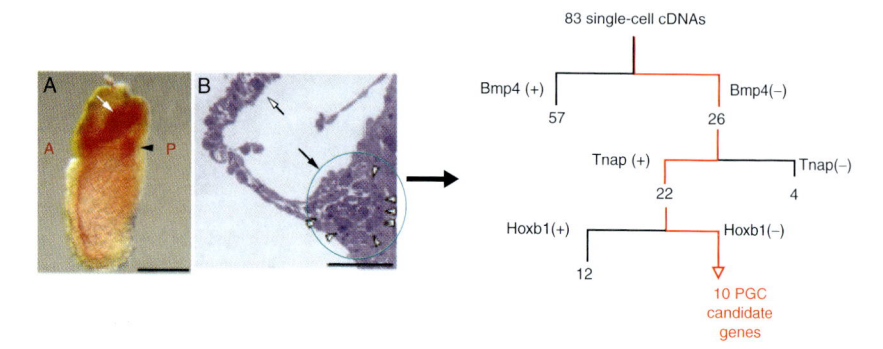

Durcova-Hills and Blanche Capel, Figure 6.4 To screen for new genes expressed during the specification of PGCs, 83 single cells were picked from the TNAP-positive region of the epiblast proximal to the extraembryonic ectoderm. A cDNA library was made from each cell, and 10 candidate PGC libraries were selected based on the observation that they did not express *Bmp4* (26), did express TNAP (22), and did not express *Hoxb1* (10). Differential expression analysis revealed two new PGC genes common to these libraries and not present in others: *Fragilis* and *Stella*. Used by permission (Saitou *et al.*, 2002).

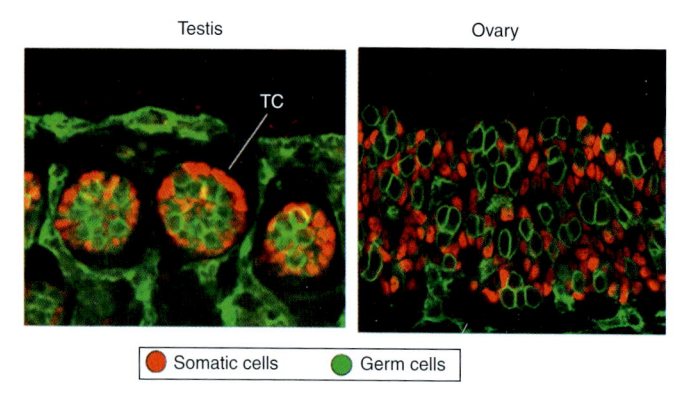

Durcova-Hills and Blanche Capel, Figure 6.7 In the testis, germ cells (round, green) are enclosed by Sertoli cells (red) inside testis cords (TC) by E12.5. In the ovary, germ cells (round, green) remain scattered between somatic cells (red) until follicles form soon after birth. The marker for germ cells is PECAM1, which also labels vasculature, most evident in the interstitial space between cords in the testis.